GIS and the Social Sciences

GIS and the Social Sciences offers a uniquely social science approach on the theory and application of Geographic Information Systems (GIS) with a range of modern examples. It explores how human geography can engage with a variety of important policy issues through linking together GIS and spatial analysis, and demonstrates the importance of applied GIS and spatial analysis for solving real-world problems in both the public and private sector.

The book introduces basic theoretical material from a social science perspective and discusses how data are handled in GIS, what the standard commands within GIS packages are and what they can offer in terms of spatial analysis. It covers the range of applications for which GIS has been primarily used in the social sciences, offering a global perspective of examples at a range of spatial scales. The book explores the use of GIS in crime science, health, education, retail location, urban planning, transport, geodemographics, emergency planning and poverty/income inequalities. It is supplemented with practical activities and data sets that are linked to the content of each chapter and provided on an eResource page. The examples are written using ArcGIS to show how the user can access data and put the theory in the textbook to applied use using proprietary GIS software.

This book serves as a useful guide to a social science approach to GIS techniques and applications. It provides a range of modern applications of GIS with associated practicals to work through, and demonstrates how researchers and policy makers alike can use GIS to plan services more effectively. It will prove to be of great interest to geographers, as well as the broader social sciences, such as sociology, crime science, health, business and marketing.

Dimitris Ballas is Professor of Economic Geography at the University of Groningen, the Netherlands. He has published widely in the fields of Social and Economic Geography, Regional Science and Geoinformatics in the Social Sciences. His recent books include *The Human Atlas of Europe: A Continent United in Diversity* (co-authored with Danny Dorling and Benjamin Hennig).

Graham Clarke is Professor of Business Geography at the University of Leeds, UK. He specialises in the application of GIS for service analysis and planning, particularly within the context of health and retailing.

Rachel S. Franklin is Associate Director of the Spatial Structures in the Social Sciences (S4) initiative and Associate Professor (Research) of Population Studies at Brown University, USA.

Andy Newing is a Lecturer in Retail Geography at the School of Geography, University of Leeds, UK. Andy contributes extensively to undergraduate and master's level GIS teaching and student supervision within the social sciences.

GIS and the Social Sciences

Theory and Applications

**Dimitris Ballas,
Graham Clarke,
Rachel S. Franklin and
Andy Newing**

Routledge
Taylor & Francis Group

LONDON AND NEW YORK

First published 2018
by Routledge
2 Park Square, Milton Park, Abingdon, Oxon OX14 4RN

and by Routledge
711 Third Avenue, New York, NY 10017

Routledge is an imprint of the Taylor & Francis Group, an informa business

British Library Cataloguing-in-Publication Data
A catalogue record for this book is available from the British Library

Library of Congress Cataloging-in-Publication Data
Names: Ballas, Dimitris, author. | Clarke, Graham, 1960- author. | Franklin, Rachel S., author. | Newing, Andy, author.
Title: GIS and the social sciences : theory and applications / Dimitris Ballas, Graham Clarke, Rachel S. Franklin and Andy Newing.
Description: New York : Routledge, 2018. | Includes bibliographical references and indexes.
Identifiers: LCCN 2017017403| ISBN 9781138785137 (hardback : alk. paper) | ISBN 9781138785120 (paperback : alk. paper) | ISBN 9781315759326 (eBook)
Subjects: LCSH: Geographic information systems. | Geographic information systems—Social aspects. | Social sciences—Methodology. | Social sciences—Data processing. | Spatial analysis (Statistics)
Classification: LCC G70.212 .B35 2018 | DDC 300.285—dc23
LC record available at https://lccn.loc.gov/2017017403

ISBN: 978-1-138-78513-7 (hbk)
ISBN: 978-1-138-78512-0 (pbk)
ISBN: 978-1-315-75932-6 (ebk)

Typeset in Minion
by Keystroke, Neville Lodge, Tettenhall, Wolverhampton

Visit the eResources: www.routledge.com/9781138785120

Contents

Figures

Tables

Preface

There are a number of excellent books on the market which provide an introduction to Geographic Information Systems (GIS) and how to get started using GIS (e.g. Heywood *et al.*, 2011; Clarke, 2011). However, as teachers of GIS and spatial analysis for a number of years we have noted the paucity of material that brings together, in textbook mode, the very different areas of GIS application, especially in the fields of human geography and related social sciences. The aim of this book is, therefore, to provide such a textbook which, in turn, we believe helps to emphasise the applied nature of GIS and its usefulness to a wide range of public and private organisations outside the traditional academic environment. There has been no such systematic review of GIS applications in human geography and related disciplines since Longley *et al.* (1999). The first four chapters of this book provide the basic ingredients that underpin much of the analyses within a GIS framework: mapping, buffer, overlay, network analysis, etc. Although we recognise that the reader could get this material elsewhere, it is important to set out the basics in these first few chapters. This helps the book to be 'self-contained' in that students and researchers new to GIS can get the mixture of theory and application in one publication. The remaining nine chapters take the reader through how these basic techniques are used in geodemographics and deprivation analysis, the study of income and well-being, crime analysis and prevention, retail analysis, health care analysis, emergency planning, education planning, transport planning and environmental justice analysis. Although we concentrate on GIS itself we inevitably have to link GIS to other spatial analysis techniques

from time to time, techniques that are not necessarily in standard GIS packages (so called soft coupling). This occurs especially in areas such as retail planning, where the competitive nature of the business requires more sophistication than the normal GIS functions can routinely offer. However, there are examples too in other fields.

In addition to the review chapters we offer a series of hands on practical activities for you to undertake using ESRI's ArcGIS software. These practicals support understanding and development of the techniques that we introduce and the application areas we discuss. We could have chosen many alternative packages for these practicals, including a number of open-source GIS. However, ArcGIS (Desktop) is one of the major GIS systems used within academia, governmental and commercial sectors around the world. It is a versatile and popular GIS package that incorporates data storage, manipulation, visualisation and analysis capabilities. We assume no previous experience in using this GIS software and, while we do not seek to provide a comprehensive guide to ArcGIS itself, readers will become confident in handling spatial data in ArcGIS and undertaking analytical tasks employed by social sciences. Teachers who use other packages should be able to translate the instructions from ArcGIS into other formats – or indeed re-write the instructions – while keeping the valuable data sets and suggestions for analysis that are provided.

The focus on social science applications is explicit – other texts and online resources provide a more comprehensive introduction to the software itself and some of its analysis tools, often in the context of broader application areas and examples not always suited to social science. Although we inevitably have more applications from Europe and the US (given our backgrounds and the widespread availability of freely available data in these contexts) we hope the examples will encourage the reader to explore data sets for their own city or region in order to replicate the examples we offer here.

The reader can apply the techniques for themselves through a number of case studies, related to application areas considered in Chapters 5–13 and built for different geographical areas. Each practical is linked to a specific chapter and draws on the techniques and application areas discussed within the text. Readers are encouraged to begin by working through Practicals A–D which, in conjunction with Chapters 1–4, introduce students to the ArcGIS interface and functionality for applying core GIS techniques. Readers can then pick and choose from Practicals 1–9, which are linked to individual application chapters and which gradually introduce new data sets, techniques and approaches. Each practical contains a comprehensive introduction outlining the topic area, spatial scale/context and the specific GIS tools and techniques covered, enabling readers to identify practicals most suited to their needs. All practicals are accompanied by comprehensive step by step instructions that assist readers in applying and understanding the techniques used. The accompanying data sets have all been derived from freely available open data sources, including governmental open data repositories, documented fully within each practical. The availability of comprehensive, timely and well-maintained spatial data from such sources is invaluable for teaching and learning GIS, enabling readers to gain familiarity with a broad range of data sets. Data sets incorporated with the practicals are broad and include small-area classifications and deprivation indices, crowd-sourced social media data related to natural hazards, road and public transport networks and incidences of reported crime, to name but a few.

Inevitably we cannot cover all areas of GIS development. In the chapters that follow, we tend not to focus too much on which GIS platform is used, other than providing the practical exercises in ArcGIS. Thus developments, for example, in open-source coding and

online, web-based GIS packages can be explored elsewhere (i.e. Mitchell, 2005; Vatsavai *et al.*, 2011; Neteler and Mitasova, 2013; Huang *et al.*, 2001; Peng and Tsou, 2003; Dragicevic, 2004). Similarly, we acknowledge that there are a large number of interesting books on 'spatial decision support systems' or 'planning support systems' which often incorporate GIS and various spatial analysis routines in one package (Carver *et al.*, 2001; Geertman and Stillwell, 2012; Leung, 2012). Again, a detailed analysis of these works takes us beyond the remit for this book. There are also a number of excellent books on 'geographical information science' which explores the links between GIS and traditional spatial analysis techniques (Longley *et al.*, 2015; Fotheringham and Rogerson, 2013; de Smith *et al.*, 2007; Longley and Batty, 1996, 2003; Stillwell and Clarke, 2004). We hope the interested readers will move on to these complementary texts once they have grasped the basics of what GIS can (and cannot) offer.

Finally we note that there is increasing interest in GIS and social media as a growing application area (cf. Sui and Goodchild, 2011). This potentially includes both the use of online interactive mapping packages for interested parties to add information directly (Google maps and OpenStreetMaps, for example) and the increasing interest in the insight that can be gained from geo-located social media data sources such as Twitter. We will speculate more on this growing application area in the conclusions when we speculate more on the future of GIS applications.

References

Carver, S., Evans, A., Kingston, R., & Turton, I. (2001) Public participation, GIS, and cyberdemocracy: evaluating on-line spatial decision support systems. *Environment and Planning B: Planning and Design*, 28(6), 907–921.

Clarke, K. (2011) *Getting Started with GIS* (5th edition), Prentice Hall, Upper Saddle River, NJ.

de Smith, M. J., Goodchild, M. F., & Longley, P. (2007) *Geospatial Analysis: A Comprehensive Guide to Principles, Techniques and Software Tools*, Troubador Publishing Ltd, Leicester.

Dragicevic, S. (2004) The potential of web-based GIS. *Journal of Geographical Systems*, 6(2), 79–81.

Fotheringham, S., & Rogerson, P. (eds, 2013) *Spatial Analysis and GIS*, CRC Press, Boca Raton, FL.

Geertman, S., & Stillwell, J. (eds, 2012) *Planning Support Systems in Practice*, Springer Science & Business Media, Berlin.

Heywood, I., Cornelius, S., & Carver, S. (2011) *An Introduction to Geographical Information Systems*, Pearson, Harlow.

Huang, B., Jiang, B., & Li, H. (2001) An integration of GIS, virtual reality and the Internet for visualization, analysis and exploration of spatial data. *International Journal of Geographical Information Science*, 15(5), 439–456.

Leung, Y. (2012) *Intelligent Spatial Decision Support Systems*, Springer Science & Business Media, Berlin/Heidelberg.

Longley, P. A., & Batty, M. (1996) *Spatial Analysis: Modelling in a GIS Environment*, Wiley, Chichester.

Longley, P., & Batty, M. (2003) *Advanced Spatial Analysis: The CASA Book of GIS*, ESRI, Inc., Redlands, CA.

Longley, P. A., Goodchild, M. F., Maguire, D. J., & Rhind, D. W. (1999) *Geographical Information Systems, Volume 2: Management Issues and Applications*, Wiley, Chichester.

Longley, P. A., Goodchild, M. F., Maguire, D. J., & Rhind, D. W. (2015) *Geographic Information System and Science*, Wiley, Chichester.

Mitchell, T. (2005) *Web Mapping Illustrated: Using Open Source GIS Toolkits*, O'Reilly Media, Inc., Sebastopol, CA.

Neteler, M., & Mitasova, H. (2013) *Open Source GIS: A GRASS GIS Approach* (Vol. 689), Springer Science & Business Media, Berlin.

Peng, Z. R., & Tsou, M. H. (2003) *Internet GIS: Distributed Geographic Information Services for the Internet and Wireless Networks*, Wiley, Chichester.

Stillwell, J., & Clarke, G. P. (eds, 2004) *Applied GIS and Spatial Analysis*, Wiley, Chichester.

Sui, D., & Goodchild, M. (2011) The convergence of GIS and social media: challenges for GIScience. *International Journal of Geographical Information Science*, 25(11), 1737–1748.

Vatsavai, R. R., Burk, T. E., Lime, S., Hugentobler, M., Neumann, A., & Strobl, C. (2011) Open-source GIS, in W. Kresse and D. Danko (eds) *The Springer Handbook of Geographic Information*, Springer, Berlin/Heidelberg, 579–595.

Part I
General concepts

1 An introduction to GIS

LEARNING OBJECTIVES

- What a GIS is and how it can be used
- Basic spatial concepts
- What map projections are and why they matter
- What a spatial data model is and how it is important

Background

The use of Geographic Information Systems, or GIS, is now longstanding within geography. Over the past few years, its usefulness and popularity have begun to diffuse into other social sciences, where both practitioners and students are increasingly expected to produce and consume information that presumes familiarity with GIS analysis. In many fields, such as health or crime, spatial analysis and GIS-based research have become almost ubiquitous. The teaching of GIS methods, however, often remains firmly lodged inside geography departments, and instruction can be stubbornly divorced from applied topics and research questions. This textbook is not aimed at those proposing to become GIS developers and specialists, as those individuals will require much more in the way of theory and background information. Rather, its purpose is to help those who are experts in a variety of social science disciplines or topical areas to become conversant

and capable with common GIS tools and concepts. Its entire point is to enable researchers to use GIS in a range of disciplinary settings.

Geographers often argue that geography is what geographers do. That is, research becomes geographical when performed by someone in that field. This tautology aside, a common thread that runs through most spatial research is a belief in the importance of place and space. How this importance is captured varies – both qualitative and quantitative approaches are common. In this book, we emphasise the utility of a GIS for quantitative analysis, with an emphasis on learning tools and methods in an applied setting. We take frequently used GIS tools and show their applicability across the social sciences and then come at the subject from the other direction, highlighting areas of research and how spatial questions and GIS analysis can contribute to new knowledge creation. Thus, the meat of this book commences with Chapter 2 and continues in the subsequent chapters in Part I, which delve into the range of tools most likely to be of interest to social scientists. Part II of the textbook explores specific GIS applications.

The goals of this first chapter are twofold: first, to show how GIS tools open new windows of opportunity and knowledge and, second, to cover important basic concepts that are fundamental to working with spatial data, as well as structuring and answering spatial questions in the social sciences. In our combined teaching experience, we have found that, naturally, everyone wants the fun part: the analytical tools. Many of the roadblocks encountered in research or in reliability of results, however, stem from inadequate attention to the very basics. So, read on!

Before jumping into the applications – the valuable and interesting things a GIS can do – and the fundamental but important material – the basics of working with spatial data – we should be clear what we mean when we refer to a GIS. A GIS, is a combination of many components, the most prominent of which is the actual piece of software being used for analysis. The GIS software, examples of which are Environmental Systems Research Institute's (ESRI) ArcGIS, open-source QGIS or MapInfo, is used to make maps but also for what is called 'spatial analysis'. Spatial analysis takes many forms, from computation of spatial variables, such as the distance from households to the nearest hospital, to measurements of area or density calculations, to more sophisticated statistical and/or mathematical modelling. A GIS can also be used purely for data management and organisation, in particular if one is collecting a variety of types of data on a particular place and wants to be able to query the data based on location.

In conversations regarding methods, one will hear phrases such as, 'I used GIS to create this map' or 'I learned GIS last year' as if the entirety of GIS is contained within a software application. In actuality, the software is but one piece of the GIS puzzle (Figure 1.1). Other key elements are the data being used, the training of the individual working with the software and the computer hardware itself. It might seem as if these are minor points. However, one quickly learns that the quality of the analytical output is only as good as the input data and the training of the user. More importantly, because spatial data can be complex and because analytical procedures tend to be computationally intensive, the processing speed and

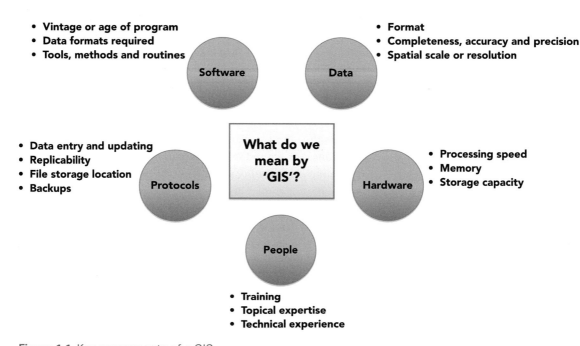

Figure 1.1 Key components of a GIS

memory of the computer can matter enormously. As with any technical or methodological subfield, the skills of software developers, GIScientists and computer scientists are indispensable. These are the academics and practitioners who not only implement the algorithms and tools underpinning the software applications, but also develop the very methods and theories that applied researchers depend upon in their own research. Although many researchers use GIS, the majority engages with it at the levels of data, tools and software. For some specialists, though, especially in geography or computer science, GIS is an object of research in itself. Issues to do with development and implementation of methods, as well as more abstract but important questions, such as how space is measured or data organised, all fall increasingly under the rubric of Geographic Information Science, or GISci.

The capabilities we take for granted in a GIS are the culmination of several disparate advances – in technology, data collection and availability, and in accumulated knowledge. Learning GIS, especially the sort of applied GIS skills emphasised in this text, is much easier than it was even ten years ago. To some extent this is due to the ubiquity today of geographical information: most of us have been exposed to spatial thinking through Google Maps or navigational software in cars and feel moderately comfortable with maps and spatial information. The accessibility of GIS and spatial analysis tools for the more casual or inexpert user – what, in essence, makes it possible to learn GIS with a book such as this – is the result of longer trends. However, two stand out in particular. First, the development of faster desktop GIS has further separated GIS from the realm of programming and complex syntax and made it much more user friendly. Most GIS tools can be reached from pull-down menus and straightforward interfaces, which are sophisticated enough for most users. A growing range of GIS tools are increasingly embedded within web-based data visualisation tools, some of which (such as ESRI's 'ArcGIS Online') have increased analytic and data handling functionality.

Second, increases over time in the sheer availability of georeferenced data have made it possible to ask and answer questions that could only have been dreamed of a few decades ago. Some of these data are the outcome of efforts and investment in the collection of baseline geographic data on the part of national governments (e.g. master address files and physical geography characteristics), while non-traditional data collection and remotely sensed data, collectively thought of as 'big' data, have also increased in quality and quantity. The result is that the average researcher or student, working alone and without a big budget, can hope to find a great deal of decent data easily and free of cost for many locations in the world. Further examples of this can be found in the practical exercises that accompany this book.

What can a GIS and spatial analysis do?

In the mid-19th century, when population growth in London was far outstripping the capacity of existing infrastructure (and when germ theory was still in its infancy), cholera outbreaks were a frequent phenomenon and response was hampered by lack of understanding regarding the mode of transmission. John Snow, a physician today considered the father of modern epidemiology, conceived of mapping cholera deaths along with the locations of water pumps. This map, seen in Figure 1.2, is often considered one of the first examples of 'spatial analysis' and even today highlights advantages of visualising data on a map, as well as considering spatial interconnections between multiple types of information (in this case, cholera deaths and water pumps). The innovation is not so much in the underlying data collection process, although Snow was thorough in interviewing affected households and querying their water collection behaviour – after all, mid-19th-century England was good at collecting statistics. The fact remains, though, that evaluating these data in tabular format, versus on a neighbourhood map, allows very different conclusions. The former tells us the typical age and sex of those affected, for example, while the latter gives a clear picture of the spatial clustering in the illness that existed (these are the black bars lining the streets in the map). The innovation that elevates this figure from simply a useful map to an excellent spatial analysis example, however,

Figure 1.2 John Snow's map of cholera mortality and pump location in a London neighbourhood

Source: Snow (1854). Image retrieved from http://commons.wikimedia.org/wiki/File:Snow-cholera-map-1.jpg (last accessed 9 May 2017)

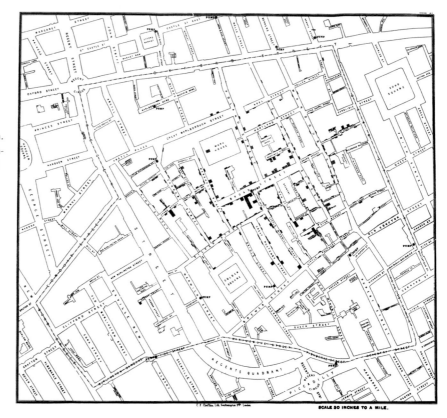

is the juxtaposition of water pump locations and cholera mortality. This allowed Snow – and subsequent generations of viewers – to hypothesise about possible connections between water quality and cholera spread. In a fitting conclusion to a fascinating tale, Snow was successful in arguing for the pump handle to be removed from the Broad Street pump, which was a common source of water for many of those afflicted.

Many current examples abound of how spatial methods or thinking, and especially GIS, can inform research in the social sciences (see Table 1.1). In general, they share an emphasis on the importance of space in mediating, affecting or determining some outcome. Applications can range from vulnerability research in health care or environmental studies to optimal locations for bus routes or school locations to the identification of food deserts. Climate vulnerability studies combine data on areal footprints of climatic events – sea level rise or the paths of hurricanes, for example – and data on the characteristics of the people or property in those areas affected to draw conclusions about the likely impacts of the event. This is an example of a GIS overlay and it lends itself well to visualisation as well as the generation of variables that then feed into more sophisticated models. In public health research, a common question regarding vulnerability is the extent to which distance from

Table 1.1 Major GIS application areas

- Geodemographics and marketing (Chapters 5 and 6)
- Crime (Chapter 7)
- Retail location and analysis (Chapter 8)
- Health (Chapter 9)
- Emergency planning (Chapter 10)
- Education (Chapter 11)
- Transportation (Chapter 12)
- Environmental justice and climate change (Chapter 13)

pollution sources can explain a variety of health outcomes, from cardiovascular problems to early childbirth. In these cases, the GIS is used less for visualisation than for calculating distance between individuals and the pollution source of interest.

Spatial optimisation or location theory occupies an important niche in quantitative geography. Its impacts are felt across the social sciences, in any research that investigates coverage or accessibility. These types of analyses might formally address the ideal or best set of locations for public infrastructure – for example, where should primary schools be located if the goal is to build as few schools as possible (as schools are expensive to construct and maintain) but also to allow children to walk as short a distance as possible? Or they might incorporate questions of fairness or equity, which tend to play out spatially in many cases. The recent popularity of food desert research is an excellent example of how a GIS allows one to assess questions of societal inequality in ways no other technology or set of techniques can. The basic idea of a food desert, that some individuals or neighbourhoods have few options nearby for purchasing food, especially fresh fruits and vegetables, is inherently spatial. Similar to vulnerability research, identifying food deserts requires the spatial combination of information about both people and the location of grocery stores. A variety of GIS techniques can be used to compare access to stores across neighbourhoods, but they all share some acknowledgement of the importance of density or distance, both discussed in more depth below. Given the necessary information, it is quick work for the GIS to compute how far away the closest store is, or how many stores lie within a given distance of a particular neighbourhood. You will make use of GIS tools in this context in the linked practical accompanying Chapter 8 where we discuss the application of GIS in retail analysis in more detail.

Another way to approach GIS capability is to consider its most common uses. A GIS can do a lot, but most applications fall into just a few categories: visualisation, data manipulation and creation and spatial analysis. Visualisation includes cartography – the making of maps – as well as other figures or diagrams that help elicit the spatial structure or characteristics of

data. All GIS software will be well equipped for the creation and export of all usual types of maps. Maps are powerful communication devices that can be used to illustrate spatial patterns of interest, locations of phenomena of interest or even just study sites. And they are useful in both publications as well as presentations. The value of a good map should never be underestimated. The visualisation of data is also important at the front and back ends of the analysis, for both exploring the data available as well as assessing the results. In sum, a map is helpful at multiple stages of the research process: from exploring data, to developing maps of study site context, to evaluating analytical results. We introduce basic visualisation techniques in the linked practical accompanying Chapter 3, with many of the accompanying practicals providing opportunities to visualise a range of data sets and GIS outputs.

Data manipulation and creation in a GIS is similar to procedures used in other quantitative arenas in the social sciences – baseline data are combined across units or variables to create new information. The difference with a GIS is that information created is extracted from the underlying spatial aspects of the data. For example, in the vulnerability overlay example on page 6, one research product might be a map showing areas under water. A more complex approach would be to generate a variable indicating whether a parcel or area is under water, or perhaps variables showing the depth of water or the speed of winds. The variables can be exported from the GIS and used in a traditional statistical software package. That is, the overall methodological approach has not changed: the GIS has been used to generate new variables. Other standard data manipulation exercises in the social sciences involve calculating distances between types of information (e.g. neighbourhoods and stores or fast food restaurants) or obtaining counts per area of some variable of interest (e.g. reported crimes or playgrounds), many of which are outlined in the accompanying practical activities.

Other common operations take place entirely within the GIS environment. These methods together are often referred to as spatial analysis (or sometimes geoprocessing) and they involve more complicated tools or sets of tools, although the finished product might very well be an intermediate step in creating information to be used in a traditional statistical package. Descriptive spatial statistics, which are used to locate 'hot spots' or density measures, which take a set of sample points as input and generate a surface as output are examples of spatial analysis: so are procedures that find potential locations for new businesses or public infrastructure. A researcher hoping to expand access to urban community gardens might use GIS to identify areas of the city without existing gardens (or without sufficient gardening space within a given distance). If she/he then seeks to create new gardens to increase access, though, she/he will need to find locations that meet certain criteria: no existing gardens; available lots or parcels with correct soil and areal characteristics; proximity to public transportation; and zoning that allows for that type of usage. The GIS, with its ability to query both spatial and non-spatial attributes of the data, can easily return a set of potential locations for new gardens. We shall explore GIS and spatial analysis in more detail in Chapter 2.

Some basic spatial concepts

Across the social sciences, much GIS research leverages insight provided by just a few key spatial concepts. The list of concepts provided below is not intended to be exhaustive, but rather to illustrate a few ways in which a GIS can be used to generate spatial information:

■ **Distance** is perhaps the most common spatial characteristic a GIS is prompted to elicit from data. This is partly because the strength of the relationship between a pair of objects is hypothesised to decrease or attenuate as distance increases. This expectation is captured in Tobler's First Law of Geography (Tobler, 1970), which states that, 'Everything is related to everything else, but near things are more related than distant things'. Thus research questions across the social sciences – from effects of pollution on pre-term births to disproportionate impacts of hurricanes on sub-populations – carry the weight of Tobler's Law.

Distance is also important because it often serves as a proxy for cost or time. While we might not always observe how much it costs an individual to get from Point A to Point B, if we know how far apart these two points are, we can make an educated guess at cost.

When measuring distance, we distinguish between so-called Manhattan distance and straight-line, 'Euclidian' or 'as the crow flies' distance. Manhattan distance can be thought of as the actual distance covered when traversing a street network. Although straight-line distance is substantially easier to calculate in a GIS, it does not always give an accurate sense of the true distance or cost of getting from one place to another. On the other hand, some variables, such as air pollution, are better captured with a straight-line distance. Thankfully, a GIS is comfortable with both types of calculation.

■ **Density** will already be familiar to many. It is the statistic used to normalise the population of a place by its area. Density, or the count of units found inside some larger area, can act as a proxy for choice or can be used to generate *per capita* statistics. Standard population density requires no GIS, as counts of people are often tabulated by administrative area as a given. Other times, however, the determination of density requires the marriage of areal, or polygon data, with other types of data: roads or parks or schools, for example.

■ **Proximity** is similar to distance but can also combine elements of density discussed above. We might capture proximity via distance but we might also seek to identify the name of the closest facility of some kind. In that case, the distance is less relevant; we simply want the unit for which distance is minimised. On other occasions we might want to know how many facilities or similar fall within a given distance. If we know that people are generally unwilling to walk more than a mile to shop, for example, we might use a GIS to find all shops that lie within a mile of some location.

■ **Accessibility** captures a more complex relationship between pairs of objects and in general seeks to answer questions about how far away the closest object is, how many are within a particular distance or how difficult it is to reach the object (this latter could be captured with distance but also with the number of transportation modes required). Accessibility is a trickier concept to handle, though, as it is by nature both spatial and aspatial. If one can conceptualise the spatial component of accessibility, whether density- or distance-based, the GIS can deliver those measurements easily. What the GIS cannot do, though, is contribute an estimate of non-spatial aspects of accessibility: access to information about availability of services driven by opening hours for instance, or willingness to use existing services.

- **Coverage** is complementary to accessibility in the sense that accessibility focuses on the experience of the entity needing access while the other, coverage, can be viewed as the area or population within reach of a particular place. So, while accessibility might measure how many doctors are located within a given neighbourhood or within some distance of the neighbourhood, coverage assesses what share of people or the characteristics of those individuals are within a certain distance of a particular clinic or hospital. Especially for public services and facilities, such as libraries or hospitals, the goal is to ensure that coverage is equitable, related to 'need' and within acceptable distance or accessibility thresholds. Again, once the spatial logic of coverage has been defined in a research project, a GIS can make quick work of the analysis.

- **Colocation** is another way of thinking of the overlay illustrations presented above. Although like density it involves the combination of at least two layers, it does not attempt to provide a count or frequency measure for one variable's occurrence within the other. Rather, colocation seeks to generate new information that is the combination of both layers. Where are the places where both x and y occur? Or, in a given place, do we find both x and y? Colocation is a feature of many types of spatial analysis and is often employed in combination with other types of tools.

Using a combination of these concepts, a GIS can help answer the following types of questions:

- Which values fall inside an area and what is their average value?
- How big or long is it? That is, what is its area or the distance of its length?
- How far away is each of a set of objects?
- Who are an object's neighbours?
- Where are high and low values?
- Are values clustered or dispersed? What might explain the pattern observed?
- How many objects are within a given distance of a location?

Conceptualising space

To get from spatially oriented research question to GIS implementation requires an understanding of how space is conceptualised. As with all quantitative research, a jump must be made from real world to abstract world. Of course, we would like to understand how some phenomenon operates in the real world, but the models we develop must somehow be simplified, lest the model itself become cumbersome and weighed down by our intrinsic inability to put the entire world into a computer. Just as variables in data sets stand in for true characteristics of groups or objects, their location in space must also be somehow abstracted. This abstraction carries with it certain considerations. First, space is considered to be either continuous or discrete. Which it will be depends on some combination of the study topic, how interactions are hypothesised to occur and the type of data available. *Continuous* space treats the study area as just that: continuous. Every location in the area will have a value and it is not possible to have an absence of value (zero being also a value, of course). Examples of phenomena that operate most logically in continuous space are precipitation and temperature. It is not possible to be somewhere on the surface of the Earth and not have a temperature reading. *Discrete* space makes more sense if we are locating objects in space. Schools or countries or rivers typically exist in discrete space (i.e. they are not ubiquitous across space). We do not require a data set that registers the presence of a school, yes or no, for every location in an area. Rather, we simply want to record the locations of the objects where they occur and perhaps allow for the possibility that additional future objects might someday also be included.

Whether discrete or continuous, space is also partitioned. This subdivision of space is referred to as *spatial scale* and it plays an enormous role in how research questions are developed and what sorts of data are required. For example, questions related to the amount of green space in a city can be asked at multiple spatial scales, employing discrete or continuous data. Even if green space is represented in continuous space, rather than assigning a green space value to each parcel or neighbourhood, the scale at which that information is captured will impact the types of questions it can be used to answer. If the purpose is to compare the amount of grass in people's gardens rather than across a set of cities, the spatial scale will need to reflect this difference. Figure 1.3 shows the distribution of tree canopy, in 30 by 30 metre pixels that measure the percentage of the area covered by trees. This scale of measurement for tree canopy is clearly adequate for even sub-neighbourhood-level analysis: parts of some neighbourhoods are much greener than others. A comparison of front and back garden tree canopy, however, would not be possible: a 30-metre square pixel is likely to have averaged out any variation occurring at that scale.

Similarly, the appropriate spatial scale of roads or cities, for instance, will depend on the type of questions that can be answered. The underlying data about the cities could be the same, but whether the spatial extent of the city is conceptualised as a point in space (appropriate for calculating distances or flows between cities) or an area (invaluable if the end goal is to measure the impact a change in the city has on neighbouring suburbs) will, in the end, dictate the sorts of research that can be accomplished. The best advice, often repeated in GIS and geography, is that the data used should match the hypothesised spatial process at hand.

Hand in hand with considerations of spatial scale or resolution are issues of data accuracy and precision. *Accuracy*, which refers to the correctness of the data, and *precision*, which captures how finely measured information is, are important for all sorts of quantitative data. In the context of a GIS, they become important because both concepts relate to spatial characteristics as well as data characteristics. Thus, locations for houses, say, might be accurate to within a few metres or several hundreds of metres. The location of each house, whether accurate or not, might be precise to the tenth of a metre or more. Although in general the preference is, of course, for better data, the extent to which such accuracy matters depends on the research question and the type of analysis. Small amounts of error in accuracy can shift the count of numbers of doctors or food stores in a neighbourhood. Those working in emergency planning might have a much lower tolerance for housing units that are not very accurately or precisely located. In other cases, there is a great deal more flexibility in how accurate the data need to be. The important takeaway message is that one should be aware of the quality of the spatial data used in a project and that this quality should be suited to the purpose for which it will be employed.

Spatial abstraction of reality involves both absolute and relative location – not only where an object is (e.g. its x, y location) but how its location is related to the locations of other objects. For example, for a set of city streets we need to not only locate each street in space but also include the 'real life details' an actual street network possesses: which streets intersect and which are overpasses? Or if we are talking about neighbourhoods within the city, and these neighbourhoods are presumed to cover the entire city, we not only want to specify the boundaries of each area, but also that boundaries are shared; that is, it should not be possible to step out of a neighbourhood and not immediately find oneself in the adjacent area. And if we layer our streets and neighbourhoods, we hope that, if a street forms the edge of a neighbourhood, then border and street are coincident in our conceptualisation of the city space. These relationships across objects are what is referred to as *topology* in a GIS setting. It is unusual for an entry level GIS user to have to work with or adjust topology. However, when issues come up with connectivity or adjacency, faulty topology is often the culprit.

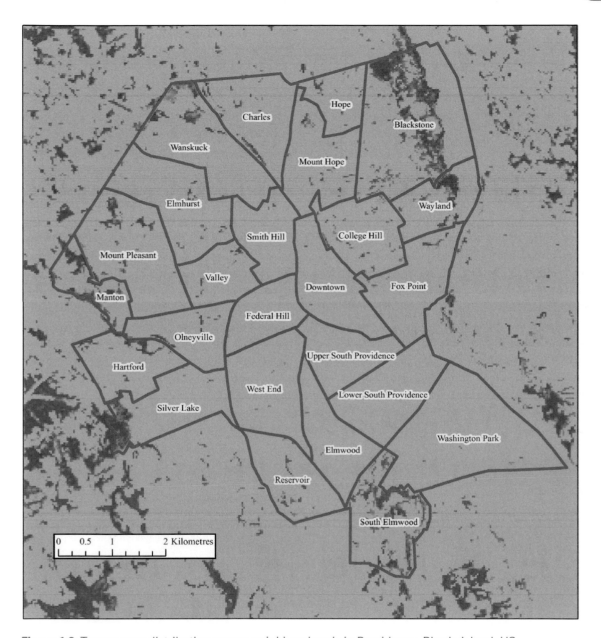

Figure 1.3 Tree canopy distribution across neighbourhoods in Providence, Rhode Island, US

The spatial concepts described above are important first considerations when embarking on work with a GIS. On the whole, the goal is to abstract space in such a way that interactions, connections and locations are represented in ways that capture whatever elements of reality are most important for the type of research being conducted. Two helpful guidelines to keep in mind are those of generalisation and parsimony. Highly accurate data at a fine spatial scale might appear at first glance to be the gold standard of data, but this information will often be difficult to come by, expensive and perhaps more complicated to work with than necessary. Although we want a reasonable abstraction of reality, we do not require a surfeit of information: this is parsimony. Generalisation reflects the idea that although objects have a true geographical footprint in terms of location, area and shape, we do not usually require a perfect representation of the object. All of these aspects of spatial data and space are relevant when it comes to the actual data used in a GIS, as discussed below.

From conceptualisation to operationalisation

Depending whether our spatial information is continuous or discrete, the GIS will store and handle the data differently. The way information is conceptualised and stored in a GIS is called the *spatial data model*. The average user will engage with the spatial data model concept at the level of file formats and the types of analysis that are conducted. The two most common spatial data models are the vector and the raster data models. Both have distinct characteristics, along with strengths and weaknesses.

The *vector model* treats space as discrete, with objects – such as streets or cities – located within space

(Figure 1.4a). In the vector model, all of reality is collapsed into points, lines or polygons, and each of these is generated from solitary points or collections of connected points, known as nodes and vertices. Figure 1.5 shows a map of the East Side of Providence, Rhode Island. Information has been captured with a map image; to extract useful information from the map, all features of interest – parks, streets or schools – would need to be translated into points, lines or polygons. Most file formats in current use keep track of the x, y locations of nodes and vertices and the type of geometry of the layer, so whether a collection of connected line segments form a longer line or a closed polygon, in ways that are not immediately visible to the average user. We choose the elements of reality most germane to our research and the spatial layers are then organised together, with the shared underlying location, within the GIS. As discussed above, spatial scale and the research questions being asked will determine whether, for example, cities are best treated as points – with no area – or polygons. Each layer will contain only points or lines or polygons.

Each layer, or slice of reality, is accompanied by a data table, which contains information about the

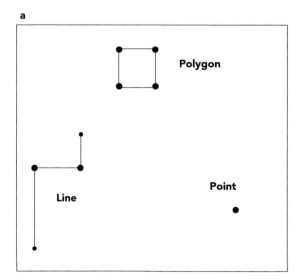
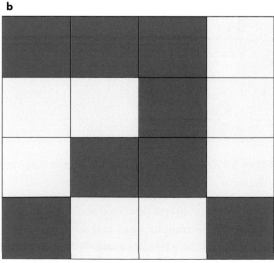

Figure 1.4 The vector (a) versus the raster (b) spatial data model. Both models abstract reality, but the result is different for each and has important implications for how research is conducted and which types of analysis can be done.

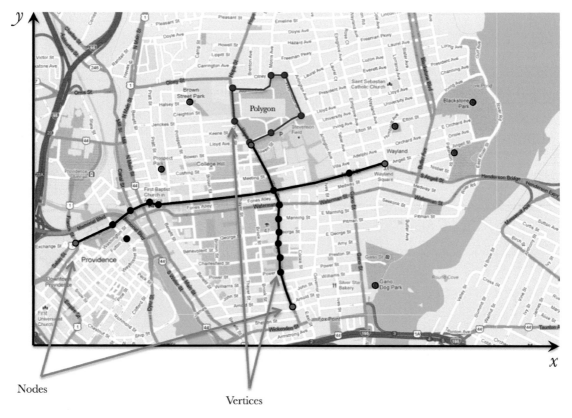

Figure 1.5 Vector representation of features in Providence, Rhode Island, US

Source: Google Maps

layer, including a unique ID and the type of feature (Figure 1.6). This is the information used to create a map or to conduct analysis. In the vector model, each element of geography is a unit of observation; some file formats will handle multipart units differently, either recognising that a region is comprised of, for example, several islands in one unit or treating each shape as a separate observation. If cities are treated as points, what is visible on the screen is simply an array of dots in space. The data table will contain a row for each observation, or city in this case, and information about that place. Figure 1.6 shows how features and their attribute tables are connected: querying features results in observations selected in both the map and the table. Subsequent analysis with these layers might work with the spatial location of objects, either with regard to other features in the same layer or different

layers, but could also involve analysis of the variables in the data, or attribute, table – and usually a combination of both. The linked practical for this chapter uses vector data and explores the link between spatial and attribute data.

The *raster model* subdivides a study area into pixels or cells, generally squares, and because each pixel is identical in size, only the x, y location of the upper left corner of the image must be recorded in order for locations of all other pixels to be determined. Space is treated as continuous, with each pixel assigned one value (Figure 1.4b) – which may be a categorical or continuous variable. Raster data are often treated as synonymous with remotely sensed data. As the term suggests, remote sensing imagery is information that has been recorded remotely, either from a satellite or an aeroplane. Satellites, for example, record the wavelength

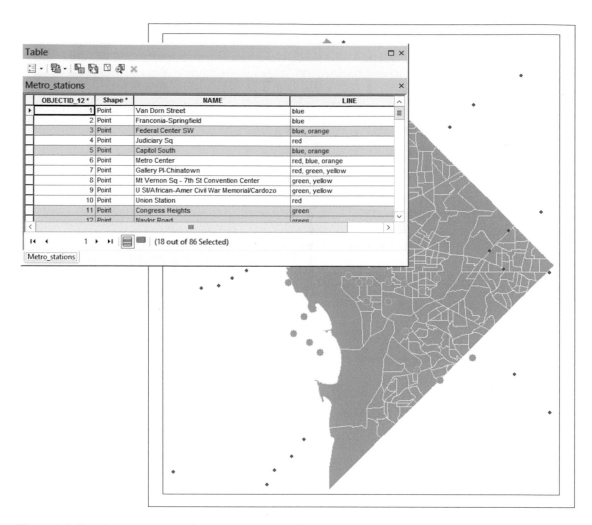

Figure 1.6 The dynamic connection between vector features and their attribute table

of reflected light from the surface of the Earth and this reflected light can be matched to spectral signatures of, say, different types of vegetation. Raster data are used quite often for environmental data, which is best thought of as varying continuously across space, without regard for administrative boundaries. Unlike vector-format data, raster information does not have an associated data table. Instead, each pixel contains just the one value, whether for precipitation or temperature or land use.

The pixel size of the layer or raster is a key characteristic of the data. Since only one reading is recorded for each cell, this value is essentially an average for all locations within the cell. A 30-metre square cell size, the resolution of Landsat satellite imagery, may be sufficient to identify variations in land use across a city (see example in Figure 1.3), but will not be able to indicate the share of a parcel that is built on. It is important to remember that raster data record a value for each and every pixel in a layer, such that even the absence of a phenomenon is noted. If the cell resolution is small, the result can be very large file sizes.

Each data model has its strengths and certain types of analysis require a given data type as input. Figure 1.7 provides an example of vector and raster data, the former for administrative characteristics of Rhode Island

a

b

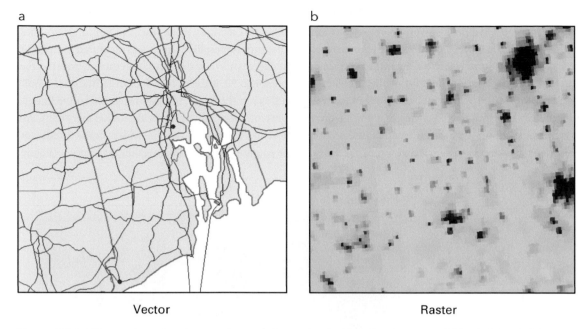

Vector   Raster

Figure 1.7 (a) Vector data showing roads, counties and point data for the state of Rhode Island; (b) raster data of population density for a portion of the United States

and the latter for population density. Either type of information could be conceptualised as either raster or vector. We could imagine a raster that, for a given cell size, notes the presence of 'road' or 'no road'. This is virtually nonsensical. Population density could be shown for administrative areas. Indeed this is often the case. However, such visualisations will miss that density does not always follow administrative borders and is characterised by clustering and variation in values.

In general, the vector model is better at precise location of objects and will generate better cartographic products, especially if the units being mapped are administrative areas. Raster data with large cell sizes will produce blocky maps for small areas, essentially squares of colour that follow no on-the-ground conception of reality. Raster data also usually have larger file sizes, a characteristic that is less of a challenge than in the past but still not a negligible consideration, as most analyses will require several rasters for a particular area. In addition, certain types of information are stored by default in one format or the other. Administrative data, for provinces, states or countries, or for streets, schools or hospitals, will usually be in

vector format. Information on land use, vegetation or other environmental characteristics will tend to come in raster format. It is normal to work across data types, combining vector and raster data.

Location in space: coordinate systems and projections

For data to be spatial, they must contain locational information. Someone, somewhere, had a way of attaching location to the entities for which data were collected. The good news for novice users is that, in many cases, but not all, data in a spatial format will already be georeferenced and the properties of the data will indicate how location is defined. In a few cases – spreadsheets containing latitude and longitude coordinates, for example – the researcher will need to know the spatial properties of the data in order to import them into a GIS. There are two aspects of location that are important in a GIS: coordinate system and projection. An accurate and appealing map, to say nothing of reliable analytical results, requires an understanding of both.

Information is located on the surface of the Earth using a *coordinate system*. Key words here are 'location' and 'surface of the Earth'. Coordinate systems are used to locate objects somewhere on the round surface of the sphere. The location portion of the problem is handled with an *x, y* location, usually latitude and longitude, which uniquely identifies that point – there is only one place on the Earth where that intersection of coordinates occurs. The subdivision of the world by latitude and longitude, for example, breaks the globe into sections from North to South called lines of latitude or parallels. Latitude is measured in terms of the degrees north and south of the equator, from 0 at the equator to 90 at the poles. Lines of latitude are parallel slices of the Earth; the distance between the lines is constant. Lines of longitude, also called meridians, are more like slices of an orange: widest in the middle (at the equator) and coming together at the poles (Figure

b Cylindrical

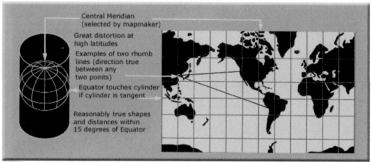

a Spherical

c Conic

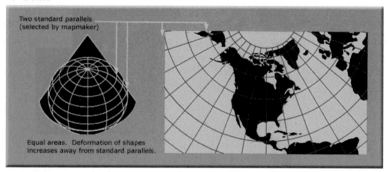

d Azimuthal or planar

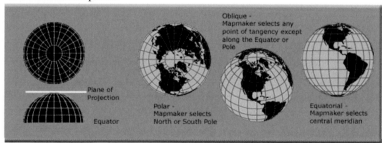

Figure 1.8 (a) A sphere or globe, with lines of latitude and longitude; (b) a cylindrical projection; (c) a conic projection; and (d) an azimuthal projection

Source: USGS: http://egsc.usgs.gov/isb//pubs/MapProjections/projections.html

1.8a). Longitude is measured from East to West, relative to the Prime Meridian, which is 0 degrees and passes through Greenwich, England. The opposite meridian, at 180 degrees, is the theoretical international date line (in practice, national governments around the world will adjust local time to match local political, social and economic needs). Both latitude and longitude are measured in terms of degrees, minutes and seconds, with 60 minutes to a degree and 60 seconds to a minute. This allows for a very good location of points on the surface of the Earth.

The surface of the Earth is a bit trickier to deal with, however, as the shape of the Earth itself is estimated. It is definitely round, but it is not a perfect sphere. Indeed, it is more like a giant beach ball that has been sat on, with the circumference around the equator longer than that going through the poles. The shape of the Earth is referred to as an oblate spheroid. Measurements of the Earth's shape are now quite decent, but as various estimates have dominated over time and many estimates tend to be more reliable for particular parts of the globe, the shape of the Earth – or the datum – used to locate x, y coordinates is not consistent. Latitude and longitude points in the United States are often collected using the North American Datum of 1983, or NAD 83. Other sets of coordinates use the World Geodetic System 1984, or WGS 1984. You will encounter both during the accompanying practicals.

In order to map geographical information or conduct analysis with it, data are projected. *Projection* is the process of going from information on a round sphere to information on a flat surface, whether a page or a computer screen. Spatial data are often not projected and must be before they can be used for research. This is a common task, though, and a GIS will either project the data on the fly, not changing the nature of the original information but only displaying it differently, or can create new versions of the data that are projected. To get from round to flat involves a fair amount of mathematics and, if you can imagine flattening an orange or a beach ball, some necessary distortion; it is impossible to go from round to flat without some loss of geographical information. Data are translated from round location to flat location by projecting the information onto a flat surface, either a cylinder, a cone

or a plane (see Figure 1.8). The cylinder or cone can then be unrolled and the information is there, on a flat surface.

If one imagines this process of 'projection' it is easy to see how some locational information must be lost or distorted. A planar projection offers perfect one-to-one correspondence at the point where the plane touches the surface of the sphere, but information becomes more distorted away from that point of contact. Every projection distorts either shape, area, distance or direction. Especially for larger areas, there is no such thing as a perfect projection, only a good projection for the type of analysis at hand. For example, if the goal is to estimate some measure of density, a priority is an area-preserving, or equal area or equivalent, projection. Other types of projections are conformal, or shape-preserving.

In an ideal world, one searches for GIS data using the internet or a database and finds exactly what is needed immediately. Using keywords such as 'GIS data' or 'shapefile' (the most common type of vector GIS data) along with other search terms can often isolate spatial data resources. Depending on the scale of analysis and the study site, a great deal of baseline GIS information is available to the public for free. Many larger cities, including London, Chicago, Washington, DC, and San Francisco, have data portals where both spatial and non-spatial information can be accessed. Some countries, such as the United States, make a great deal of data available, but across a variety of government agencies. All US states have websites called 'GIS Clearinghouses' where spatial data for the state can be accessed.

Getting data into a GIS

Finding required spatial data in a GIS-ready format does occasionally happen. More often, though, intermediate steps are required in order to be able to work with data in a GIS environment. A few of the more common strategies are covered briefly below, including geocoding, tabular joins, importing x, y coordinates, and georeferencing. Geocoding and tabular joins will be covered in deeper detail in Chapter 2. Note that these processes incorporate elements of the GIS background covered above: in particular spatial data models and coordinate systems.

If data include an address (especially with ZIP code/postcode), place name or even latitude/longitude coordinates, this information – spatial in spirit if not in file format – can be used to import a data file into a GIS. The process of locating points in space using an address is referred to as geocoding. *Geocoding* locates points in space, and creates new spatial data by comparing a spreadsheet with address information to a spatial reference file with street information and address ranges. This procedure is similar to the way in which internet mapping applications locate an address on a map. For a GIS to successfully geocode addresses, the reference file must be of good quality, complete and up-to-date. In practice this means that geocoding is easier in places that have collected, and made publicly available, good street data.

Where geocoding requires access to additional data, in the form of the reference layer, no extra spatial data is required for a GIS to locate x, y locations in space. That is, if a data file contains latitude and longitude information for the spatial units, the GIS can easily locate these in space. Locational accuracy will depend on the accuracy with which data were collected (some handheld GPS units are more accurate than others) but will also depend on the researcher's ability to identify the underlying coordinate system that was used when the points were collected. Once points have been located in a GIS, locations can be compared to spatial data for which locations are known, to make sure the results are satisfactory.

Other times spatial data are easily found, but they lack the range of variables required for analysis. This is a frequent occurrence when working with administrative or public infrastructure data, such as schools or counties. In many cities in the US, for instance, layers exist that show the locations of primary schools. These data are in GIS format and with a few clicks, one can visualise the entire set of schools for an area. The files do not, however, generally contain much in the way of information about the schools, such as age of building, number of pupils or their characteristics. To bring this sort of additional information into the GIS,

one solution is to merge, or join, the attribute table of the spatial data with an external table of data for the same spatial units. To do this, both tables of information must have a common variable that allows the tables to be matched exactly. This common variable must be unique to each observation and its format and content must be identical in both tables. Suppose the goal were a map of EU countries by average income. A layer or shapefile with country locations might be easy to come by but would not contain any information about the countries, just their shape on a map. A separate table with interesting information for each country can then be tied to the attribute table by country name. However, if one table refers to Spain and the other to España, the computer will not recognise the match. For this reason, most administrative data in most countries have numerical codes that are used to uniquely identify areas. These ease the merging of tables considerably. Once the tables are merged in the GIS, variables can be mapped and subsequent analysis completed.

Increase in demand for GIS-compatible data has led to wider availability of spatial data for many parts of the world and for many topics. Often, however, researchers are embarking on projects for which data do not exist. This could be because little or no data have been collected for a study site or because one wishes to make use of historical data, or for some other reason. One strategy is to conduct fieldwork and collect the data oneself. Another is to make use of existing cartographic projects. Researchers, civil servants, explorers and private individuals have been creating maps for centuries. The enormous amount of information contained in printed maps makes them a good resource for information, but one that requires extracting the relevant information from a paper or digital product. For recent map products, it is often worth the time to contact the author to check whether underlying GIS data might be available to be shared. In other cases, the map will be georeferenced. *Georeferencing* is the process of telling the GIS where locations on the map are in the real world. If not already in digital format, maps are scanned and then added to the GIS. The scanned image of the map, which is simply a picture, is then tagged at several locations to a basemap for which real-world locations are known. The connections between digital map and known locations help to 'line up' the map and place it where it should be. The result is mapped information that has been tied to its actual location in the world. It is, however, still only a picture. The next step is to use the scanned image as a base layer along with other data or to manually create new layers by tracing the information on the map, similar to what is seen on the left of Figure 1.5. Digitising features from a georeferenced map is time-consuming but the result is new spatial information that was not previously accessible. You will undertake some basic digitisation in the linked practical accompanying Chapter 12.

Concluding comments

The information presented in this chapter is fundamental but might seem a bit abstract. The accompanying practical should help you to begin to understand how basic spatial concepts, data models and even projections are important building blocks for any research project employing a GIS.

Below we have summarised some key ideas that might help in the development of GIS projects:

1 Identify your research question and how it is 'spatial'. Review existing research in your field that uses a GIS or spatial analysis. This will help give you a better sense of how GIS can be useful to you.
2 Use the basic spatial concepts discussed above to frame your questions as this will help when it comes to operationalising the analysis in a GIS setting.
3 What are your data requirements? This means identifying the spatial units you need information for, as well as their characteristics.
4 Do your data already exist in a GIS-ready format? If not, consider how you will get your data into the GIS.
5 Identify the best projection for your location and spatial scale of analysis and make sure all data are in this projection.

6 Determine what sorts of analysis will be needed to answer your questions. The chapters and practicals that follow will help with that.

Accompanying practical

The introductory practical (Practical A: Introduction to ArcGIS and spatial data) assumes no prior hands-on experience with a GIS and walks you through the process to launch ArcGIS Desktop, one of the most popular and versatile proprietary GIS software packages. Using exemplar data sets related to transport provision in the US city of Chicago, you will gain familiarity with loading and handling basic vector data sets.

Further reading

Iliffe, J., & Lott, R. (2008) *Datums and Map Projections: For Remote Sensing, GIS and Surveying* (2nd edition), Whittles Publishing, Caithness, Scotland. Basic information about projections and coordinate systems.

Longley, P. A., Goodchild, M. F., Maguire, D. J., & Rhind, D. W. (2015) *Geographic Information System and Science*, Wiley, Chichester. Introductory information regarding GIS (less on applications in the social sciences, but more nuts and bolts of a GIS).

United States Geological Survey (USGS) http://egsc.usgs.gov/isb//pubs/MapProjections/projections.html. Examples of projections and assistance selecting the most appropriate.

References

Snow, J. (1854) *On the Mode of Communication of Cholera*, C.F. Cheffins, London (2nd edition 1855, John Churchill, London).

Tobler, W. (1970) A computer movie simulating urban growth in the Detroit region. *Economic Geography*, 46(2), 234–240.

2 Data querying and spatial analysis in GIS

LEARNING OBJECTIVES

- Attribute and digital boundary data
- Attribute (*aspatial*) and spatial queries
- Buffer analysis, point-in-polygon and spatial overlay
- GIS-based spatial analysis in the social sciences

Introduction

In the previous chapter we introduced the reader to some basic terminology and concepts that are important to understand in order to get started. In this chapter we provide an introduction to key GIS methods that can be used to address social science research questions that have a spatial nature. As noted in the previous chapter, GIS-based spatial analysis covers a very large number of methods ranging from computation of spatial variables, such as the distance from households to the nearest hospital, to measurements of area or density calculations. In this chapter we give more specific and detailed examples of how GIS-based spatial analysis can be conducted to address simple but powerful queries relating to social science issues and problems. In particular, we discuss ways in which GIS and spatial analysis techniques can be employed to provide information and intelligence in relation to the importance and role of geography in the actions and performance of private and public sector organisations. The following questions represent examples of the types of issues and problems typically faced by organisations in the private sector (questions 1 and 2) and the public sector (questions 3 and 4):

1   Retailers wish to use socio-economic data available for different areas to assess the likely demand for their products if they open or expand an outlet; how can these areas be classified?

2   The same retailers collect information on movements of shoppers from residential zones to stores. Can we build models of such flows? Can we predict changes in such flows if we expand an outlet or open a new one?

3   Local council policy makers want to know what the spatial impacts of a council policy would be – could they use small-area data for socio-economic impact assessment?

4   National government departments need information on geographical distributions of deprivation. How can deprivation be measured, mapped and analysed?

These are typical examples of social science issues that can be analysed with GIS. They also give a small flavour of the applicability of GIS in the social sciences and of the topics and case studies that are presented and discussed in Part II of this book.

Some of these questions will be addressed using appropriate and publicly available data throughout this chapter. In particular, in this chapter we show how GIS and spatial analysis can be used to perform simple but important queries and generate relevant information and intelligence that can be used to address the above issues, using publicly available data. We begin by considering some practical issues pertaining to data availability. We then discuss examples of attribute (*aspatial*) *queries* which can be addressed with simple descriptive statistics (alongside spatial thematic mapping). Asking, for example, how many people live in each region of a country is an *aspatial* type of query that does not necessarily require GIS capability to address it, but which can be easily answered using any database management system or spreadsheet software. We then move on to *explicitly spatial queries* which make more use of the geo-analytical capability of GIS and tools such as *buffer analysis*, *point-in-polygon* and *spatial overlay*, that take into account *geographical distance*, *adjacency*, *containment* and *intersection* (see again Chapter 1).

Getting started

Before we begin, it is useful to consider again the four general questions/social science examples posed above in the context of the checklist presented in Chapter 1. Taking each item in the list in turn:

1   Identify your research question and *how it is 'spatial'*. Review existing research in your field that uses a GIS or spatial analysis. This will help give you a better sense of how GIS can be useful to you.

Looking at the spatial nature of the question of *likely demand for the products of retailers*, we note that there is a combination of socio-economic and (geo) demographic factors such as age, lifestyle and household disposable income, but also where they live and how far from the location of different retail outlets.

Similarly, with regard to the question of travel-to-shop flows we might have a combination of socio-economic and demographic factors similar to the first question, coupled with actual geographical information on the shopping trips made (derived from surveys for example).

With regard to the question about the possible spatial impacts of local city council policy we can look at specific examples such as the identification of food deserts (also briefly discussed in Chapter 1) or the decision to invest in social housing in particular areas of a city or to reconsider local tax bands. Again, there is a combination of socio-economic/demographic factors (e.g. numbers of low-income households) with geographical factors (where do they live in the city? what is the level of accessibility that they have to local council services?).

Question 4, pertaining to national government, also relates to socio-economic factors associated with the likelihood of social exclusion and of lacking access to (or being deprived of) what are considered to be basic necessities in society (such as not having access to a car or living on income below the poverty line), but at the same time geographical factors such as distance from the delivery points of public services (e.g. schools, health centres).

2   Use the *basic spatial concepts* discussed above to frame your questions as this will help when it comes to operationalising the analysis in a GIS setting.

All issues and questions are related to *distance, density, accessibility, coverage* and *colocation*, discussed in Chapter 1.

3   What are your data requirements? This means identifying the *spatial units* you need information for, as well as their *characteristics*.

First, with regard to the appropriate *spatial data model*, the *spatial units* that we need to operationalise include *points* (for features such as retail outlets, schools, hospitals, etc.), *lines* (for features such as the road network that can be used to estimate 'accessibility') and *polygons* (e.g. administrative boundaries for local council districts and electoral wards and smaller areas for which population data might be available). Therefore, the *vector model* described in Chapter 1 is the most appropriate here (and this is also generally the case for most social science applications), as applied throughout the accompanying practicals.

The next step is to identify the *characteristics* (or attributes) that need to be joined to the *spatial units*. For example, for retail store units (point data) this information can include floor-space, brand, opening hours, number of employees, number of parking spaces, etc. The road network *line* data can be enriched with additional characteristics regarding the type and capacity of road, traffic levels, etc. Finally, the area boundary (*polygon*) data can be joined with demographic and socio-economic data such as total number of residents, population by age-group, social class, car ownership, etc.

4   Do your data already exist in a GIS-ready format? If not, consider how you will get your data into the GIS.

There is a wide range of sources of digital spatial data as well as secondary social survey data that can be used to add characteristics (attributes) to the spatial data.

It is worth noting that there is a rapidly increasing number of open data sources that can be used to obtain *free digital geographical data*. A good example is the Ordnance Survey in the UK which offers open digital map data for a number of geographical features including rivers, street and road networks as well as administrative and electoral boundaries. There are similar open data resources worldwide (see Appendix 2.1 for more details). Another excellent resource for open data is OpenStreetMap, which can also be used to obtain data on the location of retail outlets and other features relevant to the issues discussed in this chapter.

Similarly, there are several suitable sources for data on characteristics or attributes that can be joined with digital map data. It should be noted that one of the key sources of data on demographic and socio-economic

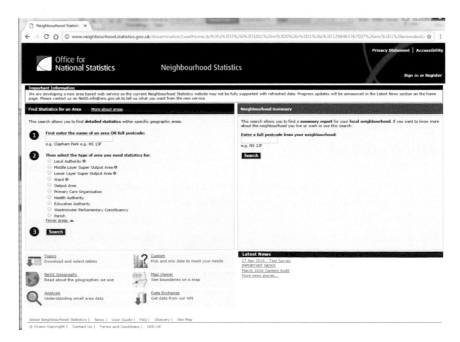

Figure 2.1

Neighbourhood Statistics topics from the UK Office for National Statistics characteristics are the censuses of population, which record demographic and socio-economic information at a single point in time and are normally carried out every five or ten years (Rees *et al.*, 2002). Census data sets describe the state of the whole national population and are extremely relevant for the analysis of a wide range of socio-economic issues and related policies. In addition to the census there is a wide range of administrative and private sector sources of suitable socio-economic data. A good example of a resource that brought together socio-economic and demographic data was the Neighbourhood Statistics service in the UK (see Figure 2.1). Neighbourhood Statistics offered free access to a wide range of social and economic data, which included census small-area data as well as data provided by local government authorities and other providers such as the Home Office (crime data) and the Land Registry (house prices). These data included information on the socio-economic characteristics of the people living in the area, as well as information on housing and crime. Neighbourhood Statistics was discontinued in 2017 but all the data previously obtained via this portal are available from other sources, many of which are listed in Appendix 2.1.

In addition, there are ongoing developments leading to new methods of geographic data collection, such as Voluntary Geographical Information (VGI) provided by 'citizens as sensors' (Goodchild, 2007) and crowdsourcing (Sui *et al.*, 2013), involving non-expert people (the 'crowd') in data production with the use of smartphones and other mobile devices (Brovelli *et al.*, 2016). There is also an increasing amount of geographical data generated via social media and great potential exists to integrate and analyse with the use of GIS (Sui and Goodchild, 2011; Croitoru *et al.*, 2013; Huang and Xu, 2014; Zhai *et al.*, 2015; Kim and Koh, 2016).

Using GIS to perform attribute (*aspatial*) queries

As briefly discussed in the introductory section of this chapter, GIS queries can be distinguished between *aspatial* attribute queries, which do not require knowledge of topological features (i.e. information on the location of spatial units and also where they are in relation to each other) and spatial, which need to take distance into account and/or involve the combination of different geographical layers. The following are examples of *attribute* queries that can also be described as *aspatial* queries:

■ What are the top ten and bottom ten regions in Europe in terms of university degree holders as a percentage of the total population aged 25–64?

■ Which areas in a city have unemployment rates that are above the national average?

■ Which are the five most and five least affluent neighbourhoods in a city?

Although these types of questions can be addressed using information systems and computer programs such as spreadsheet software and statistics packages, the use of GIS enables tabular summaries in tandem with visual analysis and mapping (as will also be discussed and explained in more detail in the next chapter). The first question posed above can be addressed with the use of any spreadsheet-based system, just by obtaining the relevant data from a suitable statistical source such as EUROSTAT and then sorting the regions in descending order of the variable in question to create Tables 2.1 and 2.2.

Table 2.1 Top ten regions: total number of 25- to 64-year-olds having completed tertiary education as a percentage of all population aged 25–64

Region	%
Inner London (UK)	60.7
Oslo og Akershus (Norway)	53.4
Prov. Brabant Wallon (Belgium)	51.9
Berkshire, Buckinghamshire and Oxfordshire (UK)	51.0
Eastern Scotland (UK)	50.4
Helsinki-Uusimaa (Finland)	50.2
Outer London (UK)	48.7
Hovedstaden (Denmark)	48.1
Zurich (Switzerland)	47.7
Stockholm (Sweden)	47.6

Source: http://ec.europa.eu/eurostat

Table 2.2 Bottom ten regions: total number of 25- to 64-year-olds having completed tertiary education as a percentage of all population aged 25–64

Region	%
Campania (Italy)	14.3
Nord-Vest (Romania)	14.2
Puglia (Italy)	14.0
Vest (Romania)	13.5
Sicilia (Italy)	13.3
Sardegna (Italy)	13.1
Regiao Autonoma dos Acores (Portugal)	13.0
Nord-Est (Romania)	12.0
Sud-Est (Romania)	12.0
Sud-Muntenia (Romania)	11.4

Source: http://ec.europa.eu/eurostat

However, these data can be integrated into GIS for mapping and further analysis. As discussed in Chapter 1, this type of attribute/tabular data can be joined with suitable geographical data (in this case that would be European region digital boundary data) in a GIS system. This would enable us to produce tables such as the above but to also provide spatial information by showing *where* these regions are and what type of geographical patterns might exist. This can be achieved by creating a thematic map of higher education graduates across all regions (see Figure 2.2).

The second and third questions posed above are also very relevant to the wider issues and example questions discussed in the previous sections. For example, it would be generally reasonable to expect the likely demand for retail products is going to be higher in more affluent areas. Similarly, knowing which areas have the highest

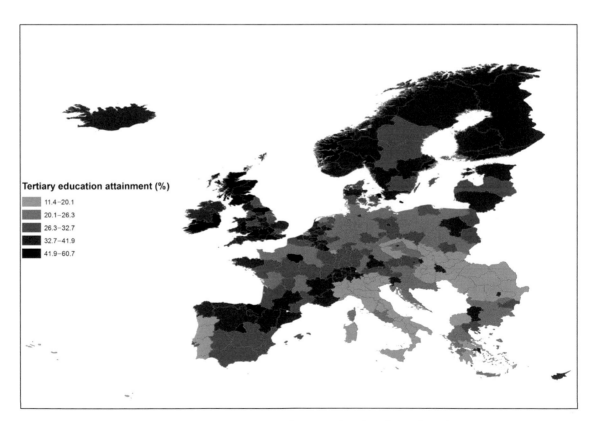

Figure 2.2 Total number of 25- to 64-year-olds having completed tertiary education as a percentage of all population aged 25–64

Source: Constructed by the authors using data from http://ec.europa.eu/eurostat

Figure 2.3 The city of Sheffield, UK, electoral wards

unemployment rates can inform discussions pertaining to local and national government policies.

Looking at the data sources briefly discussed in the previous section, we can identify relevant variables that can be used to address these attribute queries. In particular, from the publicly and freely available online data resources (see Appendix 2.1) it is possible to obtain socio-economic and demographic data from the census of population and other sources. For instance, it is possible to obtain information that could be used to address the attribute queries about unemployment, poverty and wealth. We will now give some examples of this type of attribute queries, using the city of Sheffield in the UK as an example. As we will be using the city of Sheffield as an example study area in other chapters of this book it is useful to also provide a map of the city here. Figure 2.3 shows a map of the city of Sheffield and the areas within it, according to its electoral ward geography, a UK local administrative geographical

Table 2.3 Electoral wards in the city of Sheffield, UK, with unemployment rates more than the national UK average (8.3%)

Electoral ward	Unemployment rate (%)
Manor	15.05
Burngreave	14.61
Castle	14.34
Southey Green	12.48
Firth Park	11.83
Park	11.48
Sharrow	11.01
Norton	9.79
Netherthorpe	9.77
Nether Shire	9.48
Darnall	9.28

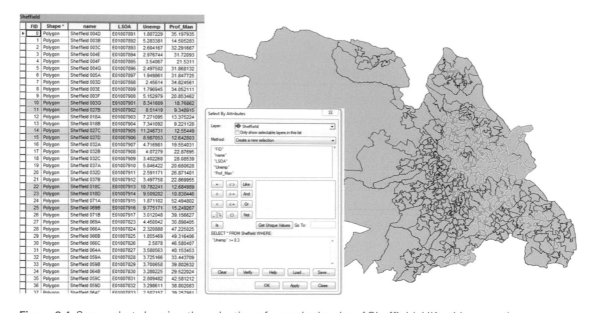

Figure 2.4 Screenshot showing the selection of areas in the city of Sheffield, UK, with unemployment rates more than the national UK average (8.3%), 2011

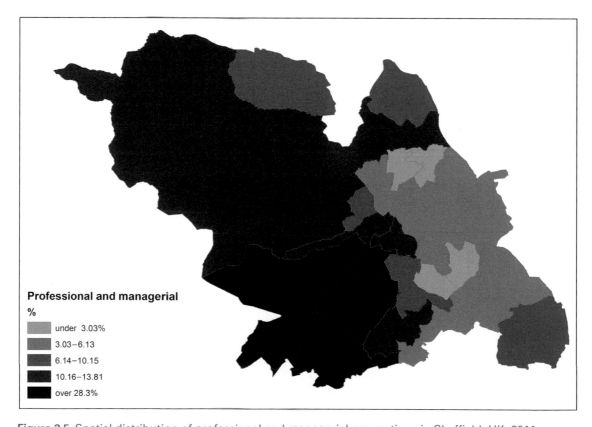

Professional and managerial %

- under 3.03%
- 3.03–6.13
- 6.14–10.15
- 10.16–13.81
- over 28.3%

Figure 2.5 Spatial distribution of professional and managerial occupations in Sheffield, UK, 2011

Table 2.4 Top five electoral wards with highest percentages of professional and managerial occupations in the city of Sheffield, UK

Electoral ward	Percentage of professional and managerial
Ecclesall	28.3
Broomhill	25.1
Hallam	25.1
Nether Edge	19.6
Dore	19.1

unit for which there is a wide range of data publicly available.

So, using the city of Sheffield as an example, Table 2.3 uses data from the census of UK population to address the unemployment question, whereas Table 2.4 utilises census data on socio-economic class to help

address the question regarding the most affluent areas in the city.

Digital boundary data enable us to display this form of information on the map (you can find out more information on how attribute data can be joined with digital boundary data throughout the linked practicals that accompany this textbook). Using the query tools described in Practical B it is possible to identify and highlight neighbourhoods in Sheffield that meet criteria of interest (Figure 2.4) or in order to explore spatial patterns (Figure 2.5).

Spatial queries: buffers and overlay operations

We can now move on to examples of spatial queries that can only be answered by processing information

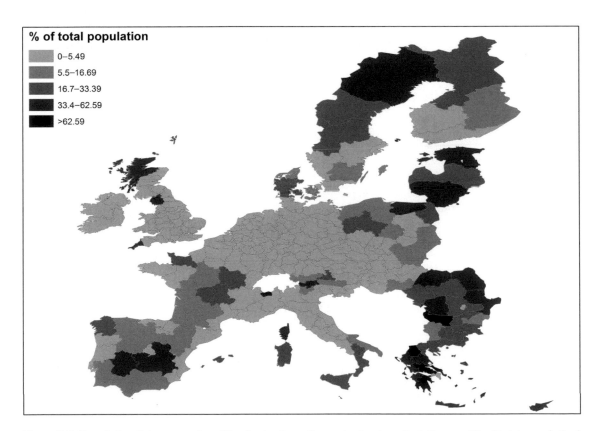

Figure 2.6 Population living more than 60 minutes from the nearest university in Europe (% of total population)

Source: Constructed by the authors using data from Annoni and Kozovska (2010)

on the *actual location* and *topographic relationships* of geographical features. For example, and with regard to the tertiary students example briefly discussed in the previous section, we could consider how far European populations live from their nearest university. Figure 2.6 addresses such a spatial query and is the output of spatial analysis involving a GIS-based calculation of how far populations in different parts of Europe live from a university in terms of travel time. This calculation also involves a consideration of the road network and relevant analysis which is discussed in more detail in Chapter 4.

We can also consider spatial queries in relation to the issues of retail demand analysis and service accessibility, building on the previous section's discussion of *attribute (aspatial) queries*. Using a function known as a *spatial join*, it is possible to combine the characteristics associated with individual grocery stores with the characteristics associated with the neighbourhoods that they serve. Table 2.5 gives an extract of the output of such a spatial join. Using this information it is

also possible to produce summary data on the average unemployment rate and social class profile of the immediate neighbourhood served by each grocery outlet.

It is also possible to use an overlay operation known as *point-in-polygon* in order to determine which (and calculate how many) grocery stores fall within each neighbourhood (see Figure 2.7).

This analysis and discussion is also very relevant to the issues of food deserts which was briefly discussed in the introductory chapter. Table 2.5 gives an indication of the types of areas and accessibility (from a socio-economic affordability view point) in relation to different supermarkets. This analysis can be further enhanced by considering the actual distance between each area and a grocery outlet in order to provide a proxy for *geographical accessibility* (these issues are discussed in greater detail in Chapters 8, 9, 10 and 11).

It is also possible to use spatial overlay techniques in order to analyse and map the distance between each neighbourhood and its nearest grocery store. We give an example here using a layer of geographical units

Table 2.5 Spatial join of grocery stores with electoral ward data

Store name	Neighbourhood		Neighbourhood	Professional and managerial (%)	Routine and semi-routine occupations (%)	Unemployment rate (%)	Never worked (%)
M&S	Castle	*Join*	Castle	4.6	14.1	14.3	8.7
Tesco	Netherthorpe	*fields*	Netherthorpe	10.8	5.5	9.8	5.5
Sainsbury's	Sharrow		Sharrow	8.9	5.9	11.0	8.3
Co-op	Hallam		Hallam	25.1	2.6	2.5	0.6

Output

Store name	Neighbourhood	Professional and managerial occupations (%)	Routine and semi-routine occupations (%)	Unemployment rate (%)	Never worked (%)
M&S	Castle	4.6	14.1	14.3	8.7
Tesco	Netherthorpe	10.8	5.5	9.8	5.5
Sainsbury's	Sharrow	8.9	5.9	11.0	8.3
Co-op	Hallam	25.1	2.6	2.5	0.6

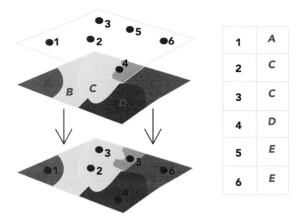

1	A
2	C
3	C
4	D
5	E
6	E

Figure 2.7 Point-in-polygon overlay

in the UK known as Lower Layer Super Output Areas (LSOAs). Using the ArcGIS ArcTool Box overlay and spatial join this layer was combined with the groceries store layer in order to calculate the distance to the nearest grocery from each geographical area. The output of this operation is shown in Figure 2.8. Nevertheless, it is useful to highlight that there are alternative (and conceptually equivalent) software approaches to buffer and distance computation. For instance, if there is no need to visualise buffers then the ArcGIS 'compute distances' function might be a faster alternative to the 'buffer distance' approach. The accompanying practical of this chapter (Practical B) illustrates the 'select by distance' approach in ArcGIS, which involves a relatively simpler series of steps to implement, compared to alternative approaches.

Such analysis can provide the basis for a more sophisticated investigation involving more complex buffer and overlays (Cheng *et al.*, 2007; Benoit and

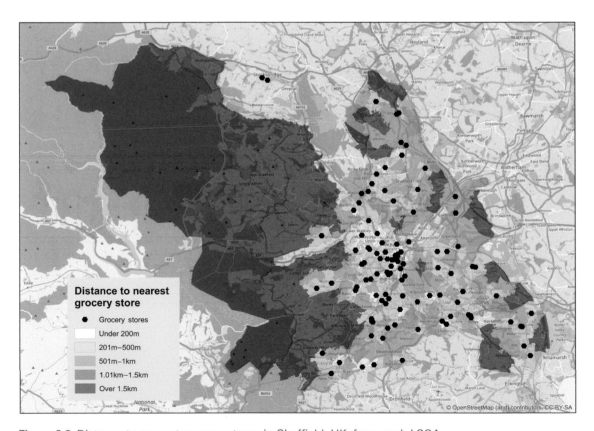

Figure 2.8 Distance to nearest grocery stores in Sheffield, UK, from each LSOA

Clarke, 1997). For example, Cheng *et al.* (2007) considered a similar but more extensive set of layers (see Table 2.6) in order to identify the best locations for the development of new shopping malls in Hong Kong.

Using spatial overlay they estimated the average distance from potential major demand points (see Table 2.7) as well as demand and total household income for existing shopping malls (see Table 2.8).

Table 2.6 Geographical layers considered in a GIS approach to shopping mall selection

Layer and sub-layer	Description of the features	Symbolisation
City map	Draw an outline of the city	Coastlines covering the whole city
District areas	Divide the city into ten district areas	Ten different colours representing the 10 districts
Roads	Draw roads on the city map	Lines with brown colour
Streets	Draw streets on the city map	Lines with red colour
Railways	Draw railways on the city map	Lines with dark green colour
Existing shopping malls	Locate the existing shopping malls on the city map	Points with labels in alphabet and number as E1, E2, etc.
Potential locations	Locate the potential sites on the city map	Points with labels in alphabet and number as S1, S2, etc.
Household incomes in each district	Add up the average monthly household incomes generated in each district	Points at the centres of districts with labels in alphabet and number as H1, H2, etc.
Demand size in each district	Add up the average monthly demands (spending) in each district	Points at the centres of districts with labels in alphabet and number as A1, A2, etc.
Household incomes in each estate	Add up the average monthly household incomes generated in each estate point	Points with labels in alphabet and number as M1, M2, etc.
Demand size in each density point (demand point)	Add up the average monthly demands (spending) in each density point	Points with labels in alphabet and number as D1, D2, etc.

Source: Cheng *et al.* (2007)

Table 2.7 Average distance from major demand points

Potential location	Total distance (kilometre)	Average distance (kilometre)
Aberdeen	258	11.73
Central	210	9.54
Causeway Bay	219	9.96
Kennedy Town	232	10.54
Tsimshatsui	192	8.72
Tseung Kwan O	256	11.64
Tai Po	340	15.47
Tuen Mun	455	20.70

Source: Cheng *et al.* (2007)

Table 2.8 Estimating potential demand and household income by shopping mall coverage area

Potential location	Coverage demand area	Existing super mall	Coverage demand (mean population)	Coverage (mean household income)
Aberdeen	Aberdeen, Kennedy Town, Central, Wanchai.	None	179,817	28,400
Central	Central, Kennedy Town, Aberdeen, Wanchai, Tsimshatsui.	Admiralty (1)	200,258	25,661
Causeway Bay	Wanchai, Central, Aberdeen, Tsimshatsui, North Point.	Causeway Bay (1)	289,890	27,929
Kennedy Town	Kennedy Town, Central, Aberdeen.	None	184,041	29,201
Tsimshatsui	Tsimshatsui, Kowloon City, Shamshuipo, Kwun Tong, Central, Wanchai, North Point.	Tsimshatsui (3)	87,733	23,600
Tseung Kwan O	Tseung Kwan O, Wong Tai Sin, Kowloon City, Kwun Tong	None	429,024	18,163
Tai Po	Tai Po, Northern District, Shatin	None	412,723	18,877
Tuen Mun	Tuen Mun, Yuen Long	None	468,951	16,500

Note: Number in parenthesis denotes the number of existing super mall(s) close to the potential location.

Source: Cheng *et al.* (2007)

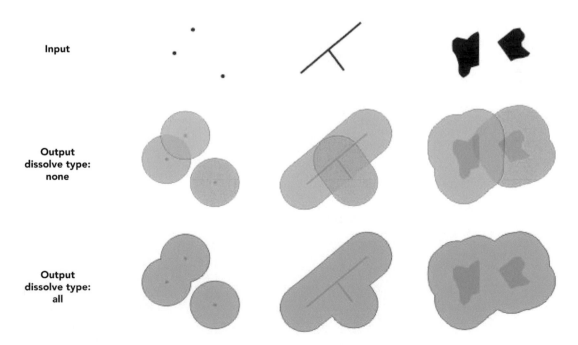

Figure 2.9 Point, line and polygon features and buffers around these features

Source: ArcGIS help system

In the discussion below we summarise some of the types of GIS operations that can be employed to address spatial queries as shown in the remainder of this section (more sophisticated approaches that involve network analysis are discussed in more detail in Chapter 4). These are all core GIS and spatial analysis techniques and feature heavily in the accompanying practicals.

Buffering

Buffering creates a polygon surrounding a feature (such as a point or line) at a specified distance. The top side of Figure 2.9 shows the features point, line and polygon, respectively. The middle part of Figure 2.9 shows the shape of the buffers associated with each in turn, and the bottom part describes the shape

of the resulting buffers when the 'dissolve' option is selected. These buffers can be useful in identifying feature space that falls within a set distance threshold (proximity) of the point, line or polygon of interest, such as a catchment area around a retail store (point feature), or an area likely to be affected by a particular pollutant (line or polygon). They can also be combined with spatial queries in order to select and identify all features that fall within or outside the boundary of the buffer.

Intersect

Intersect is used to identify features or areas that intersect (overlap) each other. It is similar to 'select by location' in that it identifies all features that fall within the intersection of two or more layers, such as

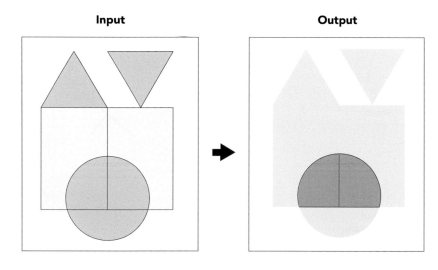

Figure 2.10 The intersect procedure

Source: ArcGIS help system

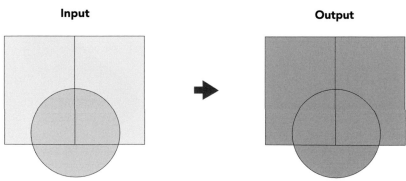

Figure 2.11 The union procedure

Source: ArcGIS help system

all properties falling within a buffer around a hazardous location. Intersect produces a new output layer containing only the features falling within the intersection. Figure 2.10 gives an illustrative example of the intersect function when applied on polygon layers.

Union

This function combines two or more layers or sets of features to create an output layer containing both sets of features. For example, this might be used to combine a layer containing major roads with a layer that represents minor roads, thus producing an output layer containing all roads (see Figure 2.11).

The Modifiable Areal Unit Problem (MAUP) and ecological fallacy

One of the issues arising from the use of GIS and spatial analysis is the so-called MAUP, arising from the use of arbitrary or artificial units of spatial reporting on continuous geographical phenomena. For example, the spatial patterns of *professional and managerial occupations* shown in Figure 2.3 and discussed earlier in this chapter are based on the use of the electoral ward geography of Sheffield. However, the occupational classification refers to individuals in unique residential locations within each electoral ward, which were then aggregated up to the ward level and divided by the total population in each area in order to compute the percentages. The summary values (e.g. whether an area has a percentage of less than 15%) are affected by the choice of geographical boundaries. For example, if instead of electoral wards, we used postal sectors, the geographical patterns shown in Figure 2.3 might have been different. In other words, by *modifying* the boundaries of the *area unit* there might be different geographical patterns as well as results when undertaking further quantitative or statistical analysis. In particular, the MAUP could introduce considerable statistical bias when the summary values are used in statistical analysis to explore geographical associations

between different variables. These issues were first identified by Gehlke and Biehl (1934) but the term MAUP was coined by Openshaw and Taylor (1979), who evaluated systematically the variability of statistical analysis results with different sets of geographical boundaries (Wong, 2009; also see Openshaw, 1984a; Openshaw and Rao, 1995; Taylor *et al.*, 2003).

Another relevant issue and concept is that of the ecological fallacy which refers to the inappropriate assignment to an individual of a property or value that has been calculated or estimated for a group. It also applies when inferences about the relationships between individual characteristics are made based on data about geographical areas (Openshaw, 1984b; Jargowsky, 2005). For example, there have been reports following the UK referendum on EU membership that there has been an increase in anti-immigrant hate crime in areas that voted to leave the EU (Stone, 2016). While it might be appropriate to suggest, based on these statistics and reports, that areas with high percentages of 'leave votes' are also likely to have relatively higher anti-immigrant hate crime rates, it would be ecological fallacy to suggest that individual people who voted 'leave' in the referendum in these areas commit these crimes.

More advanced forms of spatial analysis in GIS

The MAUP and ecological fallacy are issues that always need to be acknowledged, and if possible, addressed in GIS analysis. There are a number of advanced and sophisticated forms of spatial analysis that can be employed to that end. These include kernel density estimation, where a circular area (the kernel) of a set bandwidth is created around a geographical feature (e.g. a building or the location of a crime or health-related incident) and then the density within this area is calculated and analysed. Such advanced forms of spatial analysis are increasingly available in GIS software packages including ArcGIS (for an example of how kernel density surfaces can be created in ArcGIS using John Snow's cholera data that we also discussed in the introductory chapter see Goranson (2012)). In addition, more sophisticated approaches (which are not typically part of standard GIS software packages) include Geographically Weighted Regression (Brunsdon *et al.*, 1996; Fotheringham *et al.*, 2002; Paez *et al.*, 2002), the Local Indicators of Spatial Association (Anselin, 1995), zone design (Alvanides and Openshaw, 1999) and spatial microsimulation (Ballas *et al.*, 2005). Some of these methods (and application examples) are discussed in more detail in the second part of this book.

Concluding comments

The above spatial analysis examples and outputs give a small flavour of what is possible with GIS. Similar examples can be thought of in a wide range of social science contexts including estimating the catchment areas and performance indicators for schools, estimating likely demand for health services and the use of GIS for hospital planning and health service provision, etc. Part II of the book discusses extensively more sophisticated examples of GIS spatial analysis applications and case studies in a social science context.

Accompanying practical

This chapter is accompanied by Practical B: Spatial queries and attribute data. Continuing our introduction to ArcGIS, the practical gives you the opportunity to work further with the transport data sets introduced in Practical A and relating to the city of Chicago. You will gain experience in carrying out powerful spatial and attribute queries by exploring the relationship between different spatial data layers.

Appendix 2.1 Examples of data resources

United States Census Bureau: www.census.gov
Eurostat: http://ec.europa.eu/eurostat
United States Government Open Data: www.data.gov
European Union Open Data Portal: https://data.europa.eu/euodp/en/data
UK Government Open Data: https://data.gov.uk

Australia Government Open Data: www.data.gov.au/
Canada Open Data Government Portal: http://open.canada.ca
New Zealand Government Open Data: https://data.govt.nz
Japan City Open Data Census: http://jp-city.census.okfn.org/
GB Ordnance Survey: www.ordnancesurvey.co.uk
Open Street Map: www.openstreetmap.org/#map=7/53.041/-1.362
EDINA: www.edina.ac.uk/
ESRI ArcGIS software maps database: www.esri.com
European Regional Yearbook: http://ec.europa.eu/eurostat/statistics-explained/index.php/Eurostat_regional_yearbook
European Values Survey: www.europeanvaluesstudy.eu
ILO: www.ilo.org/global/statistics-and-databases/lang--en/index.htm
NASA MODIS sensor: http://modis.gsfc.nasa.gov
Socioeconomic Data and Applications Center (SEDAC) of the Columbia University, New York: http://sedac.ciesin.columbia.edu/data/collection/gpw-v3
The World Bank: http://data.worldbank.org
WHO: www.who.int/gho/en
Social and Spatial Inequalities group: www.sasi.group.shef.ac.uk
Worldmapper: www.worldmapper.org
GIS ESRI library website: http://training.esri.com/campus/library/index.cfm
National Trust Names: www.nationaltrustnames.org.uk/Surnames.aspx

References

All website URLs accessed 30 May 2017.

Alvanides, S., & Openshaw, S. (1999) Zone design for planning and policy analysis. *Geographical Information and Planning*, 299–315.

Annoni, P., & Kozovska, K. (2010) *EU Regional Competitiveness Index 2010*, JRC Scientific and Technical Report, EUR 24346. Publication Office of the European Union, Luxembourg. Available from: http://publications.jrc.ec.europa.eu/repository/bitstream/JRC58169/rci_eur_report.pdf.

Anselin, L. (1995) Local Indicators of Spatial Association – LISA. *Geographical Analysis*, *27*, 93–115.

Ballas, D., Rossiter, D., Thomas, B., Clarke, G. P., & Dorling, D. (2005) *Geography Matters: Simulating the Local Impacts of National Social Policies*, Joseph Rowntree Foundation contemporary research issues, Joseph Rowntree Foundation, York, ISBN 1 85935 265 0 (paperback). Free pdf copies available from: www.jrf.org.uk/file/36059/download?token=NkTWwksy&filetype=download.

Benoit, D., & Clarke, G. P. (1997) Assessing GIS for retail location planning. *Journal of Retailing and Consumer Services*, *4*(4), 239–258.

Brovelli, M. A., Minghini, M., & Zamboni, G. (2016) Public participation in GIS via mobile applications. *ISPRS Journal of Photogrammetry and Remote Sensing*, *114*, 306–315.

Brunsdon, C. A., Fotheringham, A. S., & Charlton, M. E. (1996) Geographically weighted regression: a method for exploring spatial non-stationarity. *Geographical Analysis*, *28*, 281–298.

Cheng, E. W., Li, H., & Yu, L. (2007) A GIS approach to shopping mall location selection. *Building and Environment*, *42*(2), 884–892.

Croitoru, A., Crooks, A., Radzikowski, J., & Stefanidis, A. (2013) Geosocial gauge: a system prototype for knowledge discovery from social media. *International Journal of Geographical Information Science*, *27*(12), 2483–2508.

Fotheringham, A. S., Brunsdon, C., & Charlton, M. E. (2002) *Geographically Weighted Regression: The Analysis of Spatially Varying Relationships*, Wiley, Chichester.

Gehlke, C. E., & Biehl, K. (1934) Certain effects of grouping upon the size of the correlation coefficient in census tract material. *Journal of the American Statistical Association*, *29*, 169–170.

Goodchild, M. F. (2007) Citizens as sensors: the world of volunteered geography. *GeoJournal*, *69*, 211–221.

Goranson, C. (2012) *GIS Spatial Analyst Tutorial using John Snow's Cholera Data*, YouTube video. Available from: www.youtube.com/watch?v=isVD8u6WrG4.

Huang, Q., & Xu, C. (2014) A data-driven framework for archiving and exploring social media data. *Annals of GIS*, *20*(4), 265–277.

Jargowsky, P. A. (2005) The ecological fallacy, in K. Kempf-Leonard (ed.) *Encyclopaedia of Social Measurement*, Academic Press, San Diego, CA, 715–722.

Kim, M. G., & Koh, G. H. (2016) Recent research trends for geospatial information explored by Twitter data. *Spatial Information Research*, *24*(2), 65–73.

Openshaw, S. (1984a) The modifiable areal unit problem. CATMOG 38. GeoBooks, Norwich.

Openshaw, S. (1984b) Ecological fallacies and the analysis of areal census data. *Environment and Planning A*, *16*, 17–31.

Openshaw, S., & Rao, L. (1995) Algorithms for reengineering 1991 census geography. *Environment and Planning A*, *27*, 425–446.

Openshaw, S., & Taylor, P. J. (1979) A million or so correlation coefficients: three experiments on the modifiable areal unit problem. *Statistical Applications in the Spatial Sciences*, *21*, 127–144.

Paez, A., Uchida, T., & Miyamoto, K. (2002) A general framework for estimation and inference of geographically weighted regression models: 2. Spatial association and model specification tests. *Environment and Planning A*, *34*(5), 883–904.

Rees, P., Martin, D., & Williamson, P. (2002) *The Census Data System*, Wiley, Chichester.

Stone, J. (2016) Brexit: surge in anti-immigrant hate crime in areas that voted to leave EU: police statistics show hate crimes to have tripled in some of the most eurosceptic parts of Britain. *The Independent*. Available from: www. independent.co.uk/news/uk/crime/brexit-hate-crime-racism-immigration-eu-referendum-result-what-it-means-eurosceptic-areas-a7165056.html.

Sui, D., & Goodchild, M. (2011) The convergence of GIS and social media: challenges for GIScience. *International Journal of Geographical Information Science*, *25*, 1737–1748.

Sui, D., Elwood, S., & Goodchild, M. F. (eds) (2013) *Crowdsourcing Geographic Knowledge*, Springer, Dordrecht.

Taylor, C., Gorard, S., & Fitz, J. (2003) The modifiable areal unit problem: segregation between schools and levels of analysis. *International Journal of Social Research Methodology*, *6*(1), 41–60.

Wong, D. (2009) The modifiable areal unit problem (MAUP), in A. S. Fotheringham & P. Rogerson (eds) *The SAGE Handbook of Spatial Analysis*, SAGE, Los Angeles, 105–124.

Zhai, S., Xu, X., Yang, L., Zhou, M., Zhang, L., & Qiu, B. (2015) Mapping the popularity of urban restaurants using social media data. *Applied Geography*, *63*, 113–120.

3 | Thematic mapping, GIS and geovisualisation

LEARNING OBJECTIVES:

- Thematic map making in the social sciences
- Joining geographic and associated attribute data
- Reference maps
- Creating choropleth maps
- Graduated symbol and chart-type maps
- Human-scaled visualisation and human cartography

Introduction

> You can say it, you can prove it, you can tabulate it, but it is only when you show it that it hits home.
>
> (Dorling, 2007a: 13)

One of the most powerful and creative aspects of GIS is map making and visualisation. We have already given a flavour of GIS capability in the discussion of vector and raster data models in Chapter 1 and of the examples of the outputs of spatial analysis in Chapter 2. In particular, we have already presented examples of one of the most common mapping approaches, the so-called 'choropleth' map, where areas are shaded or patterned in proportion to the measurement of a variable being displayed on the map. Most people are familiar with these types of maps as they are often used in media reports to depict the geographical distribution of socio-economic data.[1]

The creation of thematic maps involves the use of suitable contrasting symbols to portray geographic differences and is underpinned by key principles and a long tradition of cartographic theory and practice (e.g. Bertin, 2011; Brewer, 2005; Dent et al., 2008; Kraak and Ormeling, 2010; MacEachren, 2004; Monmonier, 1996; Perkins, 2003, 2004; Slocum et al., 2008). Table 3.1 summarises what can be described as key elements characterising good mapping practice.

Other issues that need to be considered when designing maps include the levels of measurement of the variables to be mapped, alternative ways of classifying data, colouring schemes, the amount and type of information to display on the map and issues of layout design. In addition key points to be considered are the overall positioning of map elements, the ways they

Table 3.1 Designing a 'good map'

Key mapping principles – what makes a good map	
Clarity	The map focuses on its purpose with as little distraction as possible.
Order	The elements of the map are organised to guide the map reader to accomplish the intended purpose.
Balance	Elements are distributed to give the map a feeling of 'evenness'.
Contrast	Important elements stand out clearly against less important ones.
Unity	The various elements come together to make the map appear as a single coherent whole.
Harmony	The elements all seem to fit together naturally.

Source: After wiki.gis.com (2016a, 2016b)

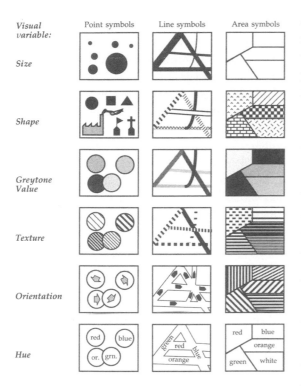

Figure 3.1 The six principal visual variables

Source: After Monmonier (1996)

classic and influential work (1996) illustrates how map symbols can differ in size, shape, greytone value, texture, orientation and hue in the form of the six principal visual variables depicted in Figure 3.1. Monmonier suggests that each of these visual variables is important for portraying one kind of geographic difference. In particular, shape, texture and hues are most effective in showing qualitative differences, for example among land uses, dominant religion or most common lifestyle group type in an area. For quantitative differences, size is more suited to showing variation in amount or count, such as the number of television or internet users by market area, or numbers of burglaries by neighbourhood. Greytone value is preferred for displaying difference in rate or intensity. Symbols varying in orientation are most appropriate for representing directional occurrences such as wind direction, migration, travel-to-work or travel-to-shop flows.

In this chapter (and accompanying practicals) we consider all these issues in a social science context. We show how thematic maps can be created and we provide an overview and examples of the most commonly used types of mapping used in the social sciences. We begin with a discussion of the different types of socioeconomic and demographic variables and data (known as 'attribute data' or 'associated tabular data') that can be joined with geographical data. We also consider the latest trends and technologies, including online mapping. Finally, we consider criticisms of these commonly used conventional ways of mapping in the social sciences and we present a case for alternative human-scaled visualisations and cartograms which are increasingly used instead of conventional maps.

can be perceived by the human eye (and map readers of different backgrounds) and the choice of symbols to represent them. To that end, Jacques Bertin (1967) identified seven main categories of visual variables: position, size, shape, value, colour, orientation and texture (wiki.gis.com, 2016a, 2016b). Bertin's work has been the basis for further modification and discussion of cartographic principles. For example, Monmonier's

Geographic and associated tabular/attribute data

As discussed in Chapter 1 there are two main data types, vector and raster. The choice of the appropriate mapping approach depends on the type of geographical data being used as well as the audience at which the map is aimed. In any case, the first step in building a data set that can be used for mapping is to combine geographical data sets (either raster or vector) with an associated attribute data table containing descriptive information about geographic objects and features, including qualitative and quantitative variables that can be mapped.

For example, a *polygon* vector layer of administrative electoral wards of a city (such as the one discussed in the previous chapter) or census tracts could be combined with an associated attribute table containing information on the total population in each area, as well as the population by age, sex, economic activity (e.g. whether in full employment, unemployed), rural/urban classification, etc. Similarly, a *point* vector data layer of supermarkets, other retail outlets or businesses could be joined with an attribute data table containing information on the size of the establishment (e.g. floor-space in squared feet), number of employees, number of parking spaces, total turnover, revenues, etc.; a *line* vector layer of the road network of a city could be joined to a data table containing information on the road type, capacity, condition and estimated travel time.

In the case of the raster model data, as discussed in Chapter 1, each pixel is assigned one value which might be a categorical or continuous variable. In the social sciences it is much more meaningful and efficient in terms of computational and storage requirements to almost always use vector data for analysis and mapping. Raster data such as satellite imagery are typically used as a background layer to thematic maps of socio-economic data. They can also be used when there is a need to analyse socio-economic data in relation to environmental indicators for which continuous data are available at small-area level (e.g. exploring the relationship between asthma cases, populations at risk and atmospheric pollution). There are also cases when socio-economic or demographic data can be available at small-area grid level, as is the case with the 'Gridded Population of the World project' (Center for International Earth Science Information Network, 2015) which provides very small-area grid level estimates of population for the world using data from censuses of housing and population (we show how this data set can be used for mapping and cartogram creation in the last section of this chapter).

Before we describe how attribute data can be mapped using GIS it is important to point out that the choice of a mapping approach depends on the type and measurement scale of each attribute for a spatial entity that is being mapped (as is also the case with statistical analysis). Attribute data are typically distinguished between *nominal*, *ordinal*, *interval* and *ratio*, following the original work of the American psychologist Stanley Smith Stevens which was published in the journal *Science* back in 1946 (Stevens, 1946). *Nominal* attributes have no quantitative value and are typically used to label spatial entities. For example, these can be the names of countries, regions or administrative units within a city (for polygon data), the name of a road (for line data) or the names of the chain to which a supermarket branch belongs (for point data). When the value of the variable being mapped can be ordered, but we do not know the exact differences between the values, then the type of attribute is known as *ordinal*. For example, regions or cities can be ranked in terms of their perceived quality of life and the area ranked number 1 is considered to be a better place to live compared to that ranked number 2, but we do not know the exact quantitative difference. Attributes are *interval* when we know not only the order but also the exact difference between the values, but when there is no meaningful value of zero. For example, temperature in Celsius or Fahrenheit is an *interval* attribute. The distance between 25 and 30 Celsius is the same as that between 30 and 35. However, in *interval* measurement ratios are not meaningful: we cannot say that 40 degrees Celsius is twice as much as 20. And the value of zero is arbitrary and artificial (zero temperature does not mean that there is no temperature). Attributes are *ratio* when differences between values make sense and at the same time there is a meaningful value of zero.

Total population, gross national product, total government consumption and total grocery floor-space are examples of ratio variables in the social sciences.

These levels of measurement are now well established in the social sciences. However, it should be noted that there have been some concerns with regard to their suitability for use in mapping and cartography (Chrisman, 1998; Forrest, 1999). For example, directional data are numerical but they cannot be ordered in a meaningful sense, due to their cyclic nature. For instance, 355 degrees is not greater or less than 5 degrees: it is just a different direction. Also, with regard to the distinction between ratio and interval scale described above, when it comes to thematic mapping (discussed below), there is not much consensus on how shading intervals might be chosen. Although the levels of measurement described above are widely adopted in geography and cartography, it is important for users of GIS in the social sciences to be aware of these issues. Chrisman (1998) presents a very interesting and comprehensive review of scales of measurement in cartography, highlighting their limitations and making a case for a broader and more suitable (for cartographic use) measurement framework.

In Chapter 2 we provided some examples of data resources for geographic and attribute data (typically socio-economic and demographic data). We now draw on some of these resources to provide examples of attribute data that can be joined with geographic data to be mapped and analysed (it is also possible for such data to be the output of original collection and processing by the GIS user). Table 3.2 shows an example of attribute data extracted for administrative areas in the city of Sheffield in the UK (similar to the data used in the previous chapter) containing information on total population, their mean age and the unemployment rate.

All data layers in a map have a field that can be used as a key in order to join data with associated attribute information (as shown in the practical exercises accompanying this chapter). Similar combinations of geographic and attribute data are possible for point and line data. Once GIS layer and associated attribute data are joined together it is possible to create thematic maps and other visual representations of the characteristics of geographic features. The remainder of this chapter discusses some of these options. We begin with the simplest form of mapping, reference maps.

Mapping location: reference maps

The simplest form of mapping that can be carried out using GIS is that of reference maps where the emphasis is on using suitable graphics to show the location of geographical entities. Reference maps include road maps, tourist maps and guidebooks. In the social sciences reference maps highlight information and the name of geographical features that are of relevance

Table 3.2 Attribute data for a selection of Sheffield, UK, electoral wards

Ward name	Population	Mean age	Unemployment rate (%)
Arbourthorne	19,133	38.9	6.9
Beauchief and Greenhill	18,815	41.6	6.3
Beighton	17,939	40.7	3.5
Birley	16,943	42.0	4.9
Broomhill	16,966	30.1	2.1
Burngreave	27,481	32.6	9.1
Central	36,412	27.0	3.8
Walkley	21,793	35.9	4.2
West Ecclesfield	17,699	43.1	3.5
Woodhouse	17,450	42.2	5.0

Figure 3.2 A reference map of Golden Gate, California, US, created using ArcGIS

Source: www.arcgis.com/home/webmap/print.html

to a social science problem (e.g. names of countries, region, location and name of grocery outlets, schools, hospitals, etc.). In addition, the decision on what and how to display depends on the audience and purpose of the map. For example, and with regard to the social science problems discussed in the previous chapters, a suitable reference map in relation to the likely demand of retail products would show the location of grocery stores, as well as the road network in a city and information on public transport (e.g. bus stops, railway stations). Creating reference maps in GIS is fairly straightforward and we have already used some illustrative examples in the previous chapters (e.g. the map shown in Figure 2.8, showing the location of grocery stores in the city of Sheffield). The emphasis in these maps is on the geographical pattern and labelling of the spatial entities being mapped and on the symbols being used to illustrate them. Proprietary GIS software has excellent capabilities to create high-quality reference maps. A simple example is shown in Figure 3.2.

The decision about what geographical features to map and which symbols to use depends on the purpose of the map and intended map users. GIS software packages have extensive libraries of symbols that can be used in reference maps.

In addition, such maps are increasingly readily available from online resources such as Google Maps and Bing. Figure 3.3 shows such a map generated via Google Maps for central London (generated by searching for 'groceries' in Google Maps). This map shows the location of all grocery stores in central London, as well as information on the road network and rail and underground train stations. The map is the product of a combination of different point, line and polygon layers.

Similarly, a suitable reference map in relation to the local council and national government policy discussed in the previous chapter could include information on the location of organisations related to a particular social policy issue. An example of such a reference map is shown in Figure 3.4, which depicts the location of charity organisations supporting the homeless in central London (again using Google Maps).

There is also a rapidly increasing number of reference maps created via crowd-sourced mapping services such as OpenStreetMap and PublicEarth, which are dynamically updated. We make use of OpenStreetMap throughout the accompanying practicals as a base map.

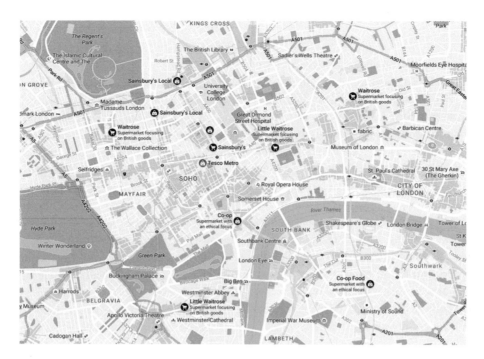

Figure 3.3 A reference map example: grocery stores in central London

Source: www.google.co.uk/maps/search/groceries/@51.5082749,-0.0985554,14.33z

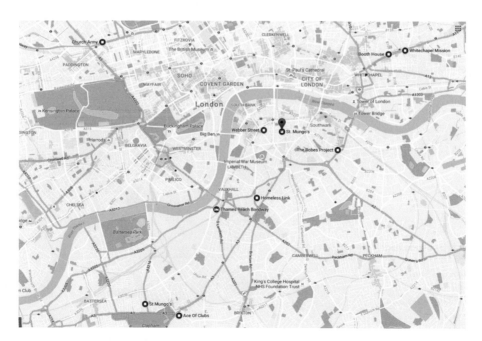

Figure 3.4 A reference map example: charities and shelters in support of the homeless in central London

Source: www.google.co.uk/maps/search/homeless+shelters/@51.5082749,-0.0985554,14z/data=!3m1!4b1

Thematic maps

One of the most powerful and attractive aspects of GIS in the social sciences is the ability to visualise geographic features such as those described in the previous section according to a particular socio-economic or political theme of interest, in order to create *thematic* maps. This involves a combination of geographic layers with associated attribute data such as those discussed in the second section of this chapter (also see accompanying practicals for examples of how such data can be joined with geographic data and mapped). There are many thematic mapping options and the choice depends on the type of layer file (vector or raster; polygon, line or point) and variable type (e.g. nominal or ratio) being mapped. In this section we present examples of the main types of thematic mapping in the social sciences.

Using graduated colours or patterns: choropleth mapping

One of the most common thematic approaches to representing the quantity of variables being mapped for polygon layers is the so-called choropleth map. The word choropleth is derived from the Greek words *choros* (meaning area or region) and *plethos* (meaning multitude). The choropleth approach involves the shading or patterning of polygons according to a categorical variable (such as geodemographic classifications) or continuous data such as an interval or ratio variable (e.g. population or income).

A simple approach involves assigning each value its own colour. This is particularly suitable for nominal or categorical data and in cases when there are only a few values. Figure 3.5 presents an example of a categorical variable being mapped for UK census Output Areas in Sheffield (OAs; areas of an average 125 households).

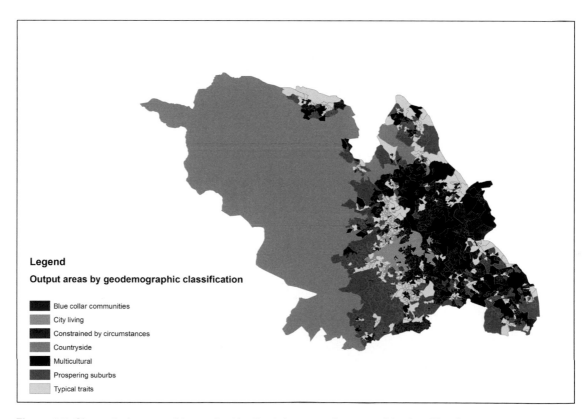

Figure 3.5 Choropleth map with nominal/ordinal data: geodemographic classifications

In particular, it presents a map of small areas classified on the basis of a combination of census data in order to label them according to the type of neighbourhood and lifestyle; this is a type of analysis of areas known as geodemographic classification (also see Chapter 5 for more details). There are 1,744 OAs in Sheffield and these are coloured in the map according to their geodemographic classification. For instance, the areas labelled as 'blue collar communities' (401 in total) are coloured in blue, whereas the areas labelled as 'countryside' (19 in total, but including very large sparsely populated areas that dominate the map) are coloured in green. The method used to create this map is based on the 'unique values, many fields' approach. There are seven geodemographic classification groups in total but it is possible for many areas to belong to the same group.

Figure 3.6 shows another example of categorical data, drawing on the Social Atlas of Europe project (europemapper.org) which we will revisit later in this chapter when discussing human cartographic

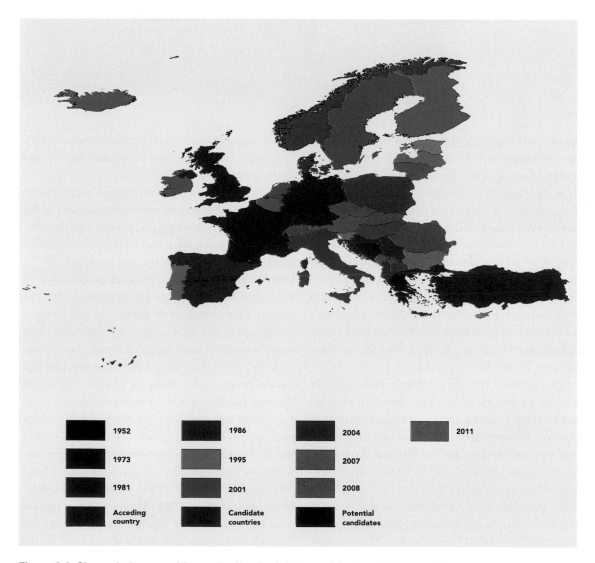

Figure 3.6 Choropleth map with nominal/ordinal data: social atlas of Europe rainbow

Source: Ballas *et al.* (2014)

Figure 3.7 Choropleth map with nominal/ ordinal data: Worldmapper regions © Worldmapper.org

Source: www. worldmapper.org/ region_map.html

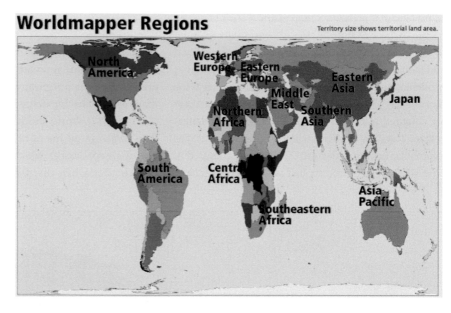

approaches to mapping in the social sciences. This time the value being mapped is the year of association of European countries with the European Union (ranging from membership to candidate and potential candidates). A rainbow scale is used to determine the colour hue for each state according to the year of association with the European Union. The more recent the formal association the nearer to the red end of the spectrum the hue is. In particular, the countries shown here are shaded using a rainbow colour scheme, starting with shades of dark red to demarcate the countries with the most recent association with the European Union and moving through to a shade of violet for the oldest members of the EU.

The rainbow colouring scheme is increasingly popular for data visualisations in the social sciences, as it is widely considered to have an attractive display. However, in some cases it can introduce visual confusion and obscure details in the data and there might be other issues with some readers. Eddins (2014) summarises the major relevant themes in the literature about the rainbow colour scheme (but this is also of wider relevance to any colouring scheme) and about the general principles of colour in scientific visualisation. It is also relevant to note here that there are a number of online platforms that can be used to create colour schemes for different types of data and map

readers such as *Colorbrewer* (http://colorbrewer2.org/), *Typebrewer* (www.typebrewer.org/) and *Indiemapper* (www.indiemapper.io/).

Figure 3.7 is a map of the world using a similar rainbow colouring scheme and produced in the context of the Worldmapper project (www.worldmapper.org) which we revisit later in the discussion of cartograms. All countries are classified into 12 world regions on the basis of development based on the United Nations Human Development Index. The regions were chosen to be geographically contiguous groups of territories that divided the world into roughly symmetrically balanced population groups, with no region containing fewer than one hundred million people. Hence there are 12 colour ranges on the maps and the shade of the colour within each range helps to identify territories and distinguish them from each other within the regions (five shades of each of the 12 regional colours are used to help differentiate territories within regions). The countries are ordered from poorest to richest by the Human Development Index published in 2004. Shades of dark red are used to demarcate the poorest territories, moving through the rainbow scale to a shade of violet for the best-off.

The choropleth approach can also be adopted to map quantitative variables (including continuous variables) such as ratio or interval. If there are relatively few values being mapped then it might be appropriate to adopt a similar approach to that described above and to assign each value a unique colour. However, in most cases with quantitative continuous variables (e.g. income, unemployment rates) there is a very wide range of values to be mapped and it is more appropriate for these to be grouped into classes. Figure 3.8 shows an example of such a map, depicting the geographical distribution of unemployment rates in the city of Sheffield in England. The particular approach adopted here is that of graduated colours, where the values of the variable being mapped are grouped into classes and each class is identified by a particular colour or shade of the same colour. There are a number of options available with regard to the number of classes and the techniques that can be used to create them. The ArcMap software default number of classes for the map shown in Figure 3.8 is five and the default classification scheme is 'natural breaks' (also see the graph in Figure 3.8). The latter is underpinned by a statistical technique that aims to create classes based on natural groups in the data distribution of the variable being mapped. In particular, the technique is aimed at minimising variance within classes and maximising the variance between classes.

The choice of classification scheme and number of classes determines which areas will fall into each class and what the map (and geographical patterns of the variable being mapped) will look like. In other words, by changing the number of classes and classification schemes you can create very different maps providing alternative messages that might inform policy-relevant decisions. For instance, the map shown in Figure 3.8 can be used to make decisions regarding the social policy and area-based priorities of local, regional and national government with regard to where the hot spots of unemployment in the city might be. There are three areas with the darkest shade in the map belonging in the top class (7.5%–9.1%). Figure 3.9 shows what happens if the same variable is mapped on the basis of a different classification method. The classification method used here (also see graph in Figure 3.9) aims to assign the same number of data values (or areas) to each class. Comparing Figures 3.8 and 3.9, we can observe that the lower end value of the top class is 6.9% in Figure 3.9 (compared to 7.5% in Figure 3.8) and there are an additional two areas in the top class coloured with the darker shade in Figure 3.9 (five in total, compared to three in Figure 3.8).

Other classification schemes that can be used to divide the geographical areas mapped include the equal interval method (dividing the range of attribute values into equal-sized sub-ranges). It is also possible to manually define the class breaks and intervals. More sophisticated methods include the geometrical interval (creating class breaks based on class intervals that have a geometrical series) and standard deviation (defining breaks on the basis of how much an area's values vary from the mean of the variable distribution). In any case, it is important to consider the differences in the geographical pattern of the variable being mapped when using alternative classification schemes and number of classes.

Using graduated/proportional symbol classes

Another approach to mapping quantitative variables is to use graduated symbols. Figure 3.10 shows how the geographical distribution of the variable of unemployment rate mapped in the previous choropleth mapping examples can be depicted using graduated symbols. As with the choropleth map example, the values of unemployment rate are grouped into classes and represented by a symbol (in this example a circle). The symbol is then resized in proportion to the magnitude of the variable being mapped. As is the case with the choropleth maps discussed above, there are similar issues regarding the classification techniques and number of classes that need to be considered when using the graduated symbol approach. In addition, a potential difficulty with the graduated symbol mapping option is that when there are too many values then differences between symbols can become indistinguishable. Further, the symbols for high value can become too large and obscure other symbols in neighbouring areas.

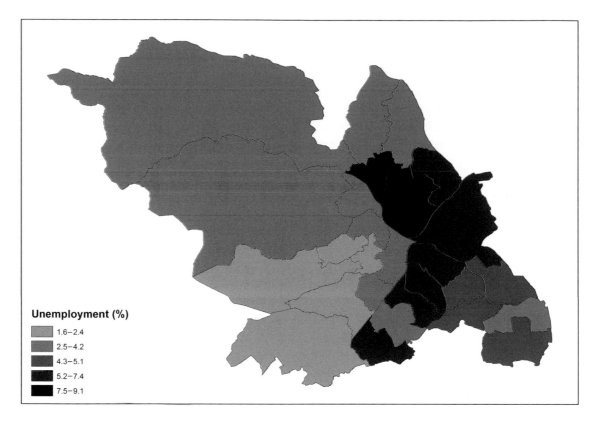

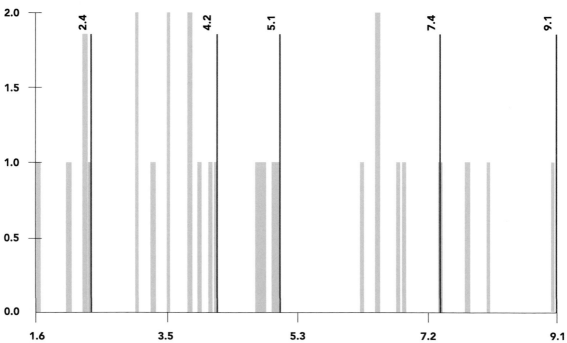

Figure 3.8 Choropleth map example for unemployment rate in Sheffield, UK, 2011 (natural breaks)

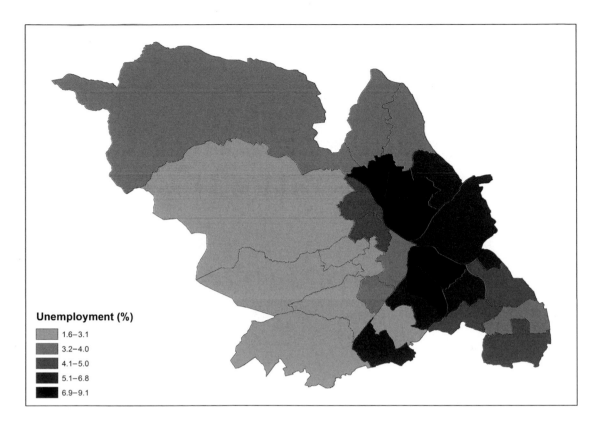

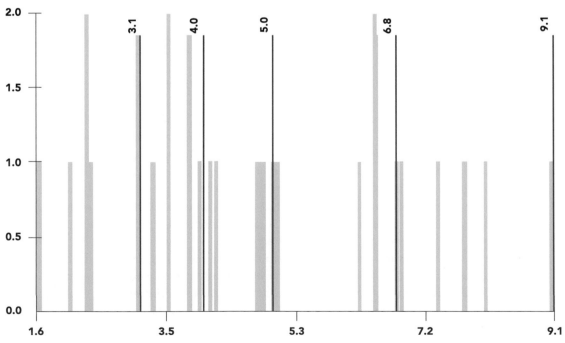

Figure 3.9 Choropleth map example for unemployment rate in Sheffield, UK, 2011 (quantiles)

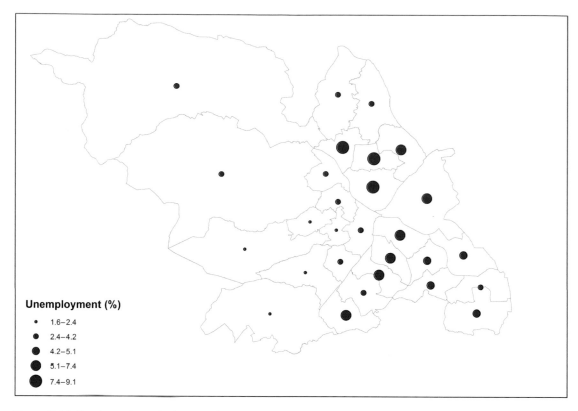

Figure 3.10 Graduated symbol example: unemployment in Sheffield, UK

Dot density

Another approach to mapping quantities is to use a dot density map. A dot density map can be used to show the amount of an attribute such as total population within an area. Figure 3.11 shows an example using population data for the city of Sheffield for electoral wards within the city.

Each dot represents a specified number of entities or incidents (e.g. number of people, number of burglaries). In the map shown in Figure 3.11 each dot represents 300 people. When creating a dot density map it is necessary to specify how many features each dot represents and how big the dots should be. As is the case with the graduated symbol map, the size of the dots must be carefully specified to ensure that they do not obscure the dots of neighbouring areas. It is also possible to map more than one variable using the dot density approach by using different colours.

Chart-type thematic maps

Another type of thematic mapping involves the use of bar and column charts, stacked bar charts and pie charts. These can be particularly suitable when mapping attributes that are meaningful to compare. For example, a suitable socio-economic variable that can be mapped in this way is occupational classification, which is derived from relevant census variables. Figure 3.12 presents an example of such a map using data on socio-economic occupational grouping obtained from the 2011 UK census of population. Figure 3.12 is a pie chart map, which can be particularly useful for illustrating the distribution of values that add up to a meaningful total. The variables being mapped in Figure 3.12 represent the number of people in each area as a percentage of all working population classified in each of the eight socio-economic occupational groupings. All values add up to 100. Figure 3.13 uses

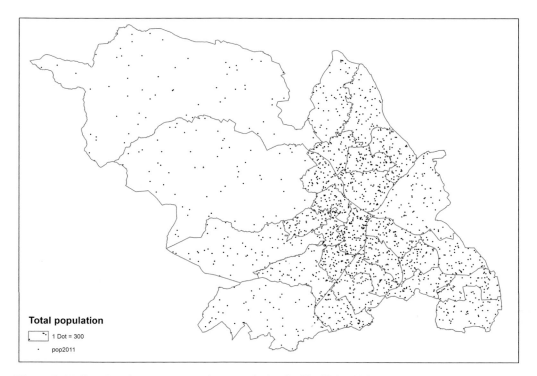

Figure 3.11 Dot density map example: population in Sheffield, UK

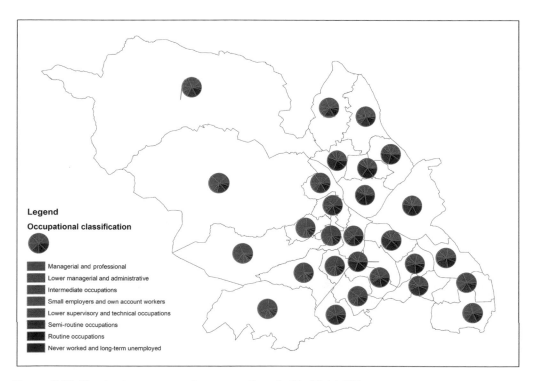

Figure 3.12 Pie chart map example: occupations in Sheffield, UK

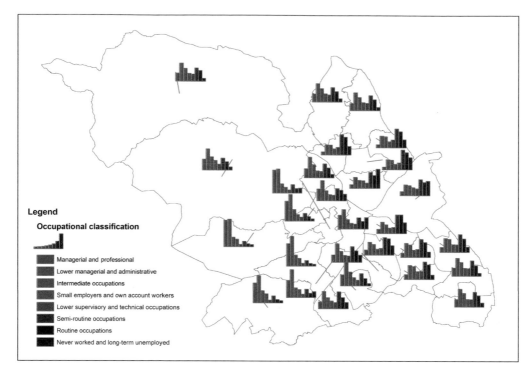

Figure 3.13 Bar chart map example: occupations in Sheffield, UK

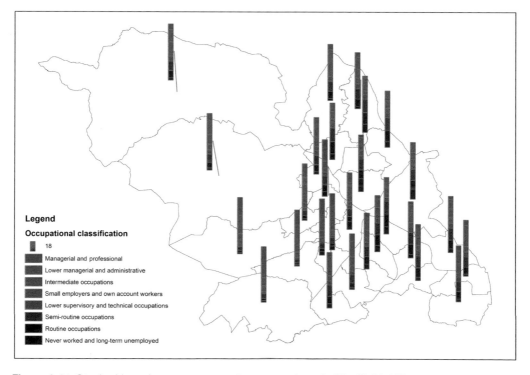

Figure 3.14 Stacked bar chart map example: occupations in Sheffield, UK

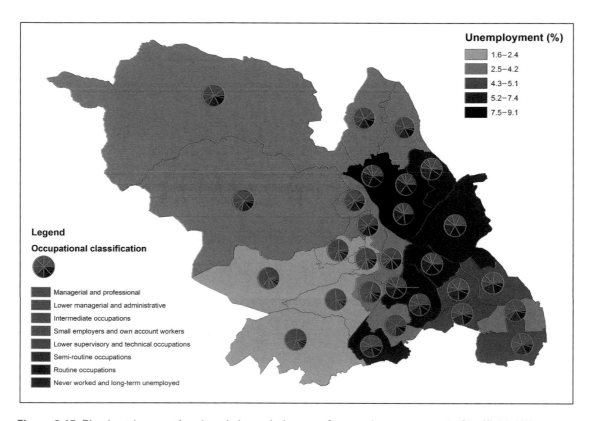

Figure 3.15 Pie chart (occupations) and choropleth map of unemployment rates in Sheffield, UK

the same data to produce a bar/column chart, while Figure 3.14 shows a stacked bar chart.

It is also possible to combine chart maps with choropleth maps. For example, Figure 3.15 combines the map shown in Figure 3.12 with the unemployment rate choropleth map shown in Figure 3.8. By combining the two types of maps it is possible to observe geographical patterns and possible associations between variables. For example, as can be seen in Figure 3.15 the three areas with the highest unemployment rates also have relatively high numbers of people in the 'never worked and long-term unemployed', 'routine occupations' and 'semi-routine occupations' social classifications. In contrast, the areas with the lowest unemployment rates have very small numbers of the least well-off social classes and higher percentages of managerial and professional occupations. Such associations and relationships between variables can be quantified and combined into new variables (such as deprivation indexes) that can

also be mapped (we will discuss examples of this type of analysis in Chapters 5 and 6).

Mapping point data

We have already presented maps of point data in the previous chapter and in the earlier discussion of reference maps in this chapter. It is possible to add thematic mapping information to such maps. For example, we can revisit the map showing the location of grocery stores in the city of Sheffield that was introduced in the previous chapter. It is possible to add thematic mapping features to this map by using different colour codes to show the chain to which each branch belongs and to also make the size of the point symbol representing each grocery outlet proportional to total floor-space (or other variables such as turnover, number of employees, parking spaces, etc.). Figure 3.16

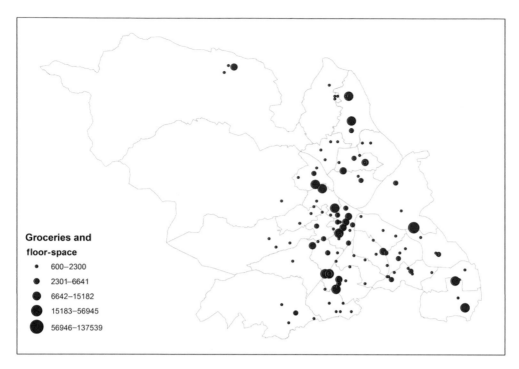

Figure 3.16 A graduated symbol point map: groceries and floor-space in Sheffield, UK

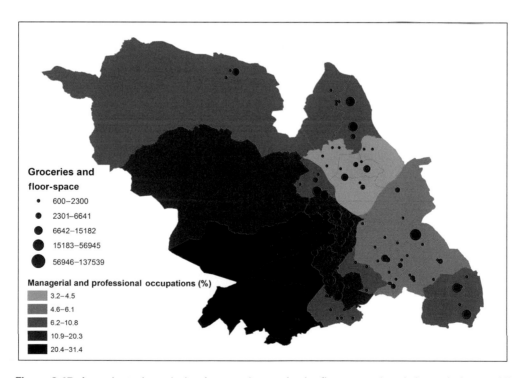

Figure 3.17 A graduated symbol point map (groceries by floor-space) and choropleth map (distribution of managerial and professional occupations) in Sheffield, UK

presents an example of such a map. It is also possible to combine this thematic information with choropleth maps showing the distribution of relevant socio-economic variables such as income or socio-economic classification. For example, it could be argued that the social geography is highly relevant to the potential demand for groceries, with some occupational groups likely to exhibit brand preferences and an overall higher spend. Figure 3.17 combines the groceries point layer with a thematic map of the numbers of managers and professionals by area (as a percentage of the total population).

Mapping line data

Thematic mapping approaches to line data typically involve displaying line features using different thickness levels and colours to represent different values. Conceptually this approach is similar to the graduated symbol maps presented above. For example, the thickness of roads can vary according to type of road or traffic volume. It is also possible to combine line mapping with choropleth mapping to visualise additional variables and highlight relevant variables. Figure 3.18 shows a mapping example of road data combined with a choropleth map of estimated travel times from different areas in London to the Department of Transport in central London at morning rush hour.

Mapping flow data

Another type of thematic mapping in the social sciences is that of movement of people, goods and services or money from one place to another. These maps typically use lines to symbolise the flow. The width of the lines can also be drawn in proportion to the size of the flow. In addition, the lines can be colour-coded to visualise further information (e.g. type of goods being transported). Flow maps can be divided into one of three categories (Akella, 2011; Prasad, 2012): radial, network and distributive. Radial maps show the link between one source and many destinations. Network maps show the quantity of the flows in an existing network. Distributive flow maps are typically used to show the distribution of commodities or some other flow that diffuses from origins to multiple destinations (Akella, 2011).

Although most proprietary GIS software packages do not include tools specifically designed to map

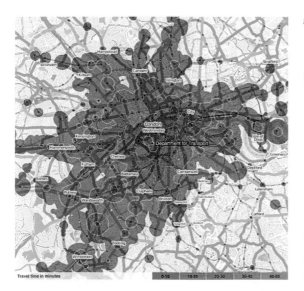

Figure 3.18 A line thematic map example combined with choropleth travel time mapping showing travel time to reach the Department of Transport (SW1) by 9 a.m. using public transport

Figure 3.19 Mapping flows example: British Coal exports in 1864

Source: https://blogs.esri.com/esri/apl/2012/09/12/generating-distributive-flow-maps-with-arcgis/

flows, there are a number of relevant applications and plug-ins. For example, Prasad (2012) presents a flow mapping tool for ArcGIS. Figure 3.19 shows how this tool was used to reproduce a well-known example of a flow map of British coal exports in 1864 by Charles Minard.

Combining different thematic map styles

As we have already seen it is possible to combine different types of mapping approaches to illustrate a wide range of data and their geographical patterns. In this section we present more sophisticated examples of how the spatial distribution of several variables can be mapped simultaneously using a combination of mapping approaches. Figure 3.20 shows a combination of chart map and choropleth mapping, using a bar chart to depict general government balance in European countries as a percentage of their GDP in selected years and choropleth mapping of the countries according

to their overall general government gross debt (as a percentage of their GDP).

Cartography and human-scaled geovisualisations

The mapping approaches presented so far in this chapter can be described as conventional. This section draws on a review by Ballas and Dorling (2011) of alternative human-scaled visualisations which includes an argument for human cartograms to be used instead of (or as well as) conventional maps in the social sciences. Most people are used to conventional maps of their regions and countries. Conventional maps appear on television in the weather reports showing geographical regions as they appear from space. However, there have long been arguments for an alternative approach to visualisation that should be used in the social sciences. Conventional maps are very good at showing where oceans lie and rivers run, for example (Ballas and Dorling, 2011). Their projections are calculated to aid

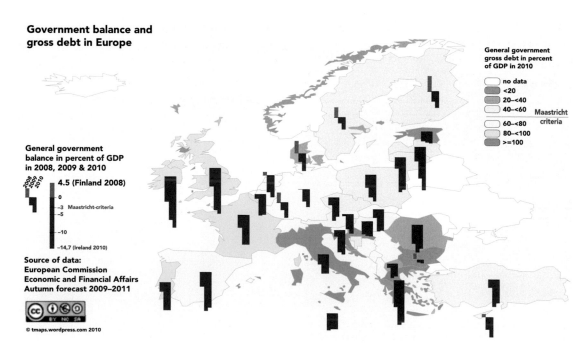

Figure 3.20 A map of government balance and gross debt in Europe

Source: Based on https://tmaps.files.wordpress.com/2010/02/defizit_en.png

navigation by compass or depict the quantity of land under crops. These maps are typically based on area projections such as that of Gerardus Mercator, developed in 1569, which was suitable as an aid for ships to sail across the oceans because it maintains all compass directions as straight lines. As we also discussed in the introductory chapter, all projections inevitably result in a degree of distortion as they transfer the area of the Earth being mapped (or the whole globe) onto a flat surface such as a piece of paper or a display unit such as a computer screen. For instance, the Mercator projection stretches the Earth's surface to the most extreme of extents and hence introduces considerable visual bias. Areas are drawn in ever expanding proportions depending on how near territory is to the poles and this results in areas such as India appearing much smaller than Greenland (when in reality India has an area more than seven times the size of Greenland). The degree to which such a distortion might be acceptable depends on the intended use of the map. There are a number of alternative projections that correspond to the actual land area size and these are much more suitable for the visualisation and mapping of environmental variables, and for pinpointing the location of physical geographical features of interest, than Mercator's map ever was (Ballas and Dorling, 2011).

However, looking at a city, region or country from space is not the best way to see their human geography. For instance, mapping the distribution of human population on a conventional map means that urban areas with large populations, but small area size, are virtually invisible to the viewer. Conversely, the large rural areas with small populations dominate such a map. When mapping data about people, it is therefore sensible to use a different spatial metaphor, one that reflects population size. Most conventional maps, regardless of the projection method that was adopted to create them, are not designed to show the spatial distributions of humans, although the single spatial distribution of people upon the surface of the globe, at one instance in time, can be shown on them. They cannot illustrate the simplest human geography of population. People are points on the map, clustered into collections of points called homes, into groups of points known as villages, towns or cities. Communities

of people are not like fields of crops. The paths through space which they follow are not long or wide rivers of water, and yet, to see anything on maps of people they must be shown as such (Ballas and Dorling, 2011).

In the remainder of this chapter we present alternative ways of mapping data in the social sciences. In particular, we draw on cartographic approaches and arguments that make the case for 'human cartograms' to be used for the geovisualisation of data in the social sciences instead of conventional thematic mapping. Such cartograms can be defined as maps in which at least one scalar aspect, such as distance or area, is deliberately distorted to be drawn in proportion to a socio-economic or demographic or any other 'human' variable of interest. Human cartograms are similar to conventional maps in that they also involve a degree of visual bias and distortion. However, unlike conventional maps, the distortion introduced by human cartograms is based on a population or social science variable of interest. In particular, the location of boundaries and size of territories of areas is redrawn on the basis of a population variable of interest. In this way the relative values of objects on a map are reflected by the size of the area and this is much easier for the human eye-brain system to assess when compared to trying to translate shades of colour into rates and then to imagine what they imply. Rescaling area to the variation in particular variables is very effective in terms of visual communication and a good example of this is the traditional homunculus used in medical science to portray the human body in terms of the degree of sensitivity: different areas of the skin are rescaled in proportion to the number of nerve endings they contain (also see Dorling, 2007a, 2007b).

Circular cartograms and Universal Data Maps

There are many options and possibilities for creating cartograms and there is a long history in human cartography, both in terms of theoretical debates as well as software development (for recent reviews see Ballas and Dorling, 2011; Hennig, 2013). Figure 3.21 presents the application of a well-used circular cartogram

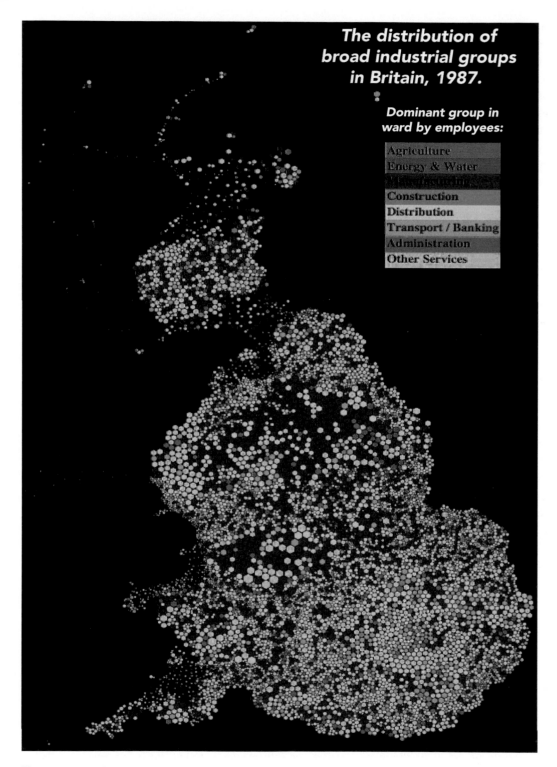

Figure 3.21 A circular (Dorling) cartogram example

Source: www.sasi.group.shef.ac.uk/thesis/small/img023.jpg

algorithm. This algorithm uses circles to depict electoral wards in Britain. The size of these circles is then drawn in proportion to their total resident population. Further, the circles are colour-coded on the basis of the dominant industry group in the area.

Another human cartographic method that is increasingly used in the social sciences is the so-called Universal Data Maps approach. This was originally developed by Durham *et al.* (2006) to build an online census atlas. This approach is suitable when mapping areas with similar population size (such as parliamentary constituencies in Britain). This approach is similar to the Dorling diagrams discussed above, but in this case all geographical features being mapped have the same size. Each geographical area being mapped is represented by a grid cell (such as a spreadsheet cell) of equal size (see Figure 3.22). Since each geographical unit contains roughly the same population, the cartogram might also be seen as a more 'democratic' view of population statistics, effectively according

each person the same space on the map (see Dorling, 2006).

This approach was further developed for different spatial units and shapes (using hexagons instead of grid cells) and applied in a wide range of contexts.

Examples of extensive use of this mapping approach include *Poverty, Wealth and Place in Britain, 1968 to 2005* (Dorling *et al.*, 2007), *The Grim Reaper's Road Map: An Atlas of Mortality in Britain* by Shaw *et al.* (2008) and *People and Places* (Dorling and Thomas, 2004).

It is often argued that a disadvantage of these types of cartograms is that they distort the original areas' real shapes and this affects the degree to which it is familiar and recognisable by a map reader. One of the ways to address this criticism is to present cartograms together with conventional maps and suitable labelling to familiarise the map reader with the cartographic approach to mapping. Figure 3.23 shows a map locator for these cartograms in relation to conventional maps.

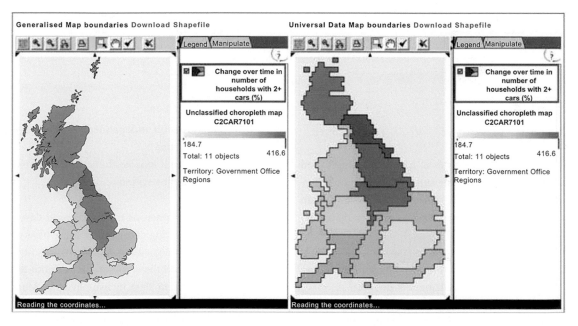

Figure 3.22 Change 1971–2001 in percentage of households with access to two or more cars

Source: Durham *et al.* (2006: 340)

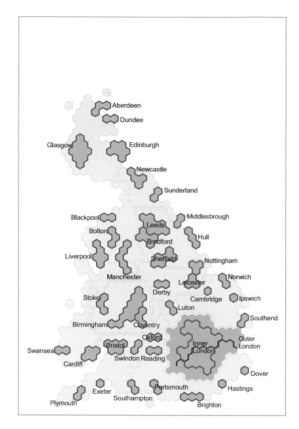

Figure 3.23 A human cartogram map locator example

Source: http://sasi.group.shef.ac.uk/publications/reaper/gr_locator_maps.pdf

Density-equalising cartograms

There are ongoing debates in human geography and cartography regarding the suitability criteria of alternative methods. Overall, the key challenges in human cartography can be summarised as follows (Ballas and Dorling, 2011):

■ Develop a method that is as simple and easy to understand and implement as possible.

■ Generate 'readable' maps by minimising the distortion of the shape of the geographical areas being mapped, while at the same time preserving accuracy and maintaining topological features.

■ Determine the cartogram projection uniquely.

■ Minimise computational speed.

■ Make the end result independent of the initial projection being used.

■ Make the end result look aesthetically acceptable.

■ Have no overlapping regions.

There have been numerous methodological developments aimed at creating cartograms on the basis of automated computer algorithms but there was little success in addressing all the above challenges until the ground-breaking work, in 2004, by two physicists, Mark Newman and Michael Gastner (Gastner and Newman, 2004). Using the diffusion of gas analogy in physics, they developed a cartogram approach that moved the borders of territories with the 'flow' of people, until density is equal everywhere. Figure 3.24 illustrates how the method works with a hypothetical example of four areas (Hennig, 2013). The size of the areas (and

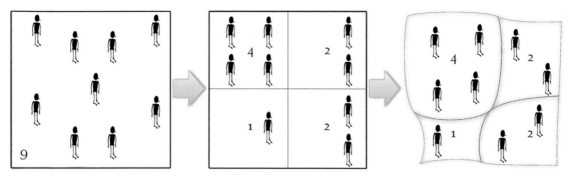

Figure 3.24 An illustration of the Gastner and Newman diffusion-based method for producing density-equalising maps

Source: After Hennig (2013)

borders) are changed until the space between the people in each area is the same everywhere (and therefore the population density in all areas is the same). The cartogram is created by 'diffusing' the people to give them an even spatial spread of population. As people diffuse, borders are moved with them until all spatial units have equal population density (Ballas and Dorling, 2011; Hennig, 2013; Gastner and Newman, 2004). The people depicted in this cartogram can represent the entire population of the area or other sub-groups such as the unemployed, the elderly, etc. It is also possible, instead of people, to use any other variable (e.g. total income) as long as it adds up to a meaningful total for all areas (in a way this type of cartogram is the geographical equivalent of a pie chart, which also only works if the data used add up to a meaningful total; also see Figure 3.26 and discussion below).

One of the first and most popular uses of this approach was to present the US presidential election results. Figure 3.25 presents one of the first applications of the method aimed at challenging the conventional approach to mapping election results. The first map shown in Figure 3.25 is a conventional choropleth thematic map of the 2004 US presidential election results (George W. Bush vs John F. Kerry). In this map each state is coloured red if more of the voters in this state voted for the Republican candidate (George W. Bush) and blue if the majority of the voters voted for the Democrat candidate (John F. Kerry). This map gives the impression that Republican 'red' states dominate the country, since they cover significantly more area than the blue ones. However, as is often the case when election results are visualised in this way, this is

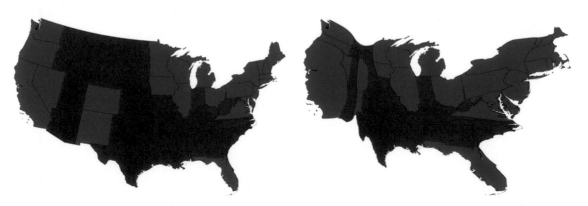

Figure 3.25 Standard versus cartogram mapping of US presidential election results, 2004

Source: Gastner *et al.* (2004)

misleading because the states where Republican voters are the majority tend to have smaller populations (and large rural areas within them), whereas the blue states might be small in area but they are large in terms of the total population (and number of voters) which is what matters in elections. This misleading effect can be corrected by using a cartogram approach that redraws each state with a size proportional to its population rather than area. The second map in Figure 3.25 shows the output of the cartogram method developed by Gastner and Newman (and which is also demonstrated in the practical exercises accompanying this chapter). In this second map the prominence of an area is given on population rather than size. For instance, in this map the state of Rhode Island, which has about a million inhabitants has twice the size of Wyoming, which has half a million inhabitants, even though Wyoming is 60 times bigger than Rhode Island in terms of topographic acreage. Using a cartogram to present the election results

paints the true picture of the situation, which is that the US was equally divided in this election.

Over the past ten years the ground-breaking diffusion-based method for producing density-equalising maps has become increasingly popular for mapping in the social sciences and it has been used extensively so far in a number of projects and examples. One of the projects that had a considerable impact in making the method widely known and used is the Worldmapper website (www.worldmapper.org) which was led by Danny Dorling and colleagues at the Social and Spatial Inequalities research group at the University of Sheffield in collaboration with Mark Newman at the University of Michigan. There is also ongoing follow-up work on this project by Benjamin Hennig at the University of Oxford (www.worldmapper.limited). One of the original aims of the Worldmapper project was to map variables in relation to the United Nations Millennium Development Goals (MDGs) using data

Figure 3.26 Bar chart of total population by Worldmapper region as a percentage of the global population.
© Worldmapper.org

Source: www.worldmapper.org

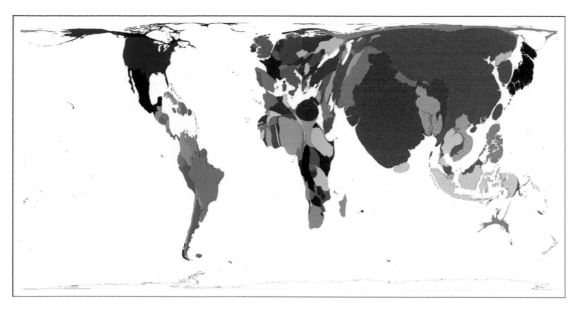

Figure 3.27 Total population (Worldmapper map 002). The size of each territory shows the relative proportion of the world's population living there. © Worldmapper.org

Source: www.worldmapper.org/display.php?selected=2. Data source: United Nations Development Programme, 2004, Human Development Report

from the World Bank, the United Nations, the World Health Organization and other sources. We have already presented a map from this project in the choropleth map section of this chapter (Figure 3.7). Figure 3.26 depicts the total population of each Worldmapper region as a percentage of the global population. It is often argued that bar charts such as that presented in Figure 3.26 are also conceptually similar to a population cartogram. It can be argued that this bar chart is a very basic and non-continuous cartogram. Every world region is sized according to a variable of interest (total population). The density-equalising cartogram method is conceptually similar to a bar or pie chart, but it maintains topology and the shape of the original territories being mapped, while at the same time it resizes them according to a variable of interest. Figure 3.27 shows a population cartogram from Worldmapper where each country is resized in proportion to each total population. Figure 3.28 shows what is perhaps one of the most influential, impactful and shocking maps of the Worldmapper project. This map was created in relation to the United Nations MDGs and shows

the distribution of all people aged 15–49 with HIV (Human Immunodeficiency Virus) worldwide, living in each country. It can be argued that when compared with conventional maps, such cartograms present a much more appropriate and powerful depiction of the magnitude of socio-economic and health issues. In addition, it can be argued that such cartograms (and especially when there is such a huge and shocking spatial disparity, as in the case of the map shown in Figure 3.28) have a much more effective and emotionally powerful visual impact compared to conventional maps or tabular descriptions of the data.

The Worldmapper project has been further refined and extended with the use of smaller area data. In particular, Hennig (2013) used data from the Gridded Population of the World (GPW) project developed by the Socioeconomic Data and Applications Center (SEDAC) of Columbia University, New York (sedac.ciesin.columbia.edu/gpw). This database includes population data and estimates from 1990 to 2015 for all countries of the world in resolutions of up to 2.5 arc minutes, leading to a population grid of 8,640 by 3,432

Figure 3.28 HIV prevalence (Worldmapper map 227). The size of each territory shows the proportion of all people aged 15–49 with HIV worldwide, living there. © Worldmapper.org

Source: www.worldmapper.org/display.php?selected=227

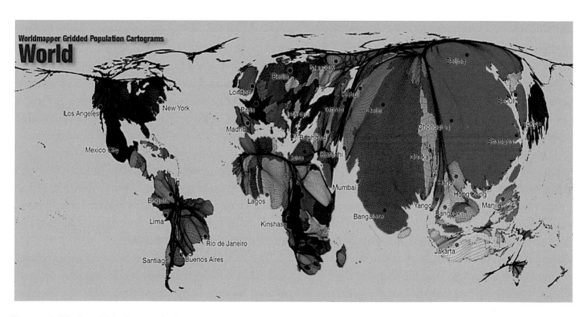

Figure 3.29 A gridded population cartogram of the world

Source: Hennig *et al.* (2010); www.esri.com/news/arcuser/0110/graphics/cartogram_2-lg.jpg

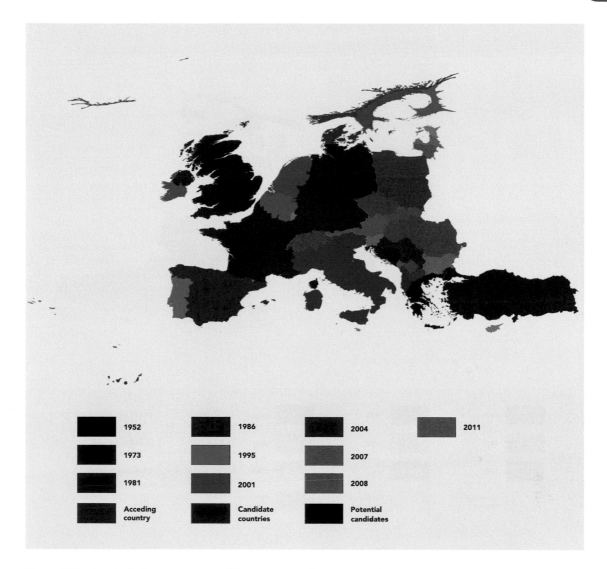

Figure 3.30 A population cartogram of Europe using Gastner and Newman's density-equalising method

pixels. The geographical database is available in raster format. Hennig and colleagues (Hennig, 2013; Hennig *et al.*, 2010) developed a method that converted this data to vector polygon format and combined it with further metadata to create a 4,096 by 2,048 pixel-sized lattice and applied the Gastner and Newman density-equalising method to resize each pixel in proportion to the number of people that live there. In particular, the process results in a contiguous gridded population cartogram (known as a Hennig Projection Gridded Population Cartogram), meaning that each new grid

cell has an area proportional to the number of people that live there, but still touches only its original eight neighbouring cells. Figure 3.29 shows the output of this method with all the grid cells coloured and shaded according to the Worldmapper colouring scheme discussed earlier.

This gridded population refinement of the Worldmapper project maps has been extensively used to provide so-called human-scaled visualisation of the world in a wide range of contexts. There are hundreds of maps of this type which have also attracted

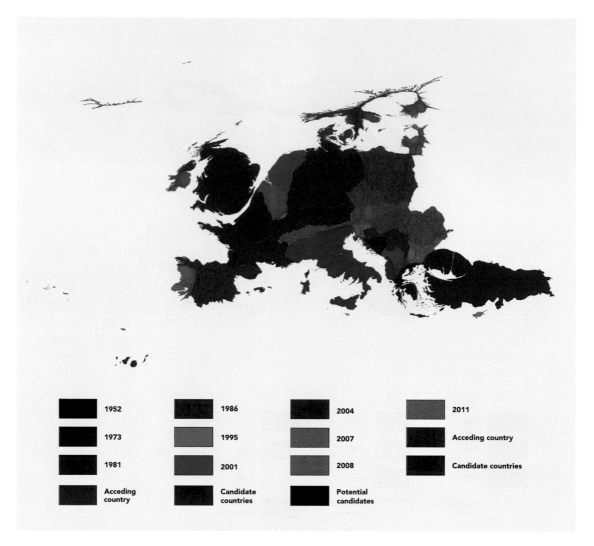

Figure 3.31 Gridded population cartogram of Europe; basemap: Hennig Projection Gridded Population Cartogram

Source: Ballas *et al.* (2014)

considerable media attention (e.g. the BBC described these as 'people-powered' maps; see Brown, 2009 for more details and numerous cartogram examples of individual countries). More examples and stories painted using this approach can be found in Hennig's original work (Hennig, 2013) as well as the accompanying website (www.viewsoftheworld.net) which is constantly updated with new data and topical themes.

A more recent application of Gastner and Newman's density-equalising method and of Hennig's refinement and gridded population cartogram was part of an effort to offer an alternative visualisation of Europe and the European integration project. The Europemapper (www.europemapper.org) project involved the creation of a Social Atlas of Europe (Ballas *et al.*, 2014) with the use of these methods to visualise Europeans living in all the states that have demonstrated a strong commitment to a common European future by being closely associated with the EU, either as current members or as official candidate states (or official potential candidates for EU accession) and/or states that are signed up to any of the following agreements: the European Economic Area, the Schengen Zone and the European Monetary Union. Figure 3.30 shows a population cartogram version of the map shown in Figure 3.6, created with the Gastner and Newman method, using the same rainbow colour scale to determine the colour hue for each state according to the year of association with the European Union. The more recent the formal association the nearer to the red end of the spectrum the hue is (also see the legend in Figure 3.6).

Further, Figure 3.31 shows a map made with the use of the Hennig gridded population cartographic projection on the basis of fine-level spatial information about where people live rather than land mass, showing even more clearly where most people are concentrated. For

example, Madrid, Paris, Istanbul and London are huge, while Scandinavia is small, whereas the Rhine-Ruhr metropolitan region in Western Europe, including the areas of Cologne, Dortmund and expanding towards the Netherlands, is much more prominent compared to on a conventional map. In addition, countries and regions that are more densely populated are more visible on the map (e.g. most of the United Kingdom, Italy, Poland, Romania) compared to large rural areas in the north of Europe.

The picture of Europe described above can be enriched further with more information, such as by employing a choropleth mapping approach to shade the gridded population cartograms according to a socio-economic or demographic theme of interest. An example is shown in Figure 3.32, where resized grid cells are coloured on the basis of publicly available data

Figure 3.32 Persons aged 25–64 with a tertiary education degree as a proportion of all people aged 25–64 living in Europe; basemap: Hennig Projection Gridded Population Cartogram

Source: Ballas *et al.* (2016)

for European statistical regions on higher education attainment of the local workforce. In particular, the shading shows the spatial pattern of the numbers of persons aged 25–64 with a tertiary education degree as a proportion of all people aged 25–64 living there. For example, the cells coloured in the darkest blue are placed within regions where more than 33% of that population has a tertiary education degree (Ballas *et al.*, 2016).

It is also possible to combine the population cartograms with physical and topographical geographical information. Figure 3.33 is the topographic version of the previous two maps, with the area being drawn scaled proportionally to population but coloured by altitude. In this way physical and human geographies can be mixed up on the map. Rather like a traditional physical geography map, upon which cities are drawn, this is a new human geography map, but one upon which mountains and valleys are also depicted.

These cartograms give just a flavour of the geovisualisation potential in the social sciences with the use of GIS and state-of-the-art human cartographic techniques. There are hundreds more examples in relation to this ongoing work (e.g. see www.worldmapper.org, www.europemapper.org and viewsoftheworld.net). One of the practical exercises accompanying this chapter shows how it is possible to create these cartograms.

Concluding comments

In this chapter we presented an overview of key cartographic issues and principles pertaining to thematic mapping in the social sciences and we introduced a number of geovisualisation approaches, ranging from so-called conventional thematic mapping (including choropleth maps, graduated symbol and chart-type maps as well as flow maps) to human-scaled visualisations and human cartograms. The latter are

Figure 3.33 Gridded population cartogram representation of the topography of Europe; basemap: Hennig Projection Gridded Population Cartogram

Source: Ballas *et al.* (2014)

increasingly appealing and particularly appropriate in the social sciences. In particular, it is increasingly and convincingly argued that conventional maps that show how cities, regions and countries appear from space, are not an appropriate way to show the spatial distributions of humans and their characteristics. The last section of this chapter provided an overview of key arguments why new innovative human cartographic methods and tools are more suitable and appropriate for the depiction of the spatial distribution of variables pertaining to human societies rather than environmental, geological or meteorological problems (which are the domains for which conventional cartography is more suitable). These methods are increasingly accessible and easy to use due to the availability of free (and user-friendly) software (as is also demonstrated in the practical accompanying this chapter). There are ongoing and exciting developments in social science cartography and geovisualisation aimed at mapping new and sometimes more complex forms of data (including flow data, voluntary geographic information user data and social media data) and involving animation and use of technologies such as *OpenStreetMap*, *Google Earth*, *Bing* and *Apple Maps*.

This chapter provided an overview of key approaches and methods, but it would have been beyond its scope to offer a comprehensive guide of what is possible and of all the issues and elements of cartographic theory and practice. The cartographic texts briefly referred to in the introductory chapter are among the excellent resources for further reading to complement what we present here in order to create social science maps and geovisualisations that are visually appealing and suitable for different audiences.

Accompanying practicals

This chapter is accompanied by two practicals (Practical C(i) and Practical C(ii)). These practicals allow you to practise generating a series of cartographic outputs using your GIS. Practical C(i): Visualisation and thematic mapping, introduces core cartographic visualisation tools in ArcGIS, using data related to small areas in the UK. You also create a human car-

togram using an external tool. Practical C(ii): Working with social media data, focuses on point data derived from a novel data source, the social media platform Twitter. You import and visualise this point data using proportional symbols, and also explore how to aggregate point data to geographic zones for subsequent thematic mapping. These skills will be applied across the range of practical activities that follow.

Note

1 E.g. see http://uk.businessinsider.com/state-unemployment-map-january-2016-2016-3.

References

All website URLs accessed 30 May 2017.

Akella, M. (2011) *Creating Radial Flow Maps with GIS*. Available from: https://blogs.esri.com/esri/arcgis/2011/09/06/creating-radial-flow-maps-with-arcgis/.

Ballas, D., & Dorling, D. (2011) Human scaled visualisations and society, in T. Nyerges, H. Couclelis, & R. McMaster (eds) *Handbook of GIS & Society Research*, Sage, London, 177–201.

Ballas, D., Dorling D., & Hennig, B. (2014) *The Social Atlas of Europe*, Policy Press, Bristol.

Ballas, D., Dorling, D., & Hennig, B. (2016) *The Human Atlas of Europe: A Continent United in Diversity*, Policy Press, Bristol.

Bertin, J. (1967) *Sémiologie graphique*, Mouton/Gauthier-Villars, Paris.

Bertin, J. (2011) General theory, from semiology of graphics, in M. Dodge, R. Kitchin, & C. Perkins (eds) *The Map Reader*, Wiley-Blackwell, Chichester, 8–16.

Brewer, C. (2005) *Designing Better Maps: A Guide for GIS Users*, ESRI Press, Redlands, CA.

Brown, A. (2009) People-powered maps. *BBC Magazine*. Available from: http://news.bbc.co.uk/1/hi/magazine/8280657.stm.

Center for International Earth Science Information Network (2015) *Socioeconomic Data and Applications Center (SEDAC): Gridded Population of the World (GPW) v3*. Available from: http://sedac.ciesin.columbia.edu/data/collection/gpw-v3.

Chrisman, N. R. (1998) Rethinking levels of measurement for cartography. *Cartography and Geographic Information Science*, 25(4), 231–242.

Dent, B. D., Torguson, J., & Hodler, T. W. (2008) *Cartography: Thematic Map Design* (6th edition), McGraw-Hill, Dubuque, IA.

Dorling, D. (2006) New maps of the world, its people and their lives. *Society of Cartographers Bulletin, 39*(1 and 2), 35–40.

Dorling, D. (2007a) Worldmapper: the human anatomy of a small planet. *PLoS Medicine, 4*(1), 13–18.

Dorling, D. (2007b) Anamorphosis: the geography of physicians, and mortality. *International Journal of Epidemiology, 36*(4), 745–750.

Dorling, D., & Thomas, B. (2004) *People and Places: A Census Atlas of the UK*, Policy Press, Bristol.

Dorling, D., Rigby, J., Wheeler, B., Ballas, D., Thomas, B., Fahmy, E., *et al.* (2007) *Poverty, Wealth and Place in Britain, 1968 to 2005*, Policy Press, Bristol.

Durham, H., Dorling, D., & Rees, P. (2006) An online census atlas for everyone. *Area, 38*, 336–341.

Eddins, S. (2014) Rainbow color map critiques: an overview and annotated bibliography, *MathWorks: Accelerating the Pace of Engineering and Science*. Available from: www.mathworks.com/tagteam/81137_92238v00_RainbowColorMap_57312.pdf.

Forrest, D. (1999) Geographic information: its nature, classification, and cartographic representation. *Cartographica, 36*, 31–53.

Gastner, M. T., & Newman, M. E. J. (2004) Diffusion-based method for producing density equalizing maps. *Proceedings of the National Academy of Sciences of the United States of America, 101*, 7499–7504.

Gastner, M., Shalizi, C., & Newman, M. (2004) *Maps and Cartograms of the 2004 US Presidential Election Results*. Available from: www-personal.umich.edu/~mejn/election.

Hennig, B. D. (2013) *Rediscovering the World*, Springer Verlag, Berlin.

Hennig, B. D., Pritchard, J., Ramsden, M., & Dorling, D. (2010) Remapping the world's population visualizing data using cartograms. *ArcUser*, Winter, 66–69. Available from: www.esri.com/news/arcuser/0110/files/cartogram.pdf.

Kraak, M., & Ormeling, F. (2010) *Cartography: Visualization of Spatial Data* (3rd edition), Pearson Education, Harlow.

MacEachren, A. M. (2004) *How Maps Work, Representation, Visualisation and Design*, Guilford Press, New York.

Monmonier, M. (1996) *How to Lie with Maps* (2nd edition), University of Chicago Press, Chicago, IL.

Perkins, C. (2003) Cartography: mapping theory. *Progress in Human Geography, 27*(3), 341–351.

Perkins, C. (2004) Cartography: cultures of mapping: power in practice. *Progress in Human Geography, 28*(3), 381–391.

Prasad, S. (2012) *Generating Distributive Flow Maps with ArcGIS*. Available from: https://blogs.esri.com/esri/apl/2012/09/12/generating-distributive-flow-maps-with-arcgis/.

Shaw, M., Davey Smith, G., Thomas, B., & Dorling, D. (2008) *The Grim Reaper's Road Map: An Atlas of Mortality in Britain*, Policy Press, Bristol.

Slocum, T. A., McMaster, R. B., & Kessler, F. C. (2008) *Thematic Cartography and Geovisualisation* (3rd edition), Pearson, Harlow.

Stevens, S. S. (1946) On the theory of scales of measurement. *Science, 103*, 677–680.

wiki.gis.com (2016a) *Visual Variables*. Available from: http://wiki.gis.com/wiki/index.php/Visual_Variables.

wiki.gis.com (2016b) *Cartographic Design*. Available from: http://wiki.gis.com/wiki/index.php/Cartographic_Design.

4 GIS and network analysis

LEARNING OBJECTIVES

■ Similarities and differences between geographical and social networks
■ Network fundamentals
■ Types of analysis relying on networks in a GIS setting

Introduction

If location is an intrinsic characteristic of geographic data, *connection* is not far behind. Connections between and across units – whether places, people, firms, water systems or streets – allow us to answer questions about relationships, flows, interactions or movements. In the case of these types of questions, knowledge about feature location is insufficient; we also need rules that capture the nature of the connections between each feature. In the case of a watershed, for example, we not only need to know whether a smaller stream joins a larger river, but also in which direction the water travels (since water travels uni-directionally in such a system). In a GIS environment, connections can be conceptualised through networks. By specifying which entities are linked to which, and what the rules of those connections are, networks are developed that can then be used in a range of social science applications.

Before delving further into the particulars of geographic networks (henceforth simply termed networks), it is important to distinguish between two different types of network analysis common across the social sciences. In this book, we are concerned with *spatial* aspects of networks, which could take the form of roads, public transportation lines or even shipping, air or rail routes. Other social science applications concern themselves with *social* network analysis: how a set of actors are connected to each other within a particular setting. Some aspects of both networks are similar. We might ask questions about information, for example, travelling through a spatial or a social network and the principles would be the same. Distance or routing in the case of a social network is not usually concerned with physical space, but rather how many people information travels through (like the child's game of Chinese whispers telephone) before reaching the destination. Increasingly, there is interest in combining spatial and social network analysis. Here, however, we limit ourselves to coverage of spatial or geographical network analysis.

Spatial networks can be interesting in and of themselves – we can learn, for instance, how many streets meet at a particular intersection – but often they will be constructed in order to feed into subsequent analysis. A street network within a particular city or region provides information about which street segments are connected to other streets, but also how long the streets are and how long it takes to traverse them. Alone, this information is not especially useful. It becomes much more powerful when used for what is broadly termed *network analysis*. Network analysis, put simply, comprises a variety of analytical techniques that employ the network, along with other data, to answer questions about routes, shortest distances, coverage/accessibility or closest facilities. Sometimes, as with routes, the network is part of the solution: a 'best' path will be routed along the network. Other times, as with service areas, the network is being used to identify locations that meet a particular distance or cost criterion; the solution will not necessarily include the underlying network but will instead be a polygon or set of polygons. And occasionally, as with distance computation,

there are multiple calculation alternatives. Other times, as in the case of most efficient route or service area, network analysis is required.

Part II of the book highlights specific applications of network analysis, and you work with network data sets in Practical D (Part I) and Practicals 5 and 6 (Part II). With an understanding of how to generate a network and what sorts of questions it can be used to answer, however, the potential range of uses of network analysis in the social sciences is enormous. The purpose of this chapter is to lay out the basics of networks, to give a general idea of the type of questions they can be used to answer and to provide an overview of the basic types of network analysis that a GIS can undertake.

So what sorts of questions can networks help answer? One of their primary uses is simply to calculate distance from Point A to Point B – from household to grocery store, doctor to patient, or individual to hazardous waste site. The locations might be known, and

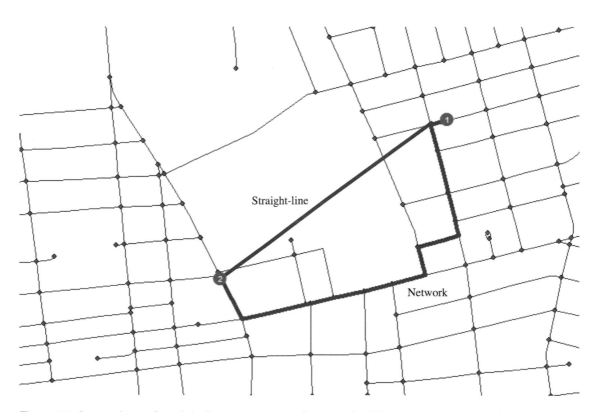

Figure 4.1 Comparison of straight-line and network distances for identical origin and destination

we only seek to quantify their distance, or we might wish to find the distance to the closest facility, whether store, doctor, or park or school. Alternatively, we can use a network to find the distance to all facilities from some starting point or points. While straight-line, or 'as the crow flies', distance is easier to calculate in a GIS as it requires no network information, depending on the application it might fail to accurately capture the true distance an individual would cover when using existing streets, footpaths or other transportation networks. As Figure 4.1 shows, not only will straight-line distance be shorter (as it must: elementary geometry tells us that the shortest distance between two points is a straight line), it might also indicate a route that is not possible in the real world. When the goal is to replicate actual routes used, then incorporation of the underlying network is helpful. This type of use of a network, usually some sort of transportation network, is common throughout the social sciences – although, given the discussion above, one encounters straight-line distance more often than one might expect. Why is that? The simple explanation is that, although a network might be the ideal choice, it is not always achievable. For some parts of the world, accurate and reliable road data simply do not exist. And even in more developed areas, some types of network data, such as for roads, are more accessible than those for rail lines or bicycle paths. Where good data do not exist, the best solution is typically to fall back on straight-line distances.

Networks contribute to answering other sorts of questions as well. Research in health or retail that seeks to measure the share of an area covered by a particular set of facilities could use buffers (see Chapter 2) but might also use a transportation network for more realistic results. Networks can help identify locations for new schools or hospitals and can also help to develop routes – delivery or transportation, for instance – that meet certain criteria. In many cases, the research question or discipline might not obviously involve a network. As in the distance examples above, what sets network analysis apart is its incorporation of a network to investigate connections and distances.

When a network can help

As previous chapters have discussed, distance (or proximity) influences a wide range of social phenomena and a strength of GIS is the estimation of these spatial relationships. GIS offers many ways to conceptualise and capture distance; it is the researcher's task to match measurement to application. Ideally, the way in which distance is measured should reflect how it matters in the real world. Because so much human interaction is governed by the constraints of transportation networks, streets in particular, it often makes sense to measure distance as traversed on these networks. Using a network can generate more accurate measurements, but also allows for considerations such as cost, time of day, intervening opportunities between origin and destination, and differences resulting from modal choice (e.g. if some patients take the bus to the doctor while others drive). Taken together, these aspects of distance can result in more realistic abstractions of reality.

We also use networks sometimes because we have to. In a GIS, many techniques, such as delineating service areas or locating new facilities, take place within a network environment because they rely on information about connectivity for computation. What this means, practically speaking, is that research using these techniques will require basic knowledge of network analysis and networks, even though the topic at hand might not appear network-dependent. We shall see, in Part II, that many applications of GIS involve knowledge of network analysis. This chapter provides introductory material that will assist in engaging more fully with those chapters.

That said, there are occasions in which networks are either overkill (too much data and work required for the result) or inappropriate to the subject at hand. Certainly, calculating straight-line distance will almost always be easier and faster. That means the decision to turn to network analysis depends on the sensitivity of the analytical results to accuracy of measurement, the availability of requisite data and a conviction that straight-line distance does not appropriately capture the spatial process under investigation. When considering distances between households and fire stations or routing ambulances from homes to hospitals, for example, accurate distance and routing is of prime importance. Longer distances – between cities in the United States – might give similar results for network or straight-line distance. Generating and then using a network for analysis requires data with certain characteristics. It also requires that these data be readily available and of good quality. Although many cities and regions in more developed countries possess good transportation data, this is often not the case for lesser-developed countries or, if the data exist, the area might be growing so quickly that data quickly become outdated. Finally, the best measurement of distance (or any other spatial variable) will depend on the research question: areas affected by a hazardous waste site might be best captured with a buffer or straight-line distance, whereas areas served by a hospital are best measured with network distance, or with travel time through that network.

Networks: the basics

A network is any representation of movement in directed space – so streets, but also utility networks or even river systems. Before performing network calculations, in general the network must be created in the GIS from raw data (the exception comes from cases in which the network was generated previously and is

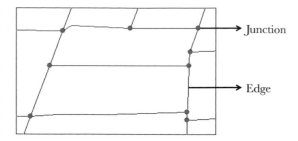

Figure 4.2 A street network represented as combinations of junctions and edges

simply being reused or borrowed). Because networks are combinations of points and lines (or *junctions* and *edges*, Figure 4.2), the starting, input data are lines. Looking at the line features, whether streets or paths or bus routes, the researcher can often discern how movement would take place along those lines. The computer, however, needs to be explicitly told which line segments connect to which and what rules govern movement or flows along the network. The network creation process takes the input lines and develops the topological rules that need to be followed. Figure 4.3 shows the difference between a street layer and a street network.

How does this work in practice? The most common input data are likely streets. The many line segments that combine to represent a city street system are used to generate the network. This process converts the line

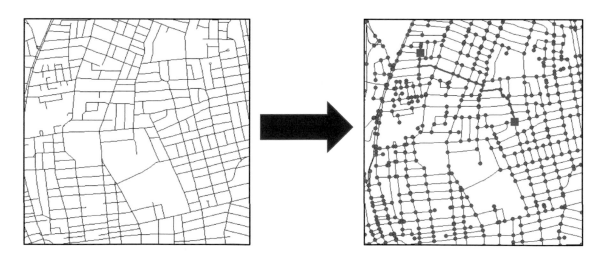

Figure 4.3 Converting streets to a street network

segments into edges – the streets carrying flows – and junctions, or intersection points. Depending on the data, the resulting street network will be of varying quality. Of paramount importance is that intersections and overpasses are accurately represented by junctions (or the lack thereof) and that connectivity in the network is complete. That is, there should be no line segments that are unreachable in the network, unless of course such a thing actually exists on the ground. We seek to be able to navigate the computer network in the same manner that we do the real network.

The network is more than its spatial component, however. Consider real-world movement on streets: this is controlled not only by where streets go and what they connect to but also by speed limits, distance and whether it is a one-way street, to name a few considerations. Traffic and toll costs might also be a factor. This is the information that lends verisimilitude to the spatial network. Some of this information – for example, distance – typically either comes with the data or can be easily calculated within a GIS. Other information such as directionality, traffic loads or speed limit must be present in the data in order to be incorporated into the network. If road type is available, it might be possible to estimate speed or traffic load based on average information for each type of road. Information about turns (and turn times) can also help provide more realistic analytical results for networks. In short, the best networks will be those containing information and variables that help it mimic real-world behaviour as closely as possible.

How does the GIS use the network to find the 'best' route? To define any sort of 'best' or ideal outcome using a network, the GIS needs decision rules and cost variables. The most common rules minimise a cost variable, such as time or distance. Rules can also maximise coverage. What rules have in common, though, is that they define best or optimal solutions. Analysis using the network – for instance, calculation of closest facility, distance or route – will also rely on one or more 'cost' or *impedance* variables. This cost is associated with travel along each link and summed to provide the total measure. In the case of closest facilities or best routes, but also other analytical tools, the GIS seeks the best of a set of options by minimising

the cost variable. This is logical in the real world as well: the preferred route will generally be the shortest or quickest or cheapest. The computer is only as smart as the input data. If no measure of distance or cost (e.g. in terms of time or money) has been included, the GIS cannot pull this information from thin air. As noted above, distance is easily calculated by adding a line segment length variable in the street attribute table, if it is not already present. Other variables such as time and monetary cost must be estimated. It is normal for connectivity checking and estimation of cost or impedance variables to require a great deal of time.

Alone, the network, no matter how perfect, does very little. It is the combination of network plus other spatial information that allows analysis to be conducted. This additional spatial information is what tells the GIS where origins and destinations are, or which facilities are being used to compute service areas. This information usually comes in the form of geo-located point data – stores, hospitals or households – that might also contain additional attribute information. That is, any network analysis incorporates information about the underlying network and data about locations. Many GIS software applications allow the user to import existing point data, but also to search for addresses or manually locate them on a map.

In real-world networks, movement is occasionally disrupted, either cut off completely or slowed. Construction might close a length of street; a flood could impact an entire neighbourhood; or a malfunctioning traffic signal might render an intersection impassable. Once a network has been constructed in a GIS it is possible to assess the impacts that changes to the network, such as those listed above, will have on movement through it. Thus, it is possible to investigate connectivity in a street system and then measure the impacts that summer roadwork, for example, might have on traffic flows (see Figure 4.4). This ability to conduct a sort of impact analysis is another capability of network analysis in a GIS framework. Changes to the network can be implemented as *barriers* – either points, lines or polygons, depending on the type of disruption envisioned – that can be turned on and off to assess changes in routes and speeds.

Figure 4.4 Impact of an areal barrier (right) on route computation

A key element of networks is that movement is occurring along prescribed lines, whether streets or some other system. This is of course different from movement in continuous space – consider the difference between birds flying from place to place and cars driving from one location to another. Even within network space, however, we can distinguish between different sorts of networks. The type of network used in this book, and, in fact, most common across the social sciences, is what is termed an *undirected* network. Within these systems, the potential paths are predetermined. A car must travel on a street to get from one place to another. However, the car driver might choose to start the journey anywhere on the network and, similarly, the destination could be any other location reached via the street network. This is different from movement or flows taking place within, say, a city's sewer system or a river system in the natural world. In those cases, flows originate at only certain locations (sources) and their subsequent route through the network is known. Imagine the confusion (and disgust) should the predictability of flows within a sewer system be upended. This type of network is called a *directed* network. The sorts of analysis done with directed networks are different from the material discussed

above. Often, the main concern is with connectivity in a directed network. If water contamination occurs in a particular location of a city, for example, which households will be affected? If maintenance on one section of the electric grid is required, which portions of the system will become temporarily unavailable? Because directed networks are encountered so rarely in social science GIS applications, they are not further discussed here.

Common types of network analysis

As discussed in the introduction of this chapter, certain types of GIS analysis are usually associated with networks and are frequently termed 'network analysis'. Some of the most common techniques are introduced below. Those seeking deeper information on any one method should see Part II of the book and also the suggestions for further reading below. It is worth noting that the algorithms and specifications for many of these methods are complex and are, in themselves, the subject of a great deal of research. This chapter provides only an introduction to the topic and is in no way meant to be exhaustive.

WHERE TO FIND ROAD DATA

- National or local governments. In some countries, such as the US, roads data are made freely available at a variety of administrative levels. This includes TIGER files at the national level but also data provided by states and cities. These data might require cleaning before being usable, however.
- OpenStreetMap (OSM, www.openstreetmap.org). This is open source, volunteered geographic information for the entire world, although data quality will vary by place and spatial scale.
- Proprietary or commercial data. ESRI, the ArcGIS software developer sells street data, as do HERE (formerly known as NAVTEQ) and Tele Atlas/TomTom.

Factors to consider in choosing data

- How old are the data? Particularly in rapidly expanding areas, data for streets can quickly become outdated as new roads are constructed.
- How complete are the data? Do you need information for smaller streets, as well as main thoroughfares? Do you need to model one-way streets? If so, your data must contain this information.
- How 'good' are the data? This includes completeness, but also accuracy – in terms of location, length of segments and even segment characteristics (e.g. number of lanes or paving type).
- Who or what is navigating your street network? Pedestrians? Cars? Bicycles or public transportation?

Routing is arguably the most frequently used network analysis tool and forms the basis of the practical accompanying this chapter. Routing in a GIS takes a set of at least two stops and finds the 'best' route, depending on the cost variable used and any barriers that might be included. Internet mapping tools that calculate directions and driving time from origin to destination use routing. A route is more than directions, though. Output is often visual (a map) and indicates which streets were used, but also includes the final cost calculation, which allows the routing option to be employed when the purpose is solely to derive a numerical result.

When more than two stops are included in a route, the GIS can be used to find not only the shortest route, but also the best ordering of stops. Figure 4.5 provides an example of a set of ten stops on a street network. In panel b, the shortest route is calculated for going to each stop *as ordered* in the list. Panel c shows the same stops: however, the GIS has ordered them in the most efficient way in order to minimise total distance travelled. The GIS will also allow the option to hold first and last stop constant or to allow all stops to be reordered.

Finding the best order for a given set of stops is known as the Travelling Salesperson Problem or TSP. The TSP is computationally intensive. Rather than the GIS having to find the shortest path for an ordered set of stops (which is already complex, given the number of routes that could be taken), it has to find the shortest path *and* find the best order of stops, given a specific cost variable (e.g. time or distance). Even with only five stops to make, ordering routes can quickly get out of hand. Should one do stop 1, 2, 3, 4 and then 5? Or 2, 1, 3, 4, 5? Or 2, 3, 4, 5, 1? Obviously, the number of options is vast (in fact, it is $n^2 - 1$, or 24, if one assumes that direction of order matters). The result is that the computer does not necessarily find the optimal solution and instead often uses heuristics to arrive at nearly optimal solutions. On the other hand, this sort of spatial problem would be difficult if not impossible to calculate without computational assistance.

Service areas comprise those areas within a given distance of a set of facilities. In this case, the necessity

Figure 4.5 Locations of (a) stops on a network; (b) a route for ordered stops; and (c) the shortest path for all stops

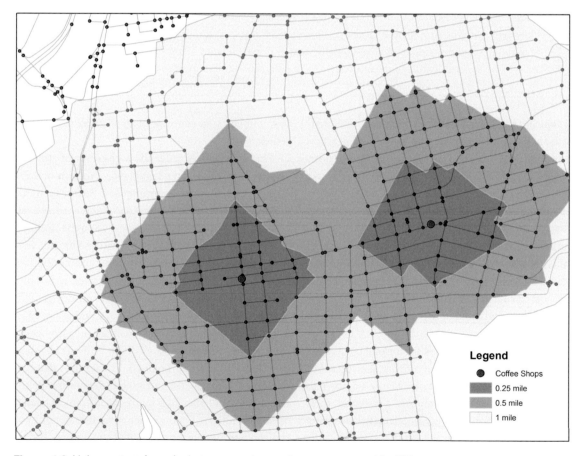

Figure 4.6 Using network analysis to compute service areas around facilities

for a network is not always apparent; one could also use buffers to identify locations close to facilities. The advantage of using the network, though, is that cost is calculated in terms of actual routes used. This advantage is made clear if time rather than distance is used as a cost variable. In that situation, 'islands' surrounding highway exits might fall into a service area, since travel time along a highway might be much faster than on smaller streets. Also the term 'service area' belies the utility of this tool: it can also be used to estimate market areas for retail and/or coverage for a set of public services, such as libraries, or accessibility in general.

The mechanics of service area computation are as follows: the GIS uses the cost variable to find all network locations that fall within the given cut-off or cut-offs. The output is either a set of polygons showing a service area or a set of lines showing all segments that meet the analytical criterion. Figure 4.6 gives a visual example of service area computation for three distances, a quarter mile, half mile and mile, for two facilities. The darkest shaded areas are those reachable within a quarter of a mile using the road network.

Service area polygons can be used in subsequent analysis to answer a range of questions, the most basic of which might be to assess the share of the population that has access to some type of facility within the given range. This coverage assessment might show whether the existing set of facilities provides coverage or access to the entire population. It might also indicate locations where additional facilities could be located, so as to increase overall coverage. The service area polygons also enable comparisons between areas inside different distance bandwidths of facilities. Perhaps the object of study is not access versus no access, but rather whether the characteristics of people or households differ according to accessibility. Service areas could thus be used to ascertain whether facilities are located equitably. Another use of service area analysis is to identify some set of locations that fall within identified polygons – for example, all the playgrounds located close to a set of schools.

Location-allocation refers to the inherently geographical challenge of locating facilities in such a way that some aspect of market or areal coverage is prioritised. The combination of *location* and *allocation*

speaks to the process by which facilities are located and demand allocated to those facilities. As a tool, location-allocation is used most often in public sector GIS applications. It is harder to use these models in retail and commercial applications as the competitive nature of these sectors makes the location criteria of distance minimisation less relevant. In retail, for example, overlapping catchments are the norm whereas location-allocation catchment areas are unique to that facility. That is the reason why they work better for sectors that are not largely in competition, such as public facilities, for example libraries, fire stations, etc. That said, Goodchild (1984) offers an interesting modification of the models to better suit retail/business applications (see also the discussion in Chapter 8).

The goal of location-allocation might vary but the initial steps are similar. First, potential locations for facilities (whether schools, stores or public libraries, for example) are decided upon. The GIS will not propose site locations from nowhere; these should have been determined from previous analysis (available land, zoning, etc.). Potential sites might include prospective locations, existing locations or some combination of the two. It is also possible to specify whether certain sites must form part of the solution. The analysis also requires information about demand – locations of households, individuals or neighbourhood centroids – which will use the sites being located. These locations might be weighted by the number of people at the demand site. The two are then compared and locations are chosen that best fit the type of problem being solved. Within this framework, the analyst has a few choices, including:

1 *Is the goal to maximise coverage?* This option seeks to locate facilities such that the furthest distance travelled is minimised or falls within a particular cut-off. Coverage maximisation is used when it is important that even the longest distance travelled between demand point and facility meet a particular threshold. For example, the average distance between fire stations and housing units is less important than ensuring the fire engines can reach the furthest location within a set amount of time.

2   *Is the goal to minimise total cost or impedance?* This option can be used to locate just one facility or several. It is referred to as the 1-median problem in the former case and the *p*-median problem in the latter – with *p* referring to the number of facilities to be located. Here, the object of the analysis is to locate facilities such that the total, weighted aggregate travel is minimised. Because this point of Minimum Aggregate Travel (MAT) minimises the total burden on demand locations – whether households or businesses – it is often used in situations in which public services such as libraries are being located.

3   *Is the goal to locate as few facilities as possible?* Perhaps building the facilities will be expensive. If so, the goal might be to locate as few facilities as possible to still meet coverage requirements. With this option, the GIS will allocate as much demand as possible to the fewest potential locations.

4   *Is the goal to secure a certain market share, given the locations of competitors?* This choice chooses locations with regard to existing competitor locations and seeks to either maximise or ensure a certain market share. Depending on competitor and facility location, a larger or smaller share of the market will be allocated to the potential facilities; the aim of the analysis is to place them such that market share is maximised (or reaches some predetermined threshold).

Closest facility tools compare a set of facilities and incidents – really just two separate sets of spatial information – and identify the closest facility to a given incident, or vice versa. The terminology employed is

Figure 4.7 Finding the closest facility for a set of 'incidents'

the same one would use to find the closest hospital to a car accident (on the assumption that the closest hospital would be the best to go to in an emergency). Figure 4.7 shows a set of incidents and facilities and demonstrates that the process finds the closest facility for each incident, and there is no requirement that each facility be used. That is, one could also use the analysis to identify underused facilities. In spite of this language, closest facility analysis is actually applied in a range of situations. This is because it serves the dual purpose of calculating cost/distance between two points but also finds the closest from a larger set. So, given a set of communities and a set of doctors, closest facility could find the closest doctor and would also estimate the distance to reach her/him. Or given locations of children and a set of primary schools, children could be allocated to a school according to the shortest distance. As with other types of network analysis, the output is visual (see Figure 4.7) but a cost variable is also generated as part of the resulting attribute table. This variable can then be used for further analysis.

Origin–destination (O–D) matrix is exactly what it sounds like. Using the underlying network, the GIS calculates the lowest cost distance between a set of origins and destinations. These locations could be a set of cities, but also doctors and patients or neighbourhoods and schools. The result is a table with locations on the margins and distances in the cells. This table can then be joined back to an attribute table for either origins or destinations and then used in further analysis. For example, most calculations of accessibility (discussed in Chapter 1 and in most of our application chapters) require distances as inputs. The O–D matrix can be used to identify locations within a particular distance threshold or to calculate the most easily reachable locations in the data set. Unlike the closest facility option above, the O–D matrix provides distances from all (and to all) origins and destinations. You will work with O–D matrices in Practical 4a (retail) and Practical 10 (emergency planning).

Location-allocation models appear in a large number of applications of GIS. We shall explore them in various chapters in Part II, and you will be able to apply a location-allocation model to maximise coverage of health services in Practical 5. You can also apply network analysis in Practical 6 to assess the provision of fire stations serving an 'at risk' population.

Concluding comments

This book is designed to introduce basic GIS concepts and functionality in Part I and provide subject-specific information in Part II. Those who would like to learn more about network analysis in the social sciences can gain more insight in the applications that follow in Part II. In Practical D we examine how to create a network in ArcMap. This helps to address the question of how to work with a network data set for routing applications. In Part II we shall explore network analysis in relation to a number of key policy-relevant questions:

Chapter 7: Crime analysis
How can we estimate the 'journey to crime', or the distance between perpetrator and victim?

Chapter 8: Retail planning
Can we identify food deserts, areas far away from grocery stores, in a particular location?

Chapter 9: Health care
Do some residents lack proper access to health care? Does distance to a hospital or a doctor partly explain health outcomes?

Chapter 10: Emergency planning
If a natural disaster impacts an area, who will be affected and what is an appropriate evacuation plan?

Chapter 11: Education planning
How can school bus routing be optimised?

We hope the reader will use these application chapters to help understand the key concepts introduced in this chapter.

Accompanying practical

This introduction to networks in a GIS is accompanied by a practical activity that gives you the opportunity to work with a network data set (Practical D: Network analysis). The practical introduces a vector data set related to tram routes in the Australian city of Melbourne. We walk you through the process to create a network data set from these tram routes before exploring the tools provided by the software in order to carry out complex routing queries.

Further reading

The papers below provide additional background regarding network theory and operationalisation in a GIS. Readers are advised to consult published literature in their field (or chapters in Part II) for applied examples of different techniques.

Boscoe, F. P., Henry, K. A., & Zdeb, M. S. (2012) A nationwide comparison of driving distance versus straight-line distance to hospitals. *The Professional Geographer*, 64(2), 188–196.

Curtin, K. M. (2007) Network analysis in geographic information science: review, assessment, and projections. *Cartography and Geographic Information Science*, 34(2), 103–111.

Fischer, M. M. (2006) GIS and network analysis. *Spatial Analysis and GeoComputation: Selected Essays*, 43–60.

Frizzelle, B. G., Evenson, K. R., Rodriguez, D. A., & Laraia, B. A. (2009) The importance of accurate road data for spatial applications in public health: customizing a road network. *International Journal of Health Geographics*, 8, 24. DOI: http://doi.org/10.1186/1476-072X-8-24.

Reference

Goodchild, M. (1984) ILACS: a location allocation model for retail site selection. *Journal of Retailing*, 60, 84–100.

Part II

Applications

5 GIS and the classification of people and areas

LEARNING OBJECTIVES

■ Theory and background of social area indicators and geodemographic classifications
■ Building area indexes of deprivation
■ Indexes of poverty and wealth
■ Indexes of quality of life
■ Building geodemographic classifications

Introduction

How much of the consumer behaviours, preferences and socio-economic characteristics of individuals are revealed by where they live? Can we really know something about who you are if we know where you live?

(Harris *et al.*, 2005: xiii)

The relationship between people and places is a key theme in GIS analysis and there has been a rapidly increasing number of relevant GIS studies that bring together multiple pieces of socio-economic and demographic data in order to create composite indicators and classifications that summarise the characteristics of residential areas. This type of research is very

relevant to the GIS functionality introduced in the previous chapters. In particular, it has almost become synonymous with GIS, especially in the geodemographics industry discussed later on in this chapter. In particular, mapping and visualisation are the main GIS functions employed, and the data are likely to be contained in the GIS where they can be manipulated relatively easily (in terms of adding/subtracting, calculating rates, etc.). As so many applications of deprivation, wealth and geodemographics are contained within a GIS platform we feel it is essential to include this within the collection of GIS applications. This chapter provides an overview of two key distinctive (but also in some ways conceptually similar) approaches to the socio-economic and demographic classification of areas and people: the social (composite) indicators approach and the geodemographic classification approach.

It can be argued that the first systematic study of social conditions of population is Friedrich Engel's *Conditions of the Working Class* (Engels, 1845). However, the first systematic attempt to classify and map socio-economic areas in this way dates back to the seminal work of Charles Booth, a British industrialist, researcher and social reformer who lived in Victorian England (between 1840 and 1916). Booth put together a team of data collectors and conducted a survey of life in the city of London. The output of this survey formed the basis for a study entitled 'The Life and Labour of the People of London' first published in 1889. This study was conducted in response to public debates at the time regarding the extent of poverty in London and reports that a quarter of the population lived below the poverty line (Vickers, 2006; Hyndman, 1911; Orford *et al.*, 2002; Simey and Simey, 1960). Charles Booth considered such reports to be exaggerated and he conducted a survey of living conditions in London aimed at proving such claims to be wrong (Vickers, 2006; Norman-Butler, 1972), but eventually concluded that the extent of poverty was actually even higher, with 30.7% of the population living below the poverty line (Simey and Simey 1960; Pfautz, 1967; Vickers, 2006).

Booth classified London's then population of four million inhabitants into seven social classes on the basis of income and produced maps of every street in London, labelling each house according to the class to which it belonged. He collected information on the living conditions of each household and each street which was then combined with information from school board visitors to establish the general socio-economic conditions in which the residents lived. This information was then used to decide to which group each street should be assigned. A colour scheme was then applied to a base map of London to graphically illustrate the general socio-economic status of the people living in each street. Although not reproduced here, examples of Booth's classification can be found in the Charles Booth Online Archive of the London School of Economics (LSE, 2015).

This work is considered to be the first attempt worldwide to classify areas and people in a systematic way and Charles Booth is widely considered as the father of area classification (Vickers, 2006; Rothman, 1989). This work was also the source of inspiration for a similar study at the end of the 19th century by Seebohm Rowntree, another British industrialist and social reformer. Rowntree carried out a detailed survey of the living standards of households in York and developed and implemented innovative methods for the measurement of poverty. He reached conclusions similar to those of Booth for London, suggesting that over one-third of the York population lived in poverty. Rowntree monitored the trends of poverty in York throughout the first half of the 20th century by undertaking surveys of the city again in 1936 (Rowntree, 1941) and in 1950 (Rowntree and Lavers, 1951).

The work of Charles Booth and Seebohm Rowntree played a significant role in the advancement of the social sciences and of area classification and poverty studies in particular. Since then there have been numerous area classification studies of poverty and of income inequalities, which have employed a wide range of new methods and alternative definitions of poverty. At the same time, there has been an increasing availability of a wide range of new socio-economic data sources in both the public and private sectors, and increased power and portability of personal computers, which have created an enabling environment for the testing and implementation of new social scientific methods and, more recently, the use of GIS. These methods are now part of a tradition in social geography and, more recently, geoinformatics and GIS, of analysing and combining different variables reflecting the circumstances and quality of life of people in order to provide suitable characterisations or profiles of neighbourhoods. These classifications of areas are routinely used in order to support decision making in the public and private sectors, ranging from allocating public resources to different neighbourhoods to customer-profiling and location-related decisions by retailers.

The remainder of this chapter provides an overview of some key studies and methods in area classification, ranging from indexes of deprivation to indexes of poverty and wealth, human development and geodemographic classifications. The accompanying practical exercise demonstrates how some of these methods can be applied using ArcMap.

Composite indicators

Indexes of deprivation

As noted above, the work of Charles Booth and Seebohm Rowntree was extremely innovative and revolutionary for their time and it has been hugely influential in the formulation of new social science research methods and the development of poverty measures and of area classification techniques. In particular, it provided the basis for the development and mapping of poverty measures and composite indicators of deprivation and quality of life. The first efforts were based on measuring absolute poverty and the extent to which the earnings of a household were sufficient in order to maintain a very basic, subsistence standard of living. This was the basis for the setting of social policy and social security benefits worldwide, especially after the end of the Second World War.

However, it has increasingly been argued that the concept of poverty constantly evolves and that the subsistence approach to the definition of poverty is inadequate. There is a need for the development of new indicators that take into account the social role of human beings (and not just the need for physical subsistence) and the extent to which they are able to meet their obligations as workers, parents, neighbours, friends and citizens; obligations that they are expected to meet and which they themselves want to meet (Gordon and Pantazis, 1997). It has been increasingly recognised that poverty is a relative concept and that it should be defined on the basis of contemporary living standards and social norms. Among the key proponents of this approach was Peter Townsend, whose work has been extremely influential in changing the policy debates and focus from *absolute* to *relative* poverty. His seminal book entitled *Poverty in the United Kingdom* (1979) provides an extensive discussion of relevant theoretical work and concepts of relative poverty and relative deprivation and argues that the poverty line should be set at a level of income below which people are excluded from participating in the norms of society in which they live. Townsend argued that people are in poverty if they additionally lack the resources

to escape deprivation (Townsend, 1987; Noble *et al.*, 2006). Townsend also distinguished between social and material deprivation and proposed ways to measure them. A key consideration in the development of such a measure is that 'double counting' should be avoided to ensure that the distribution and severity of deprivation is not misperceived and resources misallocated as a result (Townsend, 1987).

Adding a geographical dimension to this work, Townsend proposed an area-based measure of material deprivation, based on geographical areas rather than individual circumstances. This index, known as the Townsend index, is based on the following census variables (Townsend *et al.*, 1988):

1   percentage of households without access to a car or van;
2   percentage of households with more than one person per room (overcrowding);
3   percentage of households not owner-occupied (tenure);
4   percentage of economically active residents who are unemployed.

The calculation of the index involves a standardisation of the above percentages using z-scores (which can be calculated by subtracting the mean value and dividing by the standard deviation) to prevent results being excessively influenced by a high or low value for any one variable and to put each variable on the same scale, centred around zero. The four z-scores are then summed for each area to obtain a single value which is known as the Townsend deprivation index. Positive values suggest that an area has high material deprivation, whereas low values suggest that the area is relatively affluent. The Townsend index has long served as a general measure of area deprivation in academic studies in a wide range of fields including health care analysis, educational attainment and the geography of crime, but it has also been used to inform decision making in relation to resource allocation, to target areas of greatest social need.

Townsend's approach has also been the basis for the development of alternative measures of area classification. For instance, the Carstairs index of deprivation

(Carstairs and Morris, 1991; Carstairs, 1995) is derived by combining the following variables:

- *Unemployment*: unemployed male residents over 16 as a proportion of all economically active male residents aged over 16.
- *Overcrowding*: persons in households with one or more persons per room as a proportion of all residents in households.
- *Non-car ownership*: residents in households with no car as a proportion of all residents in households.
- *Social class*: residents in households with an economically active head of household in social class IV or V as a proportion of all residents in households.

As is the case with the Townsend index, the first step in calculating the Carstairs measure is to obtain percentages for the above variables and convert them to z-scores, which are then added up to give a single score for each area. Positive scores are associated with higher deprivation and negative scores with lower deprivation. The practical exercise accompanying this chapter demonstrates how the Carstairs index can be calculated and mapped using ArcMap.

An important note to be made about all area measures of deprivation is that they are based on aggregate attributes (e.g. unemployment rates) of geographic areas, rather than individual circumstances and characteristics. Not everybody in an area ranked as deprived is her/himself deprived and not all people that can be defined as disadvantaged or deprived live in deprived areas (the so-called 'ecological fallacy': see Chapter 2 and later discussion). It is also important to consider the *compositional, collective* and *contextual* dimensions to deprivation which are as follows:

- An area is deprived if it contains a large number or proportion of 'deprived' people (*compositional meaning*).
- Area effects: there is an 'area effect' which is above and beyond that which is attributable to the concentration of deprived people in the area (*collective meaning*). For example, there is evidence suggesting that where a person lives might influence

their health, even accounting for individual risk factors (Stafford and Marmot, 2003).

- Environmental: lack of facilities in the area, or some other area feature (*contextual meaning*).

There is a wide range of deprivation and other composite indicators applied at the small-area or neighbourhood level. Their suitability depends on the purpose for which they were built. Overall, the key steps for constructing an index can be summarised as follows:

- Decide the aim of the composite index (depends on the response variable; e.g. health-related applications, crime, etc.).
- Select the variables you wish to include in the index (depends on the aim).
- Decide on the form of component measure you need (e.g. percentage, z-score, ratio to mean, ranks).
- Decide whether normalisation is needed (conversion of the distribution into a normal one).
- Decide on the weighting of the components and any rescaling.

In many cases it is also useful and appropriate to assign different weights to different variables or groups of variables used to build a composite index. A good example is the official Indices of Multiple Deprivation (IMD) produced by the UK Department for Communities and Local Government. These are based on over 40 separate variables across administrative, survey and census data placed between the following 'domains' of deprivation: employment, income, health, crime, education, living environment and barriers to services (see Figure 5.1). The IMDs are produced at the UK LSOA geography (on average 1,500 residents) which are scored, ranked and mapped. The IMD was first developed and published in 2000 and was based on a combination of six separate indexes based on different domains of deprivation:[1]

- Income (25% weighting);
- Employment (25% weighting);
- Health Deprivation and Disability (15% weighting);
- Education, Skills and Training (15% weighting);
- Housing (10% weighting);
- Geographical Access to Services (10% weighting).

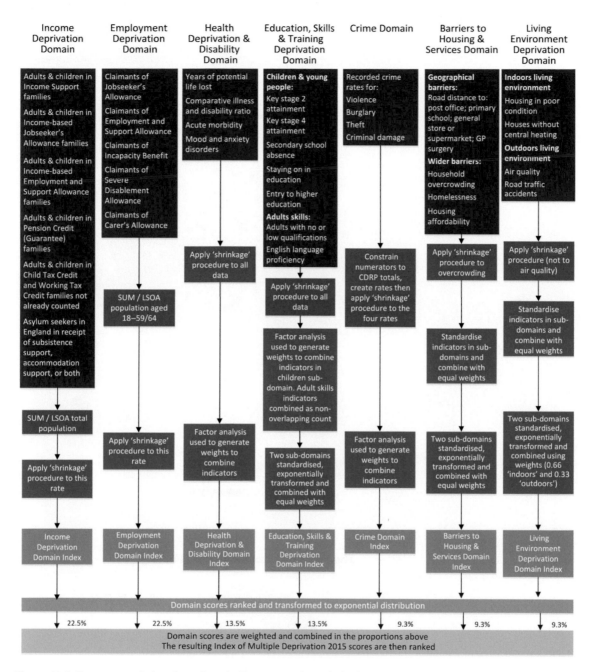

Figure 5.1 Summary of the domains, indicators and statistical methods used to create the Indices of Multiple Deprivation, 2015

Source: Smith *et al.* (2015: 19)

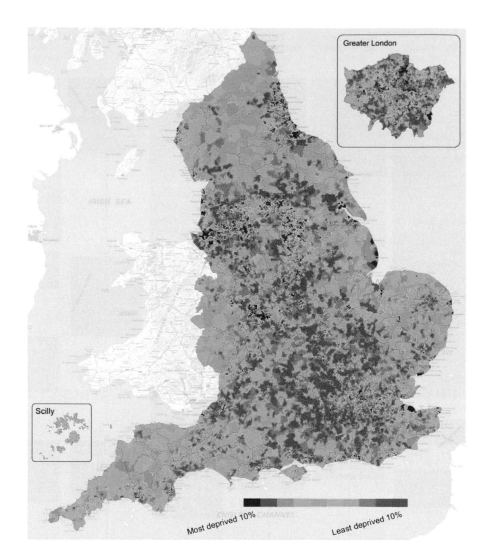

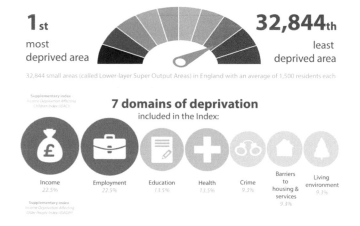

Figure 5.2 English Index of Multiple Deprivation, 2015

Source: Smith *et al.* (2015: 32). Contains Ordnance Survey data © Crown copyright 2015

The constituent variables included in the IMD have been periodically revised with new variables added and changes to weightings.[2] This means that caution is needed when comparing IMDs through time. The most recent index was released in September 2015, covering 32,844 small areas across England (see Figure 5.2) and offering a very useful insight into poverty, deprivation and social geography patterns at the local level.

Figure 5.1 summarises the data sources and method used to attach weights to different domains in the 2015 IMD. Unlike the Carstairs and Townsend index methods, the approach adopted in the IMD does not simply involve adding up the values of different variables. But how can one decide which aspects of deprivation (or 'domains') might be more important than others? Noble *et al.* (2006) ask these questions and suggest that there are at least five possible approaches to the weighting of different variables or groups of variables:

1 weighting driven by considerations emerging from the literature on multiple deprivation and social exclusion;
2 empirically driven;
3 determined by policy relevance;
4 determined by consensus; or
5 entirely arbitrary.

Figure 5.1 also provides more details on the IMD weightings.

It is also interesting to note that there are many examples of area deprivation indexes that are similar to the ones developed in Britain. For instance, Elbner and Sturm (2006) present census-tract level indices of deprivation in the US and they also examine potential links to physical and mental health outcomes. Another example is the work of Singh (2003) who developed an Area Deprivation Index and linked it to county-level mortality data in the US. Singh's index was also the basis for the development of an area deprivation index data set at the University of Wisconsin-Madison School of Medicine & Public Health (HIPxChange, 2016). Following Singh's approach they generated an index using US census block data and also developed a freely available toolkit (subject to free registration) which is described in Table 5.1.

INDEXES OF POVERTY, WEALTH AND SOCIAL EXCLUSION Other relevant approaches to area classification involve the analysis of poverty as well as wealth and the extent to which societies are geographically and socio-economically polarised and segregated. To that end a project funded by the Joseph Rowntree Foundation (Dorling *et al.*, 2007) aimed at building indicators of poverty and wealth in Britain by combining suitable national survey data with small-area geographical information from the census and other sources. A key feature of this approach was the use of the so-called *Breadline Britain* method (Pantazis *et al.*, 2006) that measures relative poverty based on the lack of what the general public (according to social surveys) considers to be a necessity of life. The Breadline Britain surveys showed that:

> Between 1983 and 1990, the number of households who lacked three or more socially perceived necessities increased by almost 50 per cent. In 1983, 14 per cent of households were living in poverty, and by 1990 this figure was 21 per cent. Poverty continued to increase during the 1990s and, by 1999, the number of households living in poverty on this definition had again increased to over 24 per cent, approximately 1 in 4 households.
> (Gordon *et al.*, 2000: 52)

The approach adopted in this project measured directly what people cannot afford that most other people think are necessities in contemporary times (e.g. being able to afford a television, a personal computer, a DVD player and, more recently, a tablet or a smartphone). In addition to producing measures of social exclusion based on relative poverty, Dorling *et al.* (2007) argued that there should also be measures of wealth, suggesting that while there has been a considerable body of work on poverty, including a wide range of approaches that attempt to measure poverty, this is not matched by the amount of literature that addresses wealth. Among the few attempts to build wealth indicators is the work of Rentoul (1987) who used Townsend's approach to suggest that the 'wealth line' can be positioned based on the amount of money needed to eliminate poverty. Thus we need to estimate the point 'in the distribution

Table 5.1 The HIPxChange Area Deprivation Index and toolkit

Variables

An Area Deprivation Index representing a geographic area-based measure of the socio-economic deprivation experienced by a neighbourhood, using the original index developed by Gopal Singh, PhD, MS, MSc which involved 17 different markers of socio-economic status from the 1990 census data. HIP has generated an updated index using 2000 census block group-level data and the original Singh coefficients from the 1990 data. The index includes the following variables:

- Percentage of the population aged 25 and older with less than 9 years of education
- Percentage of the population aged 25 and older with at least a high school diploma
- Percentage employed persons aged 16 and older in white-collar occupations
- Median family income in US dollars
- Income disparity
- Median home value in US dollars
- Median gross rent in US dollars
- Median monthly mortgage in US dollars
- Percentage of owner-occupied housing units
- Percentage of civilian labour force population aged 16 years and older who are unemployed
- Percentage of families below federal poverty level
- Percentage of the population below 150% of the federal poverty threshold
- Percentage of single-parent households with children less than 18 years of age
- Percentage of households without a motor vehicle
- Percentage of households without a telephone
- Percentage of occupied housing units without complete plumbing
- Percentage of households with more than one person per room

What does the Area Deprivation Index toolkit contain?

- A file that explains how to use the dataset
- Data sets with Area Deprivation Index scores for a variety of US Census data levels:

  - 9-digit zip codes (available in one large data set or smaller state-specific data sets)
  - Zip code tabulation areas (ZCTAs)
  - US Census block group codes
  - US County

For more information see: www.hipxchange.org/ADI

Source: www.hipxchange.org/ADI

of resources above which this could be made available'. A similar method has been employed in a study conducted as part of the United Nations Development Programme (Medeiros, 2006). Dorling *et al.* (2007) introduced the concepts of wealth-rich and income-rich; someone can be very wealthy, but have a very low income (e.g. someone living in a stately home with much in the way of accumulated resources, but little cash income), and equally someone could have a high income and little wealth (e.g. contractors with high

levels of income but insufficient stability to buy their own homes or accumulate assets).

Building on these previous relevant efforts, Dorling *et al.* assessed the changing geographies of poverty and wealth in Britain over the last three decades of the 20th century and constructed a coherent set of poverty and wealth measures for unchanging geographical areas for time periods around 1970, 1980, 1990 and 2000. The analysis involved the division and mapping of the population of each area at each time period into five groups:

1   The '**core poor**': defined theoretically according to the Taxonomy of Need (Bradshaw, 1972a, 1972b, 1994) as people suffering from a combination of normative, felt and comparative poverty, i.e. people who are simultaneously income poor, necessities/deprivation poor and subjectively poor (see Bradshaw and Finch, 2003).

2   The '**breadline poor**': under the Relative Poverty Line – defined theoretically by Townsend (1979) as the resource level that is so low that people are excluded from participating in the norms of society, and measured by the Breadline Britain Index (see Gordon and Pantazis (1997) for a detailed discussion of how the index is derived). This is the same theoretical definition of poverty used by the European Union to measure poverty and social exclusion.

3   The '**asset wealthy**': those living above an asset wealth line. Here, housing wealth data were used (developed by Thomas and Dorling, 2004) to estimate the number of wealthy households as those living above an asset wealth threshold, for time periods comparable to those used for the poverty measures. While housing wealth is a particular facet of wealth, it is most likely correlated with more general wealth.

4   The '**exclusive wealthy**': those living above a (higher) wealth line, defined as being over 'a point in the distribution of resources at which the possibility of enjoying special benefits and advantages of a private sort escalates disproportionately to any increase in resources' (Scott, 1994: 152); a resource level that is so high that people are able to exclude themselves from participating in the norms of society (if they so wish). This represented the first approach to operationalising a wealth line such as this. This was done by using the UK Family Expenditure Survey (FES) data in combination with the Households Below Average Income (HBAI) adjustments to the incomes of the very 'rich'. The HBAI adjustments account for household size and type when considering household income and are the same as those used in the Breadline Britain methodology. The adjusted FES data can then be used to discover the average band/level of income at which children go to independent schools, people use private health care, have second homes, boats, pay private club membership fees, etc.

It is assumed that the core poor are a subset of the breadline poor, and similarly that the exclusive wealthy are a subset of the asset wealthy. By defining these four groups of households, a fifth 'middle' group is also defined by default – those which are neither poor nor wealthy, counted as the remainder of households in a geographic area after accounting for groups 2 and 3 (which are presumed to contain groups 1 and 4, respectively). As described above, the availability of the data determines the exact time periods under consideration, and in the case of housing wealth data, we do not have data for 1970 or any time around then, limiting the time series with respect to this wealth measure. Table 5.2 and Figure 5.3 describe and summarise the data and methods used in this project.

Table 5.2 Poverty, wealth and place data sources and methods

Poverty and core poverty	*Asset wealth*
■ 1968/9 Household Resources & Standard of Living Survey (Townsend, 1979)	■ Tract-level data on housing asset ownership (Thomas and Dorling, 2004)
■ 1983 Poor Britain Survey (Mack and Lansley, 1985)	*Exclusive wealth*
■ 1990 Breadline Britain Survey (Gordon and Pantazis, 1997)	■ Family Expenditure Survey data for the years:

■ 1999 Poverty & Social Exclusion Survey (Gordon et al., 2000)	– 1970–72
	  – 1980–81
	  – 1990–92
	  – 2000–02

Source: Dorling *et al.* (2007)

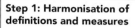

Step 1: Harmonisation of definitions and measures
a) Between surveys:
  • UK poverty surveys
  • Family Expenditure Survey
b) Between surveys and UK census (1971; 1981; 1991; 2001)

Step 5: Derivation of small-area census estimates
a) Application of (reflated) model weights to census data
b) Application of estimates to longitudinally consistent spatial units (Universal Data Mapping)

Step 2: Harmonisation of data
a) Reweighting of data to relevant UK census population distribution
b) Recalibrating of expenditure data to common time points (FES only)

Step 3: Setting thresholds for poverty and wealth
a) Poverty: 'Breadline' method (Gordon, 1995)
b) Core Poverty: A priori (Whelan *et al.*, 2001)
c) Exclusive Wealth: Cluster analysis (Ward's method)

Step 4: Modelling of poverty, core poverty and wealth
a) Logistic regression models for each census period
b) Recalibration of models to UK population estimates

Figure 5.3 Poverty, wealth and place overview of methodology

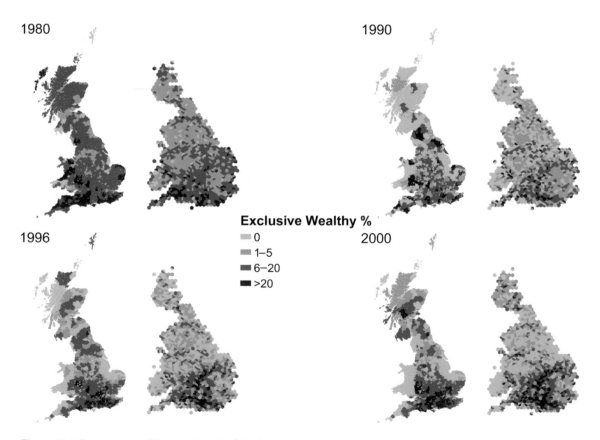

Figure 5.4 Poverty, wealth and place in Britain

Source: Dorling *et al.* (2007)

Figure 5.4 gives a flavour of the GIS mapping that was used to depict the distribution of these areas across Britain using conventional thematic mapping as well as the human cartographic approach discussed in Chapter 3. In general, mapping of these poverty and wealth data sets indicates that wealth and poverty each demonstrate similar geographical patterns at every time period. The highest wealth and lowest poverty rates tend to be clustered in the south-east of England (with the exception of most of inner London); conversely the lowest wealth and highest poverty rates are concentrated in large cities and industrialised/de-industrialising areas of Britain. However, this project revealed some interesting and substantial changes to this generalisation over time. Analyses of the degree of polarisation and spatial concentration suggest that Britain's population became less polarised with regard to area breadline poverty rates during the 1970s. However, polarisation increased through the 1980s and the 1990s. Asset wealth also became more polarised during the 1980s, but this trend reversed in the first half of the 1990s, before polarisation could be seen to be increasing again at the end of that decade.

Dorling *et al.* (2007) also conducted GIS-based spatial analysis and, in particular, spatial autocorrelation analysis, to identify clusters of poverty and wealth, or in other words to indicate the extent to which areas with high levels of poverty tend to be found near to other areas with high levels of poverty and, similarly, areas of high wealth are found near other areas of high

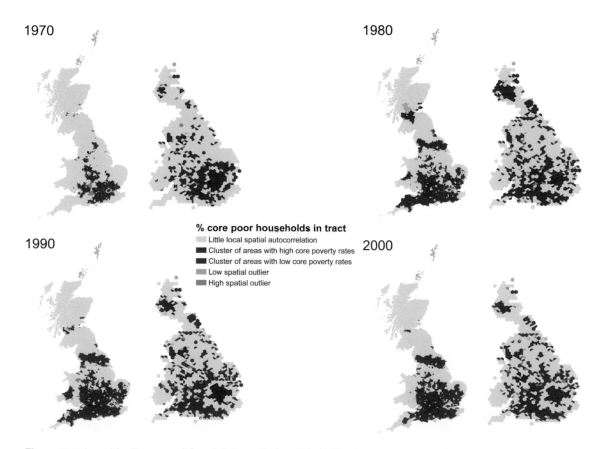

Figure 5.5 Local Indicators of Spatial Association (LISA) for the percentage of households classified as core poor at each time period

Source: Dorling *et al.* (2007)

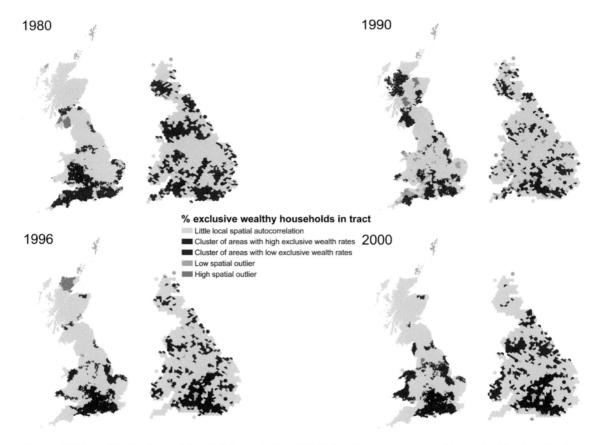

Figure 5.6 Local Indicators of Spatial Association (LISA) for the percentage of households classified as exclusive wealthy at each time period

Source: Dorling *et al.* (2007)

wealth. Spatial autocorrelation involves the calculation of measures of spatial association and they are increasingly embedded in GIS packages. For example, they are a standard feature of the Spatial Statistics toolbox of ArcGIS[3] (which is part of the 'Modelling spatial relationships' tools[4]) that enables the analysis of spatial patterns of a variable and the statistical evaluation of whether any observed patterns can be considered to be clustered, dispersed or random. Dorling *et al.* (2007) carried out spatial autocorrelation analysis to effectively measure how similar or dissimilar each area is to its neighbours, giving an indication of the degree of spatial concentration of poverty and wealth. In particular, they calculated a spatial association measure on a local/regional basis, resulting in a map of 'spatial clusters' of similar areas, building on a relevant study

by Orford (2004) that employed 'Local Indicators of Spatial Association' (LISA; Anselin, 1995) techniques to investigate the spatial concentration of poverty in 1896 and 1991 using a social class measure. Figures 5.5 and 5.6 show the results of this analysis, highlighting clusters of poverty and wealth rates, respectively (for a comprehensive discussion of these outputs and of the data and methods used see Dorling *et al.*, 2007).

INDEXES OF ANOMIE   In addition to measures of poverty and wealth there are possibilities of composite indicators that attempt to quantify concepts that relate to living conditions, quality of life and social norms and social well-being. One example is the so-called 'anomie' measure which can be seen as an index of 'local well-being' (anomie is the sociological term to describe, according to some interpretations, the feeling of 'not belonging'). Such measures can also be described as 'loneliness indices'. Ballas and Dorling (2013) and Dorling *et al.* (2008) present such an index to explore the geography of loneliness or anomie in Britain in a study commissioned by the BBC which aimed at comparing variations across BBC radio and TV regions. The index used was based on a scale and weightings which have now been widely employed in many pieces of research (Congdon, 1996). In particular, the index is calculated based on weighted sums of non-married adults, one-person households, people who have moved to the area within the last year and people renting privately. The index is equal to the sum of the following multiples in each area:

- number of non-married adults multiplied by a weight of 0.18;
- number of one-person households multiplied by a weight of 0.50;
- number of people who have moved to their current address within the last year multiplied by a weight of 0.38;
- number of people renting privately multiplied by a weight of 0.80.

The data used to calculate the index are readily available in Britain for small areas from the census of population and it can be argued that they represent a number of variables that are associated with happiness and well-being (we revisit this subject in the next

Figure 5.7 The BBC radio regions map and human cartogram

Source: After Dorling *et al.* (2008)

chapter with more sophisticated geographical approaches to small-area estimation including happiness). For instance, it has long been argued that single people appear to be on average less happy than married couples (Ballas and Dorling, 2013; Frey and Stutzer, 2002; Inglehart, 1990; Helliwell, 2003) and in general there is evidence that stable and secure intimate relationships are beneficial for happiness. In contrast, the dissolution of such relationships is damaging (Dolan *et al.*, 2007). In this context the census variables 'number of one-person households' and 'number of non-married adults' could be considered suitable to measure at the local level. Also, as noted above, length of time at current address and social networks have an impact on well-being. The census variables 'number of people who have moved to their current address within the last year' and 'number of people renting privately' capture, to some extent, the degree to which people are integrated in the local community and might feel that

they 'belong'. This variable also implicitly incorporates in the analysis the spatial process of migration (as it provides the number of in-migrants in the area within the year before the census date).

Dorling *et al.* (2008) collected these data from the British censuses for the years 1971, 1981, 1991 and 2001 to compare the anomie index levels between different regions (using the BBC radio regions as a the geographical unit of analysis). They mapped this proxy of social fragmentation or local well-being index across Britain using both conventional maps as well as human cartograms that show areas in proportion to their populations (see Figure 5.7). Figure 5.8 shows the spatial distribution of anomie index in 1971, and Figure 5.9 depicts the same variable in 2001. The gap between the index extreme values has grown over time (other than during the 1970s). Figure 5.10 shows the spatial distribution of anomie index change between 1971 and 2001.

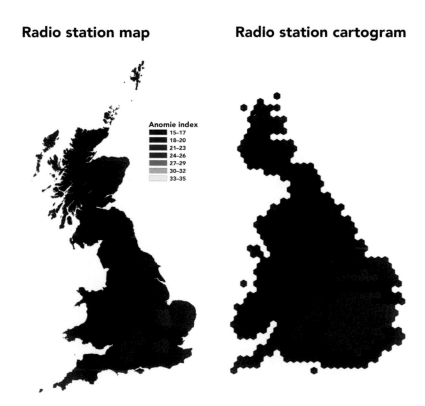

Figure 5.8 Spatial distribution of anomie index, 1971

Source: After Dorling *et al.* (2008)

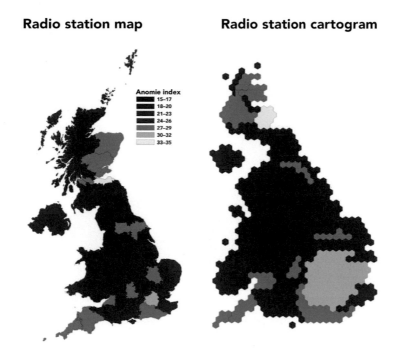

Figure 5.9 Spatial distribution of anomie index, 2001

Source: After Dorling *et al.* (2008)

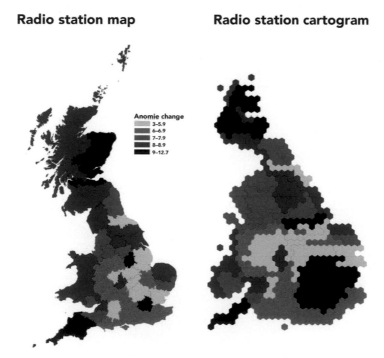

Figure 5.10 Spatial distribution of anomie index difference between 1971 and 2001

Source: After Dorling *et al.* (2008)

AREA INDEXES OF QUALITY OF LIFE, HUMAN DEVELOPMENT AND DIVERSITY Other possibilities of geographical composite indexes involve the combination of variables that relate to human development, quality of life and policy goals. A good example is the United Nations Human Development Index which measures average achievement in a series of domains (see Figure 5.11) based on the human capability approach proposed by Nobel Laureate Amartya Sen and other award-winning economists and social scientists such as Martha Nussbaum and John Rawls.

This index, which is calculated every year by the United Nations for all countries around the world, has been the basis for the development of regional and local area indexes. For instance, the European Union

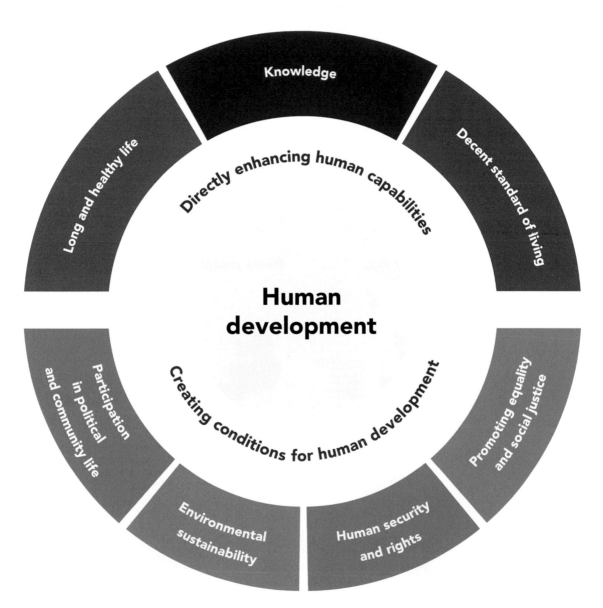

Figure 5.11 Dimensions of human development

Source: http://hdr.undp.org/sites/default/files/2015_human_development_report_1.pdf

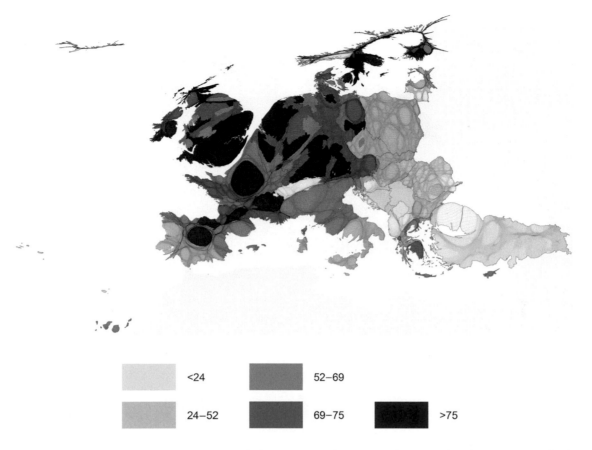

	<24		52–69		
	24–52		69–75		>75

Figure 5.12 The EU Human Development Index, 2007 (0 = low level of human development 100 = high level of human development)

Source: European Commission, 2010; based on data from Eurostat, DG REGIO (after Ballas *et al.*, 2014)

calculates the EU Human Development Index at the regional level (see Figure 5.12).[5] The index is based on life expectancy in good health, net adjusted household income per head and on levels of educational attainment of the population aged 25–64. The higher the index the better off the people of the area are said to be. Figure 5.12 is a human cartogram (Hennig projection, as discussed in Chapter 3) showing the geographical distribution of this index across the European region. The British capital city region *Inner London* has the highest value of 100, followed by the *Surrey, East and West Sussex* (93.64). The Swedish capital city region of *Stockholm* is third with an index of 93.25, followed by *Berkshire, Buckinghamshire and Oxfordshire* (92.43) in the United Kingdom and *Utrecht* (92.43)

in the Netherlands. At the other end of the range, the bottom five regions are all in Romania: *Sud-Est* with a value of zero and also *Sud-Muntenia* (0.15), *Nord-Est* (0.5), *Nord-Vest* (2.36) and *Sud-Vest Oltenia* (2.71).

Figure 5.13 shows a gridded population cartogram coloured by the geographical distribution of the so-called Lisbon Index, showing how close an EU region is to the following targets: 85% for employment rate for men aged 15–54; 64% for employment rate for women aged 15–54; 50% employment rate for people aged 55–64; 10% employment secured by early school leavers aged 18–24; 85% of people aged 20–24 attaining secondary level education as a minimum; 12.5% of lifelong learning participation among people aged

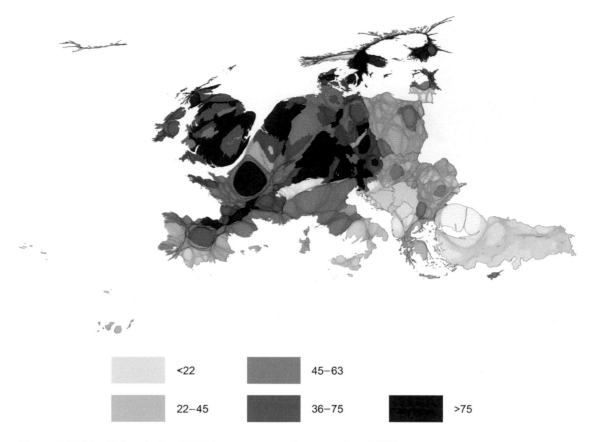

	<22		45–63		
	22–45		36–75		>75

Figure 5.13 The Lisbon Index, 2008 (average score between 0 and 100)

Source: European Commission (2010); Eurostat, DG REGIO (after Ballas *et al.*, 2014)

25–64; 2% for business expenditure in R&D as a proportion of GDP; 1% for government, higher education and non-profit expenditure in R&D as a proportion of GDP (European Commission, 2010). The index can take values ranging from 0 (for the region which is furthest away from the targets) to a maximum of 100 (meaning that all targets have been reached).

Another relevant example is the diversitydata.org project in the US. This project has produced a wide range of area indicators of diversity, opportunity, quality of life and health which are available via an interactive website (see Figure 5.14) which is aimed at allowing researchers, policy makers and community advocates to describe, profile and rank US metropolitan areas in terms of quality of life. Similarly, the diversitydatakids.org project (see Figure 5.15) focuses on child well-being and includes maps based on a

'Child Opportunity Index' (Acevedo-Garcia *et al.*, 2014).

Geodemographic classifications

> Demography is the study of population types and their dynamics therefore geodemographics may be labelled as the study of population types and their dynamics as they vary by geographical area.
>
> (Birkin and Clarke, 1998: 88)

The methods discussed so far in this chapter aim at classifying areas by a composite index in relation to different themes ranging from deprivation to poverty and wealth and human development. In this section

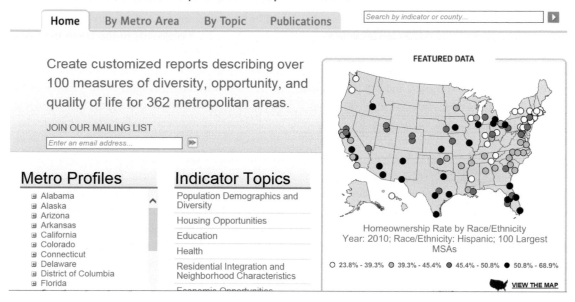

Figure 5.14 The diversitydata.org project

Source: Screenshot from www.diversitydata.org

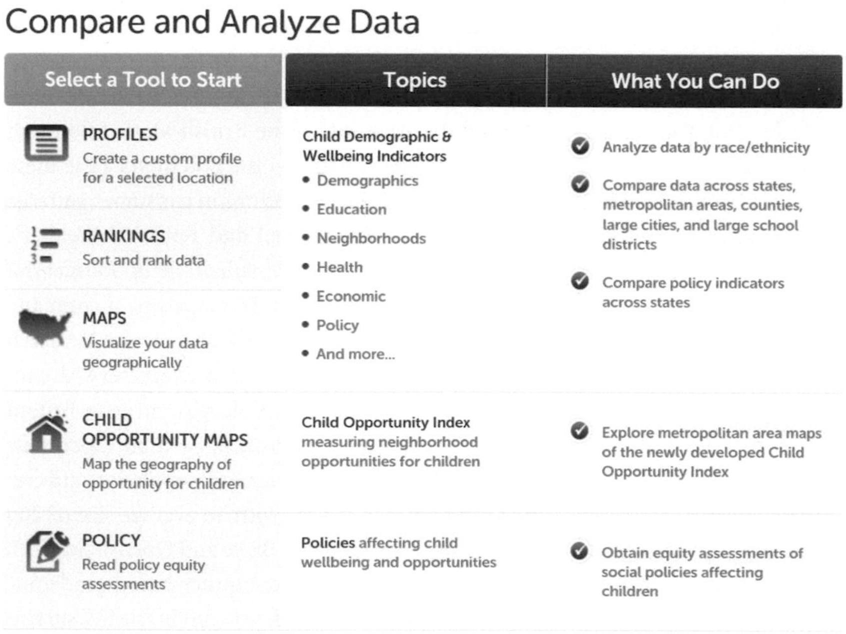

Compare and Analyze Data

Select a Tool to Start	Topics	What You Can Do
PROFILES Create a custom profile for a selected location	**Child Demographic & Wellbeing Indicators** • Demographics	✓ Analyze data by race/ethnicity
RANKINGS Sort and rank data	• Education • Neighborhoods • Health	✓ Compare data across states, metropolitan areas, counties, large cities, and large school districts
MAPS Visualize your data geographically	• Economic • Policy • And more...	✓ Compare policy indicators across states
CHILD OPPORTUNITY MAPS Map the geography of opportunity for children	**Child Opportunity Index** measuring neighborhood opportunities for children	✓ Explore metropolitan area maps of the newly developed Child Opportunity Index
POLICY Read policy equity assessments	**Policies** affecting child wellbeing and opportunities	✓ Obtain equity assessments of social policies affecting children

Figure 5.15 The diversitydatakids.org project

Source: Screenshot from www.diversitydatakids.org

we present a conceptually related but methodologically different approach to the development and application of area typologies and classifications of neighbourhoods, widely known as 'geodemographics'. As noted above, geodemographics and GIS have become almost synonymous. As is the case with composite indicators, geodemographics involves an analysis of poverty and wealth within cities and regions and the quantitative characterisation of areas (typically neighbourhoods). However, geodemographics aims to provide a more detailed and sophisticated labelling and description of places. It can be defined as the 'analysis of people by where they live' (Sleight, 1997: 16) and as the suggestion that *where* you live says something about *who* you are (Harris *et al.*, 2005) illustrated in Figure 5.16.

A key concept that underpins geodemographic classifications is that people who live close to each other tend to have similar characteristics (Harris *et al.*, 2005) which also relates to what is known as Tobler's First Law of Geography according to which, 'Everything is related to everything else, but near things are more related than those far apart' (Tobler, 1970: 236). As Vickers (2006: 16) points out, 'Geodemographics takes Tobler's law and gives it a twist, using his principle that two houses next to each other are likely to be fairly similar and contain people with comparable characteristics.' Geodemographics involves the grouping of neighbourhoods in the same locality but also similar neighbourhoods that are not geographically connected, and redefining Tobler's law as follows: 'People who live in the same neighbourhood are more similar than those who live in a different neighbourhood, but they may be just as similar to people in another neighbourhood in a different place' (Vickers, 2006: 16).

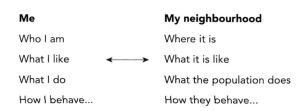

Figure 5.16 Geodemographics is about 'linking people to places'

Source: After Harris *et al.* (2005: 2)

The origins of geodemographic classification systems date back to the work of Charles Booth (which was briefly discussed in the introduction to this chapter) and by later work in social area analysis and factorial ecology by human ecologists in the US in the 1920s and 1930s (for detailed reviews of the background and history see Burrows and Gane, 2006; Harris *et al.*, 2005; Singleton and Spielman, 2014; Vickers, 2006; Birkin and Clarke, 2009). However, the first attempt to build modern computer-based geodemographic classifications can be traced back to the 1960s in Britain with the first release of the census small-area statistics in machine-readable form. This generated a lot of interest by the public sector and, in particular, local authorities that explored the possibility of using census data to classify areas by socio-economic characteristics in order to identify localities in need of investment (Vickers, 2006). One of the first analyses of these data to that end was conducted by Liverpool City Council in 1969 in their 'Social Malaise' study that was used to support decision making regarding the allocation of social services. This was followed by the work of Webber and Craig (1976) on National Classifications of areas in Britain commissioned by the UK Office of Population Censuses and Surveys (OPCS) in the 1970s and providing the foundation for the modern geodemographic industry. The OPCS classification attracted the attention of the British Market Research Bureau (BMRB) which saw the potential to use these typologies to examine variations in consumer patterns and eventually restructured and renamed Webber's classification work *A Classification of Residential Neighbourhoods* (ACORN) developed by a company called CACI[6] and launched it at the Market Research Society's 1979 conference, as a marketer's dream (Vickers, 2006). Another key development in Britain was the creation of GB Profiles by Stan Openshaw using data from the 1991 census of population and creating typologies of small (with an average size of 200 households) census tracts (Blake and Openshaw, 1995; Rees *et al.*, 2002). Similar computer-based geodemographic systems were also developed in the US, such as the work of Jonathan Robbin (Robbin, 1980).

The development of ACORN was followed by Mosaic (also led by Richard Webber) a geodemographic system developed by Experian[7] with classifications available for most of Western Europe, the US, Japan and Australia. Tables 5.3 and 5.4 provide examples and summaries of geodemographic classification systems in the US and Britain (drawing on a recent review by Singleton and Spielman, 2014), including information on the number of variables analysed, number of segment groups (level 1) and sub-groups (level 2 and micro-level) generated as well as the smallest geographical unit (taxonomic unit) at which the classification is available.

Figure 5.17 gives an example of 'pen portraits', describing geodemographic classification groups for Experian's Mosaic classification in Britain. These 'pen portraits' are created by using statistical methods, known as data reduction techniques. There are also similar geodemographic systems in the US (two examples can be seen in Figures 5.18 and 5.19).

As is the case with the composite indicator approaches, the development of geodemographics involves a classification of areas according to a set of characteristics. But unlike the composite indicator approach, which involves building a single index by combining different variables, geodemographic classification is based on a statistical method known as cluster analysis (Everitt, 1993; Everitt *et al.*, 2001). This is a classificatory technique particularly valuable and suitable for constructing typologies with data sets comprising many variables and 'cases' (which in a GIS context are geographical districts or neighbourhoods). The technique is capable of analysing large multidimensional data sets to find clusters of areas that have similar aggregate socio-demographic

Table 5.3 US general purpose area-based geodemographic classifications available in 2012

Name	Level 1	Level 2	Micro-level	Variables	Taxonomic units	Commercial	SME
Mosaic USA, Experian	19	71		~3,001	ZIP +4, block group and ZIP code	Y	N
Nielsen PRIZM	142 or 113	66		Hundreds	Block group, ZIP +4 and larger units	Y	N
Tapestry, ESRI	124 or 115	65		~60	Block group and larger units	Y	N
PSYTE Advantage, Pitney Bowes		72	400		Block group	Y	N
STI Landscape, Synergos Technologies	15	72			Block group	Y	Y
Cohorts, IXI Corporation		30				Y	N
Cameo, Callcredit Information Group	9	52			Block group	Y	N
Acxiom, Personicx	21	70				Y	N
Patchwork Nation, Jefferson Institute	12			125	County	N	N/A
American Clusters, Worldclusters	7	18			Block group	N	N/A

Note: SME = small to medium-sized enterprise.
Source: After Singleton and Spielman (2014)

Figure 5.19 The Nielsen PRIZM geodemographic classification: segment details

Source: https://segmentationsolutions.nielsen.com/mybestsegments/Default.jsp?ID=30&menuOption=segmentdetails
&pageName=Segment%Details

Table 5.5 Indicative data sets (by theme) used in the construction of Mosaic

Individual	Financial
■ Gender	■ Personal and household income and assets
■ Age	■ Shareholding value
■ Marital status	■ Outstanding mortgage
■ Surname origin	■ Directorships
■ Length of residency	■ Employment status
■ Head of household	
Property	Family
■ Age	■ Presence of children
■ Type	■ Household composition
■ Size	■ Lifestage
■ Tenure	
■ Council tax band	
■ Value	

Source: Compiled by the authors based on Experian Mosaic marketing materials

publicly available information on the data and variables used (e.g. how many age groups there are in the age variable described in Table 5.5) and methods (e.g. regarding any data transformation and the particular cluster analysis technique) used in commercial geodemographic systems such as Mosaic. This is due to commercial competition and confidentiality reasons. However, there has been an increasing number of so-called 'open geodemographic' classification systems

which do not have such limitations. A good example of such a system is the official UK Office for National Statistics geodemographic classification developed by Dan Vickers at the Universities of Leeds and Sheffield (Vickers, 2006, 2010; Vickers and Rees, 2007, 2011) using data from the 2001 census of population. Appendix 5.1 provides more details on how this system was developed, and Figure 5.20 presents maps of these outputs for South and West Yorkshire.

Building on this work Chris Gale and colleagues (2014) at University College London developed an open geodemographic classification using data from the 2011 census of population (see Figure 5.21). Such open geodemographic classification systems are free to download[8] and use for any application and the data and methods used to develop them are publicly available.

The names of the clusters described (in Figure 5.21 and Appendix 5.2) are the output of the final step in building a classification that involves a meticulous analysis in order to produce a short description by combining text and visual information such as graphs summarising statistical information, photographs of typical homes and lifestyle-related consumer products. In order to generate a suitable description the average values for each of the variables used in the cluster

analysis are calculated across all areas in each cluster in turn. These average values are then compared to the average values for the full data set to establish the characteristics of each cluster group. Clusters that have extreme values for one or more variables are generally easier to describe than clusters that have average values for all variables (Vickers and Rees, 2006). The process of identifying common characteristics and attributes that can be used to name clusters can be greatly enhanced with the use of radial plots for each cluster, as demonstrated by Vickers (2006) and Vickers and Rees (2007) in the development of the UK Office for National Statistics Census Output Area Classification discussed above. Figure 5.22 shows a radial plot cluster summary for the supergroup 'Countryside' within this classification. The radial plots represent the standardised values for each variable and the numbers on the

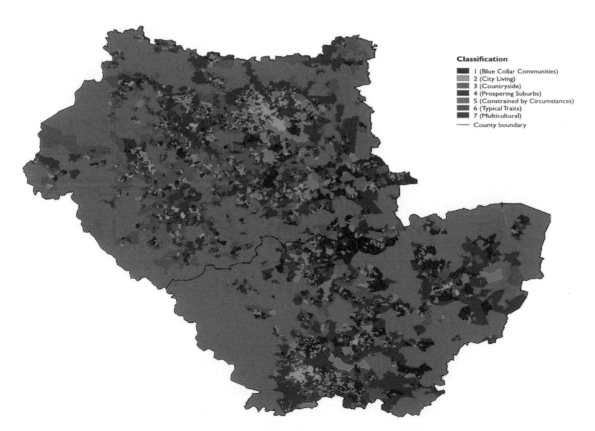

Figure 5.20 Mapping the first UK output area classification, South and West Yorkshire, UK, 2001

Source: Vickers and Rees (2006)

1 - Rural Residents
2 - Cosmopolitans
3 - Ethnicity Central
4 - Multicultural Metropolitans
5 - Urbanites
6 - Suburbanites
7 - Constrained City Dwellers
8 - Hard-Pressed Living

Southampton

0   1
Kilometre

Figure 5.21 The UK Office for National Statistics output area classification for the city of Southampton, UK, 2011

Source: Gale *et al.* (2016). Contains National Statistics data © Crown copyright and database right 2016; contains Ordnance Survey data © Crown copyright and database right 2016

scale show the difference from the mean value for that variable (i.e. the mean for all variables is 0). The value of each variable for each cluster group can be seen by the distance each point is above or below the middle (0) ring, with the outer ring representing a value of 0.5 and the inner ring a value of –0.5 (Vickers and Rees, 2007: 399).

Vickers and Rees (2007) adopted two general principles when naming their clusters: the names must not offend residents (and it is interesting to compare these cluster names with those of Booth presented in Figure 5.1) and they must not contradict other official classifications or use already established names. For instance, in their classification the terms 'rural' and 'urban' were

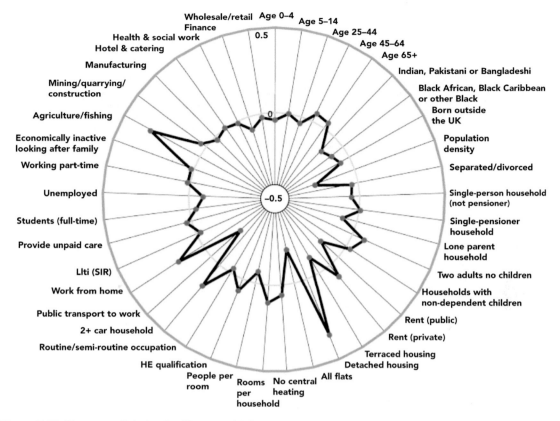

Figure 5.22 Cluster radial plot for 'Countryside' supergroup

Source: After Vickers (2006)

not used to avoid confusion with official urban-rural classification of the Office for National Statistics, whereas words such as 'affluent' were avoided due to giving potential stigma of wealth (or non-wealth) to areas. As they put it: 'Coming up with descriptive inoffensive names for some areas is easier than for others. For a pleasant area it is not such an arduous task as for areas where few would choose to live' (Vickers and Rees, 2007: 399).

Finally we note that many geodemographic suppliers are now producing pan-international classifications, in addition to classifications for different individual countries. Figures 5.23 and 5.24 for example, shows CAMEO's international classification scheme mapped for Los Angeles and Sydney, respectively. These maps pick out the highest income areas in both cities using the same worldwide classification scheme.

Concluding comments

In this chapter we have provided an overview of key methods for the classification and mapping of areas and people. We discussed the first attempts to create such classifications and the rationale for classifying areas using secondary data in the social sciences, briefly drawing on the social indicators literature and then giving examples of how indexes of deprivation and quality of life can be built on the basis of existing secondary data. We also introduced geodemographic classification methods, providing examples highlighting the key steps needed to develop geodemographic typologies of places and the issues that need to be considered. There is a growing number of new systems and further developments in this field (for a recent review see Singleton and Spielman, 2014) and ongoing research on the strengths and limitations of these approaches.

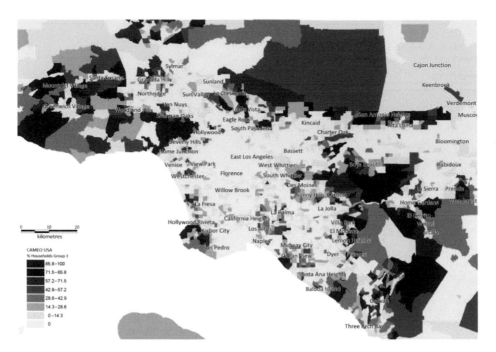

Figure 5.23 Distribution of high-income earners in Los Angeles, California, US, using geodemographics

Source: Callcredit's CAMEO International 2014

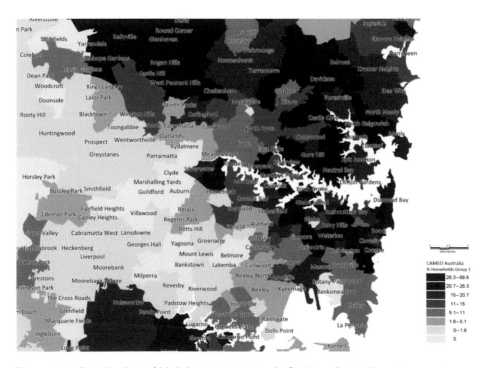

Figure 5.24 Distribution of high-income earners in Sydney, Australia, using geodemographics

Source: Callcredit's CAMEO International 2014

Nevertheless, it is also important to note that the methods discussed in this chapter have long been at the centre of lively debates and concerns pertaining to privacy, confidentiality, the production/labelling of space, ethical issues, consumer identity and stereo-typing and electronic surveillance. Among the first and most influential and comprehensive overviews of these issues is a book entitled *Ground Truth: The Social Implications of Geographic Information Systems* (Pickles, 1994) as well as an article published in the journal *Economic Geography* and entitled '"We know who you are and we know where you live": the instrumental rationality of geodemographic systems' (Goss, 1995). Also of relevance are debates about the social implications of the 'geospatial web' (Ellwood, 2010). There have been ongoing efforts to further consider and address such concerns in a wide range of contexts (e.g. see Flowerdew, 1998; Uprichard and Burrows, 2009; Beer and Burrows, 2013) and to also more generally acknowledge and highlight the societal and ethical implications of using GIS in the social sciences.

An important methodological limitation of the methods and studies discussed in this chapter is that area-level index value or geodemographic labels (as is the case with all group averages) can lead to what is known as the ecological fallacy, which was introduced and discussed in Chapter 2. This can happen when statistical information relating to an aggregated areal unit (e.g. the neighbourhood) is assumed to apply to all smaller units or individual within this areal unit (Tranmer and Steel, 1998). This issue is particularly important for areas of high levels of socio-economic and demographic diversity. For instance, it is possible to find pockets of deprivation (or single individuals or households that are considered to be deprived) in areas that are overall affluent and, similarly, pockets of affluence in areas that otherwise appear to be poor. This issue can be potentially addressed with the use of more sophisticated small-area estimation methods such as spatial microsimulation which are presented in the next chapter.

Accompanying practical

This chapter is accompanied by a linked practical (Practical 1: Area classification – creating a small-area deprivation index). We walk you through the process of downloading, preparing and importing data from the 2011 census in England. You use these data to construct an area-based deprivation index known as the Carstairs index (see above). We use the Carstairs index due to its relative simplicity, allowing you to gain experience of all the steps needed to download and process the data in order to construct and visualise the index in your GIS.

Notes

1 See http://webarchive.nationalarchives.gov.uk/2010 0410180038; http://www.communities.gov.uk/archived/ general-content/communities/indicesofdeprivation/ indicesofdeprivation/.

2 See www.gov.uk/government/collections/english-indices-of-deprivation.

3 http://pro.arcgis.com/en/pro-app/tool-reference/spatial-statistics/spatial-autocorrelation.htm.

4 http://pro.arcgis.com/en/pro-app/tool-reference/spatial-statistics/modeling-spatial-relationships.htm.

5 Also see Chapter 3 for a discussion of the cartographic method used to create this map and www.europemapper. org for more details on the geography of Europe.

6 www.caci.co.uk.

7 www.experian.com.

8 E.g. see www.opengeodemographics.com.

Further reading

Dorling, D., Rigby, J., Wheeler, B., Ballas, D., Thomas, B., Fahmy, E., *et al.* (2007) *Poverty, Wealth and Place in Britain, 1968 to 2005*, Policy Press, Bristol.

Harris, R., Sleight, P., & Webber, R. (2005) *Geodemographics, GIS and Neighbourhood Targeting*, Wiley, Chichester.

Singleton, A. D., & Spielman, S. E. (2014) The past, present, and future of geodemographic research in the United States and United Kingdom. *The Professional Geographer*, 66(4), 558–567.

Vickers, D., & Rees, P. (2007) Creating the UK national statistics 2001 output area classification. *Journal of the Royal Statistical Society: Series A. Statistics in society*, 170(2), 379–403.

Online resources (including data sources)

Consumer Data Research Centre: www.cdrc.ac.uk
Open Geodemographics: www.opengeodemographics.com
The 2001 UK Census Area Classification, Office for National Statistics Open Geodemographic Classification system: www.sasi.group.shef.ac.uk/area_classification
The 2011 UK Census Area Classification, Office for National Statistics Open Geodemographic Classification system: http://geogale.github.io/2011OAC
UK Output Area Classification User Group: www.area classification.org.uk

Appendix 5.1 Dan Vickers' Open Geodemographic Classification developed for the UK Office for National Statistics (after Vickers, 2006; Vickers and Rees, 2007)

Data: 41 census variables

Demographic

1   Age 0–4: Percentage of resident population aged 0–4

2   Age 5–14: Percentage of resident population aged 5–14

3   Age 25–44: Percentage of resident population aged 25–44

4   Age 45–64: Percentage of resident population aged 45–64

5   Age 65+: Percentage of resident population aged 65+

6   Indian, Pakistani or Bangladeshi: Percentage of people identifying as Indian, Pakistani or Bangladeshi

7   Black African, Black Caribbean or other Black: Percentage of people identifying as Black African, Black Caribbean or other Black

8   Born outside the UK: Percentage of people not born in the UK

9   Population density: Population density (number of people per hectare)

Household composition

10 Separated/divorced: Percentage of residents 16+ who are not living in a couple and are separated/divorced

11 Single-person household (not pensioner): Percentage of households with one person, who is not a pensioner

12 Single-pensioner household: Percentage of households which are single-pensioner households

13 Lone parent household: Percentage of households which are lone parent households with dependent children

14 Two adults no children: Percentage of households which are cohabiting or married couple households with no children

15 Households with non-dependent children: Percentage of households comprising one family and no others with non-dependent children living with their parents

Housing

16 Rent (public): Percentage of households that are resident in public sector rented accommodation

17 Rent (private): Percentage of households that are resident in private/other rented accommodation

18 Terraced housing: Percentage of all household spaces which are terraced

19 Detached housing: Percentage of all household spaces which are detached

20 All flats: Percentage of household spaces which are flats

21 No central heating: Percentage of occupied household spaces without central heating

22 Average house size: Average house size (rooms per household)

23 People per room: The average number of people per room

Socio-economic

24 HE qualification: Percentage of people aged 16–74 with a higher education qualification

25 Routine/semi-routine occupation: Percentage of

people aged 16–74 in employment working in routine or semi-routine occupations

26 2+ car household: Percentage of households with two or more cars

27 Public transport to work: Percentage of people aged 16–74 in employment who usually travel to work by public transport

28 Work from home: Percentage of people aged 16–74 in employment who work mainly from home

29 LLTI (SIR): Percentage of people who reported suffering from a Limiting Long Term Illness (Standardised Illness Ratio, standardised by age)

30 Provide unpaid care: Percentage of people who provide unpaid care

Employment

31 Students (full-time): Percentage of people aged 16–74 who are students

32 Unemployed: Percentage of economically active people aged 16–74 who are unemployed

33 Working part-time: Percentage of economically active people aged 16–74 who work part-time

34 Economically inactive looking after family: Percentage of economically inactive people aged 16–74 who are looking after the home

35 Agriculture/fishing employment: Percentage of all people aged 16–74 in employment working in agriculture and fishing

36 Mining/quarrying/construction employment: Percentage of all people aged 16–74 in employment working in mining, quarrying and construction

37 Manufacturing employment: Percentage of all people aged 16–74 in employment working in manufacturing

38 Hotel and catering employment: Percentage of all people aged 16–74 in employment working in hotel and catering

39 Health and social work employment: Percentage of all people aged 16–74 in employment working in health and social work

40 Financial intermediation employment: Percentage of all people aged 16–74 in employment working in financial intermediation

41 Wholesale/retail trade employment: Percentage of

all people aged 16–74 in employment working in wholesale/retail trade

How many data inputs are involved?

223,060 OAs, 41 Variables = 9,145,460 data points

Data transformation: Log transformation

■ Reduce the effect of extreme values
■ Range standardisation (0–1)
■ Problems will occur if there are differing scales or magnitudes among the variables. In general, variables with larger values and greater variation will have more impact on the final similarity measure. It is necessary, therefore, to make each variable equally represented in the distance measure by standardising the data.

Clustering method: K-means clustering

K-means is an iterative relocation algorithm based on an error sum of squares measure. The basic operation of the algorithm is to move the seeded cluster centres to see if the move would improve the sum of squared deviations within each cluster (Aldenderfer and Blashfield, 1984). The case is then assigned/re-allocated to the cluster to which it brings the greatest improvement. The next iteration occurs when all the cases have been processed. A stable classification is therefore reached when no moves occur during a complete iteration of the data. After clustering is complete, it is then possible to examine the means of each cluster for each dimension (variable) in order to assess the distinctiveness of the clusters (Everitt *et al.*, 2001).

When choosing the number of clusters to have in the classification there were three main issues which need to be considered:

■ **Issue 1:** Analysis of average distance from cluster centres for each cluster number option. The ideal solution would be the number of clusters that gives the smallest average distance from the cluster centre across all clusters.

■ **Issue 2**: Analysis of cluster size homogeneity for each cluster number option. It would be useful, where possible, to have clusters of as similar size as possible in terms of the number of members within each.

■ **Issue 3**: The number of clusters produced should be as close to the perceived ideal as possible. This means that the number of clusters needs to be of a size that is useful for further analysis.

Appendix 5.2 The 2001 ONS Census Output Area Classification groups

1: Blue Collar Communities	1a: Terraced Blue Collar	1a1: Terraced Blue Collar (1)
		1a2: Terraced Blue Collar (2)
		1a3: Terraced Blue Collar (3)
	1b: Younger Blue Collar	1b1: Younger Blue Collar (1)
		1b2: Younger Blue Collar (2)
	1c: Older Blue Collar	1c1: Older Blue Collar (1)
		1c2: Older Blue Collar (2)
		1c3: Older Blue Collar (3)
2: City Living	2a: Transient Communities	2a1: Transient Communities (1)
		2a2: Transient Communities (2)
	2b: Settled in the City	2b1: Settled in the City (1)
		2b2: Settled in the City (2)
3: Countryside	3a: Village Life	3a1: Village Life (1)
		3a2: Village Life (2)
	3b: Agricultural	3b1: Agricultural (1)
		3b2: Agricultural (2)
	3c: Accessible Countryside	3c1: Accessible Countryside (1)
		3c2: Accessible Countryside (2)
4: Prospering Suburbs	4a: Prospering Younger Families	4a1: Prospering Younger Families (1)
		4a2: Prospering Younger Families (2)
	4b: Prospering Older Families	4b1: Prospering Older Families (1)
		4b2: Prospering Older Families (2)
		4b3: Prospering Older Families (3)
		4b4: Prospering Older Families (4)
	4c: Prospering Semis	4c1: Prospering Semis (1)
		4c2: Prospering Semis (2)
		4c3: Prospering Semis (3)
	4d: Thriving Suburbs	4d1: Thriving Suburbs (1)
		4d2: Thriving Suburbs (2)
5: Constrained by Circumstances	5a: Senior Communities	5a1: Senior Communities (1)
		5a2: Senior Communities (2)
	5b: Older Workers	5b1: Older Workers (1)
		5b2: Older Workers (2)

		5b3: Older Workers (3)
		5b4: Older Workers (4)
	5c: Public Housing	5c1: Public Housing (1)
		5c2: Public Housing (2)
		5c3: Public Housing (3)
6: Typical Traits	6a: Settled Households	6a1: Settled Households (1)
		6a2: Settled Households (2)
	6b: Least Divergent	6b1: Least Divergent (1)
		6b2: Least Divergent (2)
		6b3: Least Divergent (3)
	6c: Young Families in Terraced Homes	6c1: Young Families in Terraced Homes (1)
		6c2: Young Families in Terraced Homes (2)
	6d: Aspiring Households	6d1: Aspiring Households (1)
		6d2: Aspiring Households (2)
7: Multicultural	7a: Asian Communities	7a1: Asian Communities (1)
		7a2: Asian Communities (2)
		7a3: Asian Communities (3)
	7b: Afro-Caribbean Communities	7b1: Afro-Caribbean Communities (1)
		7b2: Afro-Caribbean Communities (2)

References

All website URLs accessed 30 May 2017.

Acevedo-Garcia, D., McArdle, N., Hardy, E. F., Crisan, U. I., Romano, B., Norris, D., et al. (2014) The Child Opportunity Index: improving collaboration between community development and public health. *Health Affairs*, 33(11), 1948–1957.

Aldenderfer, M. S., & Blashfield, R. K. (1984) *Cluster Analysis*, Sage, London.

Anselin, L. (1995) Local Indicators of Spatial Association – LISA. *Geographical Analysis*, 27, 93–115.

Ballas, D., & Dorling, D. (2013) The geography of happiness, in S. David, I. Boniwell, & A. Conley Ayers (eds) *The Oxford Handbook of Happiness*, Oxford University Press, Oxford, 465–481.

Ballas, D., Dorling, D., & Hennig, B. (2014) *The Social Atlas of Europe*, Policy Press, Bristol.

Beer, D., & Burrows, R. (2013) Popular culture, digital archives and the new social life of data. *Theory, Culture and Society*, 30, 47–71.

Birkin, M., & Clarke, G. P. (1998) GIS, geodemographics, and spatial modelling in the UK financial service industry. *Journal of Housing Research*, 9, 87–111.

Birkin, M., & Clarke, G. P. (2009) Geodemographics, in R. Kitchin & N. Thrift (eds) *International Encyclopaedia of Human Geography*, Vol. 4, Elsevier, Amsterdam, 382–389.

Blake, M., & Openshaw, S. (1995) Selecting variables for small area classifications of 1991 UK census data, *Working Paper 95/2*, School of Geography, University of Leeds. Available from: www.geog.leeds.ac.uk/papers/95-2.

Bradshaw, J. R. (1972a) The concept of social need. *New Society*, 496, 640–643.

Bradshaw, J. R. (1972b) The taxonomy of social need, in G. McLachlan (ed.) *Problems and Progress in Medical Care*, Oxford University Press, Oxford, 69–82.

Bradshaw, J. R. (1994) The conceptualisation and measurement of need: a social policy perspective, in J. Popay, & G. Williams (eds) *Researching the People's Health*, Routledge, London, 45–57.

Bradshaw, J., & Finch, N. (2003) Overlaps in dimensions of poverty. *Journal of Social Policy*, 32, 513–525.

Burrows, R., & Gane, N. (2006) Geodemographics, software and class. *Sociology*, 40(5), 793–812.

Carstairs, V. (1995) Deprivation indices: their interpretation and use in relation to health. *Journal of Epidemiology and Community Health*, 49, 3–8.

Carstairs, V., & Morris, R. (1991) *Deprivation and Health in Scotland*, Aberdeen University Press, Aberdeen.

Congdon, P. (1996) Suicide and parasuicide in London: a small-area study. *Urban Studies*, 33(1), 137–158.

Dolan, P., Peasgood, T., & White, M. (2007) Do we really

know what makes us happy? A review of the literature on the factors associated with subjective well-being. *Journal of Economic Psychology*. DOI: https://doi.org/10.1016/j.joep.2007.09.001.

Dorling, D., Rigby, J., Wheeler, B., Ballas, D., Thomas, B., Fahmy, E., *et al.* (2007) *Poverty, Wealth and Place in Britain, 1968 to 2005*, Policy Press, Bristol.

Dorling, D., Vickers, D., Thomas, B., Pritchard, J., & Ballas, D. (2008) *Changing UK: The Way We Live Now*, report commissioned for the BBC. Available from: http://news.bbc.co.uk/1/shared/bsp/hi/pdfs/01_12_08_changinguk.pdf.

Elbner, C., & Sturm, R. (2006) US-based indices of area-level deprivation: results from HealthCare for Communities. *Social Science and Medicine, 62*(2), 348–359.

Ellwood, S. (2010) Geographic information science: emerging research on the societal implications of the geospatial web. *Progress in Human Geography, 34*(3), 349–357.

Engels, F. (1845) *Conditions of the Working Class in England.* Marx/Engels Internet Archive. Available from: www.marxists.org/archive/marx/works/1845/condition-working-class/index.htm.

European Commission (2010) Investing in Europe's future: fifth report on economic, social and territorial cohesion. Available from: http://ec.europa.eu/regional_policy/sources/docoffic/official/reports/cohesion5/index_en.cfm.

Everitt, B. S. (1993) *Cluster Analysis*, Edward Arnold, London.

Everitt, B. S., Landau, S., & Leese, M. (2001) *Cluster Analysis* (4th edition), Edward Arnold, London.

Flowerdew, R. (1998) Reacting to ground truth. *Environment and Planning A, 30*, 289–301.

Frey, B., & Stutzer, A. (2002) *Happiness and Economics*, Princeton University Press, Princeton, NJ.

Gale, C., Singleton, A., Bates, A. G., & Longley, P. (2014) *Creating the 2011 Area Classification for Output Areas (2011 OAC)*, JOSIS discussion forum. Available from: http://josis.net/index.php/josis/article/view/232/150.

Gale, C. G., Singleton, A. D., Bates, A. G., & Longley, P. A. (2016) Creating the 2011 area classification for output areas (2011 OAC). *Journal of Spatial Information Science, 12*, 1–27.

Gordon, D. (1995) Census based deprivation indices: their weighting and validation. *Journal of Epidemiology and Community Health, 49*(Suppl.2), S39–S44.

Gordon, D., & Pantazis, C. (eds) (1997) *Breadline Britain in the 1990s*, Ashgate, Aldershot.

Gordon, D., Adelman, A., Ashworth, K., Bradshaw, J., Levitas, R., Middleton, S., *et al.* (2000) *Poverty and Social Exclusion in Britain*, Joseph Rowntree Foundation, York.

Goss, J. (1995) 'We know who you are and we know where you live': the instrumental rationality of geodemographic systems. *Economic Geography, 71*(2), 171–198.

Harris, R. J., Sleight, P., & Webber, R. J. (2005) *Geodemographics, GIS and Neighbourhood Targeting*, Wiley, London.

Helliwell, J. F. (2003) How's life? Combining individual and national variables to explain subjective well-being. *Economic Modelling, 20*, 331–360.

HIPxChange (2016) *Area Deprivation Index*. Available from: www.hipxchange.org/ADI.

Hyndman, H. (1911) *The Record of an Adventurous Life*, Macmillan, New York.

Inglehart, R. (1990) *Culture Shift*, Princeton University Press, Princeton, NJ.

London School of Economics (LSE) (2015) Charles Booth Online Archive. Available from: http://booth.lse.ac.uk/static/a/2.html.

Mack, J., & Lansley, S. (1985) *Poor Britain*, George Allen & Unwin, London.

Medeiros, M. (2006) Poverty, Inequality and Redistribution: A Methodology to Define the Rich. International Poverty Centre, United Nations Development Programme, *Working Paper 18*, May. Available from: www.ipc-undp.org/pub/IPCWorkingPaper18.pdf.

Milligan, G. W. (1996) Clustering validation: results and implications for applied analyses, in P. Arabie, L. J. Hubert & G. De Soete (eds) *Clustering and Classification*, World Scientific, Singapore, 345–379.

Noble, M., Wright, G., & Smith, G. (2006) Measuring multiple deprivation at the small-area level. *Environment and Planning A, 38*, 169–185.

Norman-Butler, B. (1972) *Victorian Aspirations: The Life and Labour of Charles and Mary Booth*, George Allen & Unwin, London.

Orford, S. (2004) Identifying and comparing changes in the spatial concentrations of urban poverty and affluence: a case study of inner London. *Computers Environment and Urban Systems, 28*, 701–717.

Orford, S., Dorling, D., Mitchell, R., Shaw, M., & Smith, G. D. (2002) Life and death of the people of London: a historical GIS of Charles Booth's inquiry. *Health & Place, 8*(1), 25–35.

Pantazis, C., Gordon, D., & Levitas, R. (eds) (2006) *Poverty and Social Exclusion in Britain*, Policy Press, Bristol.

Pfautz, H. W. (1967) *Charles Booth on the City Physical Pattern and Social Structure*, University of Chicago Press, Chicago, IL.

Pickles, J. (1994) *Ground Truth: The Social Implications of Geographic Information Systems*, Guilford Press, New York.

Rees, P., Denham, C., Charlton, J., Openshaw, S., Blake, M., & See, L. (2002) ONS classifications and GB profiles: census typologies for researchers, in P. Rees, D. Martin, & P. Williamson (eds) *The Census Data System*, Wiley, Chichester, 149–170.

Rentoul, J. (1987) *The Rich Get Richer: The Growth of Inequality in Britain in the 1980s*, HarperCollins, London.

Robbin, J. E. (1980) Geodemographics: the new magic. *Campaigns and Elections, 1*(1), 106–125.

Rothman, J. (1989) Editorial. *Journal of the Market Research Society*, *31*(1), 1–5.

Rowntree, B. S. (1941) *Poverty and Progress: A Second Social Survey of York*, Longman, London.

Rowntree, B. S., & Lavers, G. R. (1951) *Poverty and the Welfare State*, Longman, London.

Scott, J. (1994) *Poverty and Wealth: Citizenship, Deprivation and Privilege*, Longman, London.

Simey, T., & Simey, M. (1960) *Charles Booth, Social Scientist*, Oxford University Press, London.

Singh, G. (2003) Area deprivation and widening inequalities in US mortality, 1969–1998. *American Journal of Public Health*, *93*, 1137–1143.

Singleton, A. D., & Spielman, S. E. (2014) The past, present, and future of geodemographic research in the United States and United Kingdom. *The Professional Geographer*, *66*(4), 558–567.

Sleight, P. (1997) *Targeting Customers: How to Use Geodemographic and Lifestyle Data in Your Business*, NTC Publications, Henley-on-Thames.

Smith, T., Noble, M., Noble, S., Wright, G., McLennan, D., & Plunkett, E. (2015) *The English Indices of Deprivation 2015: Research Report*, Department for Communities and Local Government. Available from: www.gov.uk/government/publications/english-indices-of-deprivation-2015-research-report.

Stafford, M. & Marmott, M. (2003) Neighbourhood deprivation and health: does it affect us all equally? *International Journal of Epidemiology*, *32*, 357–366.

Thomas, B., & Dorling, D. (2004) *Investigation Report: Know Your Place*, Shelter, London. Available from: http://scotland.shelter.org.uk/__data/assets/pdf_file/0008/48329/Knowyourplace.pdf/_nocache.

Tobler, W. (1970) A computer movie simulating urban growth in the Detroit region. *Economic Geography*, *46*, 234–240.

Townsend, P. (1979) *Poverty in the United Kingdom: A Survey of Household Resources and Standards of Living*, Penguin Books and Allen Lane, London.

Townsend, P. (1987) Deprivation. *Journal of Social Policy*, *16*, 125–146.

Townsend, P., Phillimore, P., & Beattie, A. (1988) *Health and Deprivation: Inequality and the North*, Croom Helm, London.

Tranmer, M., & Steel, D. G. (1998) Using census data to investigate the causes of the ecological fallacy. *Environment and Planning A*, *30*, 817–831.

Uprichard, E., & Burrows, R. (2009) Geodemographic code and the production of space. *Environment and Planning A*, *41*, 2823–2835.

Vickers, D. W. (2006) *Multi-Level Integrated Classifications Based on the 2001 Census*. PhD thesis, University of Leeds.

Vickers, D. W. (2010) England's changing social geology, in J. Stillwell & P. Norman (eds) *Spatial and Social Inequality*, Springer, Berlin, 37–51.

Vickers, D., & Rees, P. (2006) Introducing the area classification of output areas. *Population Trends*, *125*, 15–29.

Vickers, D., & Rees, P. H. (2007) Creating the UK national statistics 2001 output area classification. *Journal of the Royal Statistical Society: Series A. Statistics in Society*, *170*(2), 379–403.

Vickers, D., & Rees, P. H. (2011) Ground-truthing geodemographics. *Applied Spatial Analysis and Policy*, *4*(1), 3–21.

Webber, R. J., & Craig, J. (1976) Which local authorities are alike? *Population Trends*, *5*, 13–19.

Whelan, C. T., Layte, R., Maitre, B., & Nolan, B. (2001) Income, deprivation and economic strain: an analysis of the European Community Household Panel. *European Sociological Review*, *17*(4), 357–372.

6 GIS and small-area estimation of income, well-being and happiness

LEARNING OBJECTIVES

■ Combining small-area data with national social surveys to estimate small-area information
■ Identifying and studying interdependencies between variables at the local level
■ Becoming familiar with statistical modelling approaches to the analysis of small-area data
■ Becoming familiar with geosimulation and spatial microsimulation
■ Estimating small-area information on 'soft' variables such as happiness and well-being
■ Becoming familiar with a wide body of literature on small-area estimation

Introduction

In this chapter we provide an overview of GIS-related methodological tools that can be used to combine different sets of secondary data in the social sciences in order to estimate small-area information on very important policy-relevant variables that are not typically available from publicly available data sources. Examples of such variables range from household income and individual earnings (which are typically collected by governments and tax authorities, but which are often deemed to be politically and socially sensitive and are not made publicly available due to confidentiality concerns) to variables that are often not routinely available such as health and health-related information, subjective well-being and happiness. The estimation and mapping of small-area distribution of these types of variables at small-area levels has always been a very challenging exercise, but there has been a rapidly increasing number of projects that utilise GIS-related methods to build relevant geographical databases.

The best source of small-area socio-economic information in most countries is the census of population. However, in many cases the census does not provide any information on variables such as household income, wealth and taxation in order to preserve confidentiality (Marsh, 1993) and to minimise non-response even though collecting information on income would be extremely useful for policy analysis. Aspects of such information are available from various government surveys but in many cases the finest spatial scale to which these survey data are coded is the region or, in the case of surveys such as Understanding Society (formerly known as the British Household Panel Survey [BHPS]), in the UK, the local authority district. Even then such sample surveys provide incomplete geographical coverage. Despite these problems and issues, there has been considerable progress (especially over the past 20 years) in the development of small-area microdata estimation methods.

Even in the cases where small-area information on income and economic circumstances is available, there is typically a lack of additional detail in the form of small-area microdata (e.g. a list of individuals with estimated income, employment status, number of children, educational attainment, etc.) that would enable the study of interdependencies of variables at the small-area level (e.g. poverty and educational attainment or educational opportunities, or estimating the number of children living below the poverty threshold in families with unemployed parents).

This chapter (and accompanying practical) provides examples of how it is possible to combine national social survey data with geographical data in order to address the paucity of microdata at small-area level. In addition, it shows how these methods can be used in a GIS context for what-if policy analysis. We begin by providing examples of national survey data that can be combined with the types of data reviewed and discussed in previous chapters (including some of the small-area data sources presented in Chapter 2). We then discuss relatively simple methods of estimating income at the small-area level (e.g. by combining social class and income from a survey with small-area data on social class), before moving on to more sophisticated modelling methods. Finally, we provide applied examples and outputs of such methods and discuss their policy relevance and societal importance. These examples include the geographical analysis of national social and area-based policies, involving the investigation of the geographical implications of national policies, current trends in socio-economic polarisation and inequalities between and within cities.

Overall, the data and methods presented in this chapter demonstrate the huge potential for GIS to be used to estimate very important small-area informa-tion and to also address a series of important policy questions from a geographical perspective. However, as the reader will see, we also need to introduce techniques used in parallel with GIS – modelling techniques such as regression and spatial microsimulation. The GIS remains the container for the data and the core software for data visualisation, but greater ana-lytical power comes from coupling other modelling techniques.

Combining small-area with national social survey data

As we noted in Chapter 2, there is a wealth of geo-graphical data ranging from regional to small-area level that can be stored, analysed and visualised with the use of GIS. As argued in Chapter 2, one of the most important and reliable sources of small-area socio-economic data is the census of population, as it is the most authoritative and spatially comprehensive social survey and is used by governments around the world to support decisions pertaining to the allocation of public expenditure at different geographical levels. In addition, census data are very valuable commercially, as they form the basis for geodemographic analysis (as also discussed in the previous chapter) and are used in marketing analysis and retail modelling (see Chapter 8). However, the census data variables are relatively limited on cost grounds and in order to preserve con-fidentiality. For instance, the questionnaire of the last census of the UK population had 56 questions pertain-ing to themes such as work, health, national identity, passports held, ethnic background, education, second homes, language, religion and marital status.[1] However, it did not include any questions on household income or lifestyle.

In Chapter 2 we provided an overview of sources of data on a number of themes and variables not measured by the census (such as crime- and safety-related variables, health, deprivation, income and lifestyle information, etc.) but estimated on the basis of social survey or administrative data and, in many cases, by combining them with census data with the use of small-area estimation methods. Such data sources are made available to meet a very strong need for small-area data from academics, private and public sector organisations and government departments. Again our examples here are drawn from the UK – we hope the reader can search for the equivalent in their home country. The important point is how many of these variables can be mapped and analysed in a GIS framework to provide important insights into spatial patterns of policy-relevant socio-economic variables.

There is also a wealth of UK social survey data that can be used, as is demonstrated in this chapter, in combination with census and other small-area data to produce small-area information on variables that are not available from other sources as well as synthetic populations which can be analysed to study *interdependencies* between different variables at the individual level (e.g. age, health status), household level (e.g. household type, household income) and different area-level variables (e.g. availability and quality of local amenities, access to job markets, etc.).

There are numerous UK socio-economic surveys of households and individuals that can be used to study these interdependencies but with limited scope for GIS analysis. The outputs of most of these surveys are typically released at relatively coarse levels of geography (e.g. region or local authority district level geographies). Among the most widely used socio-economic surveys in the UK are the New Earnings Survey (conducted by the Office for National Statistics (ONS)), which records employment data of a relatively large sample of employees. Further, the Family Expenditure Survey (FES) has been widely used for policy analysis. The FES is a continuous survey of household expenditure and income carried out by the ONS (sample size 10,000 households). It should be noted that the FES was combined with the National Food Survey from 2001 and renamed to the Expenditure and Food Survey. It is now known as the Living Costs and Food Survey.[2] Among the survey's topics are:

- food consumption and nutrition;
- expenditure on goods and services, with considerable detail in the categories used;
- income, including details about the sources of income;

- possession of consumer durables and cars;
- housing.

Further, the Family Resources Survey[3] (FRS) is a continuous survey that was launched in October 1992 by the Department of Social Security (ONS, 2000b; Dhanecha *et al.*, 2003). The FRS has a relatively large sample size (around 26,000 households per year). The FRS includes:

- basic household and individual characteristics (tenure, ethnic origin, employment status, etc.);
- housing costs (rents, mortgages, Council Tax, water and sewerage charges, insurance);
- household income (including benefit receipt, unearned income, pensions, etc.);
- other costs (travel to work costs, childcare costs, maintenance payments, etc.);
- ownership of vehicles and consumer durables.

(ONS, 2000b; Dhanecha *et al.*, 2003)

The General Lifestyle Survey (GLS), formerly known as the General Household Survey, was conducted from 1971 to 2012 (UK Data Service, 2017a) and covered five 'core' subjects: population and family information, housing, employment, education and health. In addition, special topics were added from year to year. In 1998 these supplementary topics included smoking, drinking, hearing, contraception and day care. Questions related to elderly people were also repeated from earlier years, with the results published as a separate report 'People aged 65 and over' (UK Data Service, 2017a). The GLS can be used to explore the relationships between income, housing, economic activity, family composition, fertility, education, leisure activities, drinking, smoking and health. In addition to regular 'core' questions, certain subjects are covered periodically, such as:

- family and household formation;
- health and related topics;
- use of social services by the elderly and participation in sports and leisure activities.

Another widely used survey for labour market analysis is the Labour Force Survey (LFS) which is conducted by the ONS. The survey was biennial from 1973 to 1983. Since 1984 the LFS has been conducted annually, and since 1992 quarterly, with around 40,000 sampled households (ONS, 2015; UK Data Service, 2017b). The LFS has a panel design and every household is interviewed for five waves. The LFS asks a range of questions including:

- household composition;
- housing tenure;
- ethnicity;
- education and training;
- employment, unemployment and job search activities;
- reasons for not wanting to work;
- income;
- labour mobility;
- travel to work;
- trade union membership;
- current working conditions;

- hours of work and health (sickness, accidents and health problems/disabilities).

One of the most comprehensive surveys in Britain is the BHPS/Understanding Society,[4] which is an annual survey of the adult population of the UK, drawn from a representative sample of over 5,000 households. The aim of the survey is to deepen the understanding of social and economic change at the individual and household levels in Britain, as well as to identify, model and forecast such changes, their causes and consequences in relation to a range of socio-economic variables (Taylor *et al.*, 2001). Appendixes 6.1 and 6.2 outline the core household and individual questions asked in the BHPS questionnaires (and the basis for the Understanding Society Survey, which is also known as the UK Household Longitudinal Study). These questions have generated a wealth of socio-economic and demographic variables, which make the BHPS unique in that it contains almost all the variables contained

Figure 6.1 Geographical distribution of subjective happiness in Europe (see Figure 3.6 for details of colour scheme used)

Source: Ballas *et al.* (2014), based on data from the European Values Survey

in most other national social survey data in Britain. However, as is the case with all other large surveys, BHPS gives information at relatively coarse levels of geography.

The UK social survey data sets described above can be used in many cases to obtain data on variables such as average income or more unusual data such as life satisfaction and subjective happiness at national or regional level. For instance, Figures 6.1 and 6.2 present cartographic representations (using the methods described in Chapter 3) of national subjective life satisfaction in Europe (based on the European Union Statistics on Income and Living Conditions (EU-SILC) data) and regional disposable income (based on data from the European Values Survey).

Overall, there are very few sources of geographically detailed microdata sets, but it is increasingly possible to use GIS and related methods to combine such surveys with local area (within regions and cities) data. The remainder of this chapter provides an overview of these methods as well as new approaches that utilise online data sources. In addition, the accompanying practical provides data and a worked example of how to apply one of the methods described below.

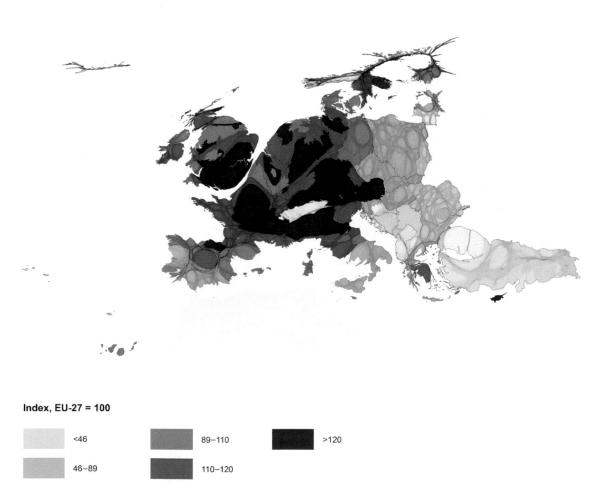

Index, EU-27 = 100

▢	<46	▢	89–110	▢	>120
▢	46–89	▢	110–120		

Figure 6.2 Net adjusted disposable income of private households (purchasing power consumption standard), 2007

Source: Ballas *et al.* (2014)

Generating indirect non-survey designed estimates

The simplest approach to combining national social survey data with small-area information is to obtain small-area total numbers from the census on variables that could be correlated with a 'target variable' to be estimated (e.g. income would be correlated with 'occupational classification'). The next step is to obtain information at the national or sometimes regional level on the same variable cross-tabulated by the census variable. For example, Figure 6.3 shows the average monthly earnings in euros (before tax) by broad occupational classification in the UK according to the EU-SILC. These data can be matched with suitable small-area data from sources such as the census of population. For instance, the UK census of population provides information on total population by broad and more detailed occupation group categories. Table 6.1 presents these broad categories.

Table 6.1 Census of population small-area broad occupational groupings

1. Managers, directors and senior officials
2. Professional occupations
3. Associate professional and technical occupations
4. Administrative and secretarial occupations
5. Skilled trades occupations
6. Caring, leisure and other service occupations
7. Sales and customer service occupations
8. Process, plant and machine operatives
9. Elementary occupations

Source: UK Census of Population Table (QS606EW) Occupation (Minor Groups) (QS606EW) obtained via https://neighbourhood.statistics.gov.uk

The data sets from the EU-SILC and the census of population illustrated in Figure 6.3 and Table 6.1, respectively can be used to generate what can be described as 'indirect non-survey designed estimates of earnings' at the small-area level. This can be achieved

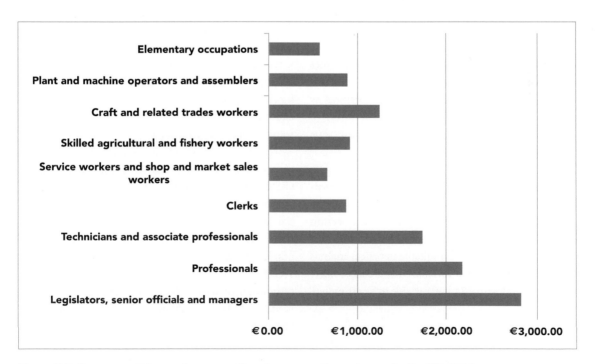

Figure 6.3 Gross monthly employee earnings by occupation category in the UK, 2011

Source: Authors' calculations using data from the EU-SILC

by multiplying the average income for each occupational group shown in Figure 6.3 with the total number of people belonging to this group (or a similar group, if there is no direct match). It should be noted that both the EU-SILC and the census of population data sets contain more detailed information on occupational groupings that could be used (more than 100 sub-groups) to provide more sophisticated estimates. In addition, when suitable data are available it is also useful to combine different variables (e.g. calculate average earnings by socio-economic occupation and by age and sex and then multiply this by the respective total numbers at the small-area level).

The simple approach described above can be adopted to estimate any other variable at the small-area level, as long as there is some association (according to relevant theories or appropriate statistical analysis) between the variable to be estimated and the relevant variables available at the small-area level. The steps to be adopted with this simple approach can be summarised as follows:

- Obtain small-area total numbers from the census on variables that could be correlated with a 'target' variable (e.g. income would be correlated with 'occupational classification').
- Obtain information at the national or sometimes regional level on the same variable cross-tabulated by the relevant census variable (e.g. earnings by occupational classification).
- Multiply the known census totals by average value for each area.

The accompanying practical for this chapter provides an example of how this method can be applied to estimate expenditure for small areas. The next section discusses more sophisticated approaches to small-area estimation that are based on suitable statistical analysis that identifies small-area variables that are most likely to be correlated with a 'target' variable to be estimated.

Statistical model-based estimates

The simple approach described above can be useful if there are limited resources, but it does not take advantage of geographical models that can be used for a more sophisticated estimation of information at the small-area level using GIS-related methods. In particular, there has been a great deal of more sophisticated modelling work conducted by geographers and regional economists who have been involved in the construction of small-area income estimation models. In contrast to the above simple approach, these modelling approaches explore the relationship between the target variable to be estimated (typically income) and a wide range of other socio-economic and demographic variables. In this section we give a flavour of these developments over the past 20 years drawing on and updating a review of relevant work (Ballas *et al.*, 2006a). Most of these developments are based on statistical analysis and modelling which are increasingly embedded in GIS proprietary software. For instance, one of the key methods is regression analysis which can be used to understand, model and predict a variable (such as income, known as the dependent variable) on the basis of the observations of explanatory (or independent) variables. For example, as also discussed in the previous section, income is correlated with variables such as educational attainment, socio-economic classification and occupational status, employment status and age. Regression analysis is typically employed to identify the causes of particular phenomena (e.g. the relationship between income inequality and crime) at different geographical levels and it is a standard tool that is now embedded in GIS software packages such as ArcGIS[5] (see Figure 6.4).

Nevertheless, it should be noted that the standard regression models have relatively limited potential for GIS-based analysis, as they do not take into account spatial dependencies between observations and potential local variations in statistical relationships. Such issues can be addressed with the use of more 'geographically aware' methods such as Geographically Weighted Regression (GWR), which involves the estimation of local regression equations (Brunsdon *et al.*, 1996; Fotheringham *et al.*, 2002) and Anselin's (1995) Local Indicators of Spatial Association (LISA). There is specialised stand-alone software that can be used to apply these methods but there are also features in proprietary GIS packages such as ArcGIS that enable GWR[6] and LISA[7] analysis. These are very

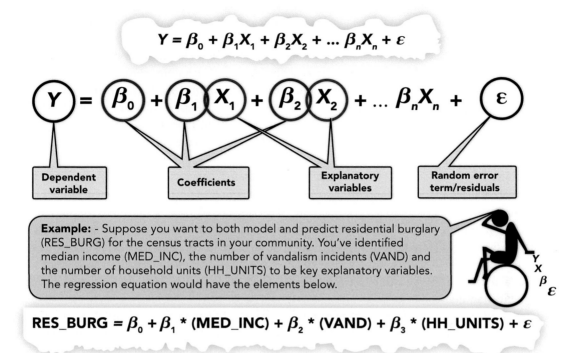

- **Dependent variable** (*Y*): this is the variable representing the process you are trying to predict or understand (e.g. residential burglary, foreclosure, rainfall). In the regression equation, it appears on the left side of the equal sign. While you can use regression to predict the dependent variable, you always start with a set of known *Y*-values and use these to build (or to calibrate) the regression model. The known *Y*-values are often referred to as *observed values*.

- **Independent/explanatory variables** (*X*): these are the variables used to model or to predict the dependent variable values. In the regression equation, they appear on the right side of the equal sign and are often referred to as *explanatory* variables. We say that the dependent variable is a function of the explanatory variables. If you are interested in predicting annual purchases for a proposed store, you might include in your model explanatory variables representing the number of potential customers, distance to competition, store visibility and local spending patterns, for example.

- **Regression coefficients** (β): coefficients are computed by the regression tool. They are values, one for each explanatory variable, that represent the strength and type of relationship the explanatory variable has to the dependent variable. Suppose you are modelling fire frequency as a function of solar radiation, vegetation, precipitation and aspect. You might expect a positive relationship between fire frequency and solar radiation (the more sun, the more frequent the fire incidents). When the relationship is positive, the sign for the associated coefficient is also positive. You might expect a negative relationship between fire frequency and precipitation (places with more rain have fewer fires). Coefficients for negative relationships have negative signs. When the relationship is a strong one, the coefficient is large. Weak relationships are associated with coefficients near zero.

    β_0 is the regression *intercept*. It represents the expected value for the dependent variable if all of the independent variables are zero.

Figure 6.4 Regression in GIS as explained by ArcMap instruction manuals

Source: http://resources.esri.com/help/9.3/arcgisdesktop/com/gp_toolref/spatial_statistics_toolbox/regression_analysis_basics.htm powerful toolkits for the study of local variations and associations between variables, but there a number of alternative approaches building on different 'flavours' of statistical modelling (including regression) that are particularly aimed at estimating small-area information and, in particular, income. These are discussed in some detail in the remainder of this section.

Among the early most sophisticated and robust small-area income estimation modelling approaches in the UK is the work of Bramley and Smart (1996) who developed a formal model capable of estimating local (at local authority district level) income distributions on the basis of census data, as well as data from the UK FES and the UK National Online Manpower Information System.[8] Their modelling approach was based on classifying FES households by type and economic activity (e.g. no household members in employment, one household member in employment, 2+ members in employment) obtained from the census at the local authority district level. To each household type they then allocated a median weekly income based on the data shown in Table 6.2. One of the key characteristics of their model was the ability to estimate proportions of households by local authority districts that have an income below a given level. Bramley and Smart (1996) applied their modelling approach and generated results for England. They also validated their results, by comparing their estimates with FES data at the national level, as well as regional

and district income data from the Inland Revenue, and found that their model estimates fitted these data reasonably well.

Bramley and Lancaster (1998) refined this model by incorporating housing tenure disaggregation. They have also modified model parameter values on the basis of the analysis of microdata specific to Scotland, before implementing the model for Scottish districts and sub-district areas (postal sectors). They then tested the predictions of the model for Scotland against various external data sets at different geographical levels. An additional important development in their model was a method of updating and projecting the estimates forward. The method was based on the utilisation of a wider range of data including the LFS, New Earnings Survey, claimant unemployment data, Scottish Office household projection data, housing stock by tenure and house prices. For instance, they used 1992 and 1995 district-level data from the LFS to calculate the rate of change for a number of indicators (economic activity rates, unemployment rates, part-time workers, industry and occupation structure). They then applied these rates to the 1991 census values in order to provide updated data for their model. Another example of small-area estimation is the work of Rusanen *et al.* (2001, 2002) who examined income data in Finland and carried out an analysis of mean taxable incomes per household throughout the 1990s at a 1 by 1 km grid cell resolution, comparing their results with

Table 6.2 Household groupings and income distributions used by Bramley and Smart (1996), based on 1991 Family Expenditure Survey

Household type	Composition (%)	Median income per week
Single elderly	14.6	£92.30
Single adult	12.8	£217.30
Lone parent + child	4.3	£109.20
Elderly couple	9.8	£177.90
Couple/2 adults	22.1	£405.70
Couple + 1 child	8.3	£392.20
Couple + 2 children	10.6	£413.10
Couple + 3+ children	5.2	£391.30
3+ adults	12.1	£517.50
All households	**100**	**£296.80**

income data at postal district and municipality levels. They suggested that the smaller the areal unit used, the greater the income differences, and vice versa, and concluded that the results of analyses based on different areal units cannot necessarily be regarded as comparable. Hamnett and Cross (1998) used GHS data and New Earnings Survey data to examine the existence and extent of income polarisation in London. Their research suggested that the evidence for polarisation was relatively weak, and that where polarisation exists it is asymmetric, with much greater growth in the size of groups at the top of the earnings distribution than at the bottom.

Another relevant (and more recent) example is the work of Longley and Tobon (2004) who presented developments in the provision and quality of digital data, arguing that these developments create new possibilities for spatial and temporal measurement of the properties of socio-economic systems at finer levels of granularity. In addition, they suggest that the 'lifestyles' data sets collected by private sector organisations in the UK and the US provide one such prospect for better inferring the structure, composition and heterogeneity of urban areas. Using a case study of the city of Bristol they compare the patterns of spatial dependence and spatial heterogeneity observed for a small-area income measure with those of the census indicators that are commonly used as surrogates for it.

Using similar 'lifestyle' data, the CACI consultancy reported findings on household wealth in the UK. In particular, CACI publishes an annual report on the 'Wealth of the Nation', examining average household income by broad geographic regions such as county or local authority and exploring local concentrations of wealth and poverty by ranking local neighbourhoods with the highest proportion of households earning a lot, or very little. CACI estimates household income distributions at geographical level down to unit postcode based on four million market research records, ACORN (their household-level geodemographic classification), national census and survey data, and statistical modelling. CACI weights household records from commercial data sets so that they match target distributions by ACORN class, postcode area and income band.

CACI's reports (2005) are typically based on PayCheck, a system that provides estimates of gross household income (including investment income and social security) right down to the level of postcode. In particular, PayCheck is CACI's estimate of household income at postcode level. It is based upon government data sources together with income data for millions of UK households collected from lifestyle surveys and guarantee card returns (CACI, 2005). The core directory provides estimates of mean gross household income by postcode and a banding of these mean incomes, which can be used for profiling or classifying postcodes. The extended directory provides more detailed information on the expected income distribution within each postcode. A similar source of household income information is the Experian database, which is available for academic use (Webber, 2004; Census Dissemination Unit, 2005). This database comprises the Mosaic UK household classification data, classifying all UK households into 11 groups, 61 types and 243 segments and is updated every year. It also comprises median household income data which are estimated on the basis of a multi-stage process that predicts personal and household income for a number of standard employment types using survey data (including MORI's Financial Tracking Survey). Nevertheless, it should be noted that a serious flaw of both CACI and Experian data is the dependence on lifestyle surveys and postcode addresses (with associated biased response rates) and a reliance on a small number of lifestyle categories to model income. McLoone (2002) provides a detailed discussion and critique of CACI's data.

Williamson and Voas (2000) pointed out that CACI small-area income estimates might be the best that have been produced due to the size of the underlying data set. Nevertheless, it is very difficult to evaluate the results and the techniques, given that these are not in the public domain (with the exception of the methods used by Experian[9]). Williamson and Voas (2000) and McLoone (2002) also stress that possible weaknesses of the CACI approach include an over-reliance on geodemographic classification and an analysis of geographical context restricted to postcode area level.

In the US, Fay and Herriot (1979) were among the first researchers to use a statistical approach to estimate income for small areas (defined as 'areas with a population of less than 1,000 residents'). More recently, Fisher (1997) presented the 'Small-area Income and Poverty Estimates' (SAIPE) programme, which aimed at estimating median household income for states and counties, as well as poverty for states, counties and school districts. These estimates were based on statistical models that used decennial census data, household survey data, administrative records data and population estimates. Gee and Fisher (2004) discuss ways of identifying the degree of error in the estimates generated by this method. Cressie (1995) also discussed geographical data on income in the US, suggesting that the data published by the US Census Bureau at the county level are often a noisy representation of the true geographic distribution of rates over the small areas, and he presented a Bayesian statistical method for smoothing raw rates.

Lynch (2003) also used US census data in combination with US Internal Revenue Service data in order to produce estimates of income levels and income inequality in the United States for 1988, 1995 and 1999. He analysed trends in national and regional income inequality and income levels between 1988 and 1999. At a smaller area level, Cloutier (1995, 1997) explored the inter-urban variation in family income distribution in the 1980s. Cloutier (1995) argued that the estimation of percentile incomes within intervals reported by the Census Bureau has often been carried out under the assumption that incomes are uniformly distributed. Using log-normal extrapolation methods Cloutier (1995) demonstrated that income might be significantly underestimated if the assumption is applied to lower level percentiles in the black family income distribution and suggested that analysis based on US census data about the level and determinants of the relative income of poor black families could be misleading.

In a more applied context, Cloutier (1997) used similar methods to demonstrate that urban development, rising female-headship, a widening educational distribution and changes in the industrial and occupational mix were major contributing factors to rising inequality in US metropolitan areas. Hammer *et al.* (2003) studied the impact of income estimates derived using the US Census Bureau SAIPE method upon the

development of a 'distressed county' index that was used by the US Appalachian Regional Commission (ARC). In particular, they evaluated the potential impact of incorporating the SAIPE estimates into the index and they suggested that these would alter the index but not to a radical degree. They also suggested that combining the SAIPE point estimate and the SAIPE upper bound estimate in the determination of distressed status would achieve the objective of using more current estimates of poverty while reducing the negative consequences of using an estimate of poverty with greater statistical variation than decennial census-derived estimates.

As can be seen thus far from the examples discussed above, there have been numerous attempts worldwide to estimate income at geographical levels that are much smaller than the official statistical units at which survey-based data are typically available.

The best way of validating small-area estimation methods is by comparing estimated or modelled data with actual census or census-style survey data. This became possible in England and Wales when the ONS tested an income question during their large-scale census rehearsal in April 1999, following a survey of census users (Rees, 1998). The question was dropped from the final version of the census form used in 2001 due to concerns about potential negative impacts on response rates. However, Williamson and Voas (2000) conducted research for the Economic and Social Research Council (ESRC) between 2000 and 2001 (Williamson, 2005) and carried out analysis of these rehearsal data in order to explore alternative small-area income imputation methods. Figure 6.5 shows the census rehearsal data income distribution.

This distribution approximates a log-normal pattern, which is what should be expected according to relevant literature in economics. Nevertheless, it is interesting to explore this distribution in different areas. Figure 6.6 shows the income distribution for enumeration districts classified on the basis of overall income. As can be seen, the pattern varies considerably. Also, as Williamson (2000) points out, there are considerable variations in households in different income bands by area. In particular, at electoral ward level, the proportion of household heads in the top income

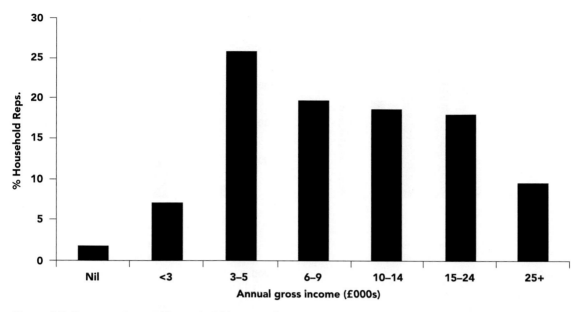

Figure 6.5 Census rehearsal household income distribution

Source: After Williamson and Voas (2000)

band is on average 9.1%, ranging from 2.8% to 21.6% (Williamson and Voas, 2000). In addition, Williamson and Voas suggest that 89% of enumeration districts contained one or more household heads who were in the top income decile of the population.

Williamson's research project made extensive use of the census rehearsal data in order to evaluate existing and proposed small-area income estimation methods. He assessed the efficacy of alternative small-area imputation strategies when applied to sub-district spatial units. In particular, he evaluated the census rehearsal data set and evaluated established small-area income estimation methods. Williamson suggested that by far the most effective simple proxy for income is the proportion of the economically active population in National Statistics Socio-economic Classification (NS-SEC) categories 1 and 2 (Managerial and professional occupations). He also suggested that this finding applies regardless of whether mean income is measured per person, per adult or per number of persons in the household. The proportion of NS-SEC 1+2 captured 74% to 81% of the observed variation in income between enumeration districts (areas with an average size 100 households). The only income measure for which a better alternative exists is mean total household

income, for which the proportion of households with no car offers better performance. Williamson also explored the association of a range of deprivation indexes (such as those discussed in Chapter 5) with income data from the census rehearsal exercise. According to his analysis, these indexes, based on the combination of a number of proxy measures, performed less well than % NS-SEC 1+2. Williamson's results suggested that only in certain circumstances can even commercial geodemographic (area) classifications match the performance of the simple indicator % NS-SEC 1+2.

Voas and Williamson (2000) and Williamson's (2005) research were evaluated by Patrick Heady of the UK ONS and his colleagues who consequently built a *regression synthetic estimation fitted using area-level covariates approach* (Heady et al., 2003).[10] Their modelled estimation method involves combining survey data with other data sources that are available on an area basis and is underpinned by the area-level relationship between the survey and auxiliary variables (usually administration data or census data). In this context, Heady et al. (2003) modelled ten variables at the small-area level: household income from the FRS, household income from the GHS, a measure of social capital, children from ethnic minorities, number of

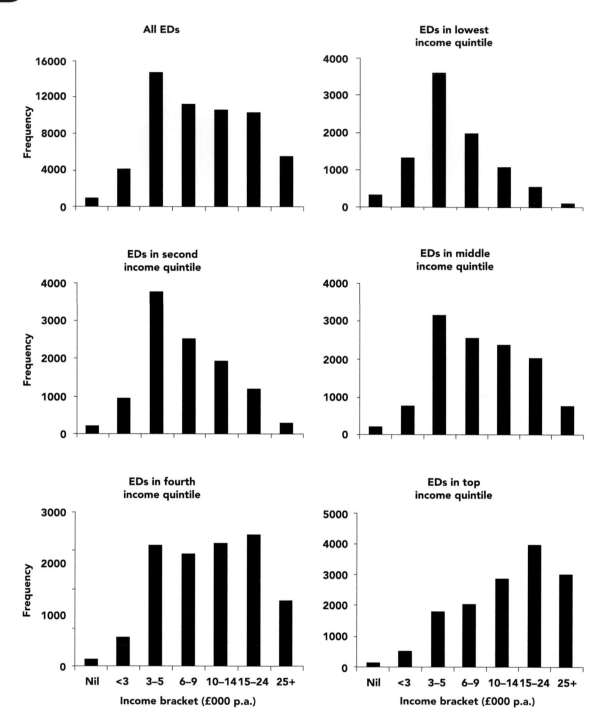

Figure 6.6 Income distribution of household representatives in different local neighbourhoods (census enumeration districts) classified by income

Source: After Williamson and Voas (2000)

people to help in a crisis, single-parent families, over-crowding and three measures of poor health. They developed a regression model that was used to estimate 'average weekly household' income at the electoral ward level in England and Wales on the basis of the following predictors:

- the social class of the ward population;
- household type/composition;
- regional/country indicators;
- the employment status of the ward population;
- the proportion of the ward population claiming DWP benefits;
- the proportion of dwellings in each of the Council Tax bands in a ward.

Their model-based approach is based on finding a relationship between weekly household income (as measured in the FRS) and covariate information (usually from census or administrative sources) for the wards that are represented in the FRS. The Small-area Estimation Project (SAEP) methodology has been applied by ONS to produce ward-level estimates of income in Britain with the latest data (at the time of writing this book) in 2011/12.[11]

Geosimulation and spatial microsimulation

In this section we discuss ways of building on the experience obtained through the studies reviewed so far, in order to produce improved and more detailed small-area estimates of variables that are not available at small-area level. So far we have described a wide range of methodologies aimed at estimating income at various geographical scales in different contexts. Most of these regression-based modelling approaches to small-area estimation have been very successful in identifying census and other variables that are useful for predicting a target variable such as average income at various scales. However, they are not suitable for estimating combinations of variables (e.g. number of low-income elderly individuals with no access to a car; or numbers of households with dependent children below the

poverty threshold) or proportions of households or individuals earning below or above particular income thresholds. Such estimates are particularly useful for the geographical analysis of the impacts of national public policies as well as area-based policies. In this section we present more sophisticated geographical approaches that can be used to produce small-area microdata including a wide range of variables (and combination of variables) that include income but also other information, and which can support the spatial analysis of national social policies as well as urban and regional policies. In particular, we present GIS-based research efforts that are aimed at creating simulation models that can be used for the estimation of the spatial impacts of social policies, as well as their socio-economic impact, in order to address the need for spatial analysis of national social policies. These geographical simulation methods are conceptually very relevant to popular life simulation computer games such as SimCity and the Sims, but, instead of using game-based rules and hypothetical imaginary data on the synthetic characters of the game, it involves the merging of small-area statistics (such as census of population data) and social survey data to simulate a population of individuals within households (or different geographical units), whose characteristics are as close to real populations as it is possible to estimate. These methods can also involve the development and use of computer agent-based models (Heard et al., 2015) and dynamic microsimulation models which involve forecasting past changes forward to produce the best estimate possible of an individual's circumstances in the future – were current trends to continue – or under different policy scenarios.

Microsimulation is a technique that has been broadly developed and used by economists and other social scientists.[12] In particular, economists have long been involved in the development of microsimulation models that are capable of modelling the impact of national government policies. The results of national (aspatial) microsimulation models are widely quoted in the media when covering the possible impact of government budget changes upon different types of households. Microsimulation models aim to build large-scale data sets on the attributes of individuals or households (and/or on the attributes of individual firms or organisations) and to analyse policy impacts on these micro-units; by permitting analyses at the level of the individual, family or household they provide the means of assessing variations in the distributional effects of different policies (e.g. see Mitton *et al.*, 2000; Redmond *et al.*, 1998). However, these models have been very limited in terms of geographical analysis capacity and this was mostly due to the lack of good quality geographical data, issues of computational power and software developments. In particular, until very recently there were very few sources of geographical socio-economic data. Even today, with a very few exceptions, there are no very small-area population microdata, which are the standard data sets used by sophisticated economic microsimulation models.

Microsimulation models become geographical when spatial information about the simulated entities is available (or estimated). Spatial microsimulation involves the creation of large-scale population microdata sets and the analysis of the impacts of any policy changes that change the attributes contained in these micro databases in some way. In other words, adding spatial detail to traditional microsimulation involves creating geographically referenced microdata that refer to a particular locality, to a geographically defined and restricted area. Since there are very few sources of geographically detailed microdata, there is a need to create these data using spatial microsimulation techniques by merging census and survey data to simulate a population of individuals within households (for different geographical units) whose characteristics are as close to the real population as it is possible to estimate.

Various types of spatial microsimulation models can be distinguished. For instance, there are *static models* that are based on simple snapshots of the current circumstances of a sample of the population at any one time, and *dynamic models* that vary or age the attributes of each micro-unit in a sample to build up a synthetic longitudinal database describing the sample members' lifetimes into the future. The main characteristic of dynamic models is that they incorporate behavioural responses under different policy scenarios.

As noted above, spatial microsimulation involves the creation of large-scale population microdata sets and the analysis of policy impacts at the micro level. *Population microdata* contain information on individuals rather than aggregate data. *Population microdata* can be divided into *individual microdata* that contain *information on individuals*, household microdata that might contain *household information* only and household microdata that might contain *individual* and *household information*. For example, the BHPS/Understanding Society data that we described earlier can be used in combination with census small-area data to estimate health-related variables at the small-area level, as well as to explore the interdependencies of these variables with socio-economic variables such as income, social class, access to health services, etc. The BHPS/Understanding Society is a representative longitudinal survey on the social situation of private households and can be presented in the format of a list of individuals within households (see Tables 6.3 and 6.4).

Spatial microsimulation techniques involve the merging of survey data such as the BHPS with census and other geographical area data to simulate a population of individuals within households (for different geographical units), whose characteristics are as close to the real population as it is possible to estimate. In other words, geographical microsimulation models simulate *virtual populations* in given geographical areas, so that the characteristics of these populations are as close as possible to their 'real-world' counterparts. One of the major advantages of microsimulation is that it can be a substitute for conducting detailed surveys to produce survey data such as the BHPS described above at the small-area level.

The spatial microsimulation method typically involves three major procedures:

- The construction of a microdata set from samples and surveys.
- *Static what-if* simulations, in which the impacts of alternative policy scenarios on the population are estimated: who would benefit from a particular local or national government policy? Which geographical areas would benefit the most?
- Dynamic modelling, to update a basic microdata set and future-oriented *what-if* simulations.

Table 6.3 The BHPS microdata format

PERSON	*HID	PID	*AGE12	SEX	*JBSTAT	...	*HLLT	*QFVOC	*TENURE	*JLSEG	...
1	1000209	10002251	91	2	4	...	1	1	6	9	...
2	1000381	10004491	28	1	3	...	2	0	7	−8	...
3	1000381	10004521	26	1	3	...	2	0	7	−8	...
4	1000667	10007857	58	2	2	...	2	1	7	−8	...
5	1001221	10014578	54	2	1	...	2	0	2	−8	...
6	1001221	10014608	57	1	2	...	2	1	2	−8	...
7	1001418	10016813	36	1	1	...	2	1	3	−8	...
8	1001418	10016848	32	2	−7	...	2	−7	3	−7	...
9	1001418	10016872	10	1	−8	...	−8	−8	3	−8	...
10	1001507	10017933	49	2	1	...	2	0	2	−8	...
11	1001507	10017968	46	1	2	...	2	0	2	−8	...
12	1001507	10017992	12	2	−8	...	−8	−8	2	−8	...

Table 6.4 Variable descriptions for Table 6.3

PERSON	Person number
*HID	Household identifier (number of household to which the listed individual belongs)
PID	Person identifier (a unique number to identify the individual)
*AGE12	Age at 1/12/ *
SEX	Sex
*JBSTAT	Current labour force status (e.g. self-employed, in paid employment, unemployed, family care, etc.) in year *
*HLLT	Health status in year *
*QFVOC	Vocational qualifications in year *
*TENURE	Tenure status in year *
*JLSEG	Socio-economic group: last job (in year *)

The first procedure can also be defined as static spatial microsimulation. This involves the reweighting of an existing microdata sample (which is only available at coarse levels of geography), so that it would fit small-area population statistics tables. For instance, an existing microdata set such as the BHPS described above can be reweighted to 'populate' small areas. The BHPS provides a detailed record for a sample of households and all of their members. Reweighting methods aim to sample from all the microdata records to find the set of household records that best matches the population described in the UK small-area statistics. First, a series of small-area tables (e.g. from the census or other sources) that describe the small area of interest must be selected. For example, a reweighting method would sample from the BHPS to find a suitable combination of households that would fit the statistical data in two hypothetical areas or neighbourhoods within cities and regions presented in Table 6.5.

The task would be to select the records of the BHPS microdata that best match these statistical descriptions using statistical matching or geographical microsimulation reweighting techniques. However, there is a vast number of possible sets of households that can be drawn from the BHPS sample. There is

Table 6.5 A hypothetical small-area statistical data set for two areas

Small-area table 1 (household type)	Small-area table 2 (economic activity of household head)	Small-area table 3 (tenure status)
Area 1	Area 1	Area 1
60 married couple households	80 employed/self-employed	60 owner occupier
20 single-person households	10 unemployed	20 Local Authority or Housing Association
20 other	20 other	20 rented privately
Area 2	Area 2	Area 2
40 married couple households	60 employed/self-employed	60 owner occupier
20 single-person households	20 unemployed	20 Local Authority or Housing Association
40 other	30 other	20 rented privately

a wide range of techniques that can be employed to find a set that fits the target tables well. There are a number of alternative static spatial microsimulation approaches that are capable of generating small-area microdata based on estimation methods ranging from Iterative Proportional Fitting-based deterministic reweighting (Ballas, 2004; Ballas et al., 2005a, 2005c) and synthetic reconstruction (Ballas and Clarke, 2000) to combinatorial optimisation heuristic tech-niques such as simulated annealing (Williamson et al., 1998; Ballas et al., 2007b). These models also include systematic validation of the simulations by compar-ing simulated outputs to published data and in some cases also involve creating Graphical User Interfaces and Planning Support GIS systems (Ballas et al., 2007b; see Figure 6.7) that allow interactive visualisation and mapping of the outputs under different scenarios. Recent reviews of the state of the art and relevant

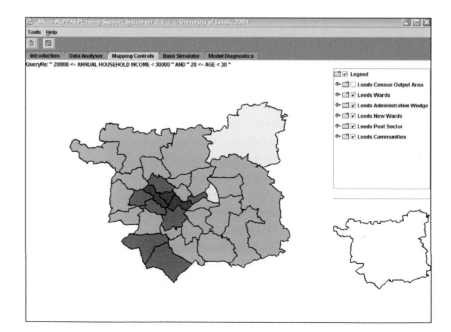

Figure 6.7 Spatial microsimulation query results

developments are presented by Hermes and Poulsen (2012) and Tanton and Edwards (2013).

Overall, these models result in the creation or synthesis of small-area population microdata which can be achieved by combining different small-area census cross-tabulations or by merging survey data such as census and other geographical area data to simulate a population of individuals within households (for different geographical units), whose characteristics are as close to the real population as it is possible to estimate. In other words, the models simulate virtual populations in given geographical areas, so that the characteristics of these populations are as close as possible to their 'real-world' counterparts. The simulation outputs can include a wide range of policy-relevant variables such as earned income, tenure status, household type, socio-economic group, consumption patterns, car ownership and so forth.

An example of a model that combines national survey data with small-area census data to estimate small-area microdata is the SimBritain model (Ballas *et al.*, 2005a, 2005c) which adopts a so-called deterministic approach to reweighting survey microdata so that they fit given small-area statistics tables. In particular, this methodology was used to estimate a wide range of non-census variables (including household income) at the small-area level. This model has been used to assess the socio-economic as well as geographical impact of a wide range of national social policy changes in the UK (Ballas *et al.*, 2007a; also discussed in more detail in the next section). Another example of such a model is the Microsimulation Modelling and Predictive Policy Analysis System (Micro-MaPPAS) (Ballas *et al.*, 2007b – an example is shown in Figure 6.7) which is open-source software implementing a geographical microsimulation model which is capable of constructing a list of approximately 715,000 individuals living within households along with their associated attributes for any point in time (past or future) at very small-area levels (down to UK OAs, an average of 100 households). The technique applies a simulated annealing combinatorial optimisation algorithm to data from the census of population and national survey microdata. There is also ongoing work and recent developments aimed at producing freely

available generic code and software that can be combined with GIS (see the Further reading and resources section at the end of this chapter).

Dynamic microsimulation involves forecasting past changes forward to produce the best estimate possible of an individual's circumstances in the future – were current trends to continue, or were they to change under different policy scenarios. Dynamic microsimulation typically involves the modelling of behavioural and second-order effects. This can be carried out on the basis of calculated probabilities for a series of event changes that occur during the lifetime of individuals. Another aim of dynamic spatial microsimulation is the analysis of household and individual reactions and behavioural changes that might result from policy changes. This adds further to the complexity of the task.

The task becomes even more difficult when there are attempts to introduce geographical detail. Spatial dynamic microsimulation involves the behavioural modelling of individuals over time and at various geographical scales. It also involves the modelling of individual decisions that have a strong geographical element, such as migration. The latter is dependent on a series of individual characteristics such as age, socio-economic background and tenure.

Spatial dynamic microsimulation involves the modelling of different types of transitions on the basis of each individual's attributes and circumstances. Nevertheless, one of the biggest problems associated with both spatial and non-spatial *dynamic* microsimulation is that they can be extremely complex and difficult to develop, implement and explain to policy practitioners who might be interested in using them. It has often been argued in the microsimulation literature that there is a need for transparency and simplicity in the construction of models. An alternative to the traditional comprehensive dynamic microsimulation models is to combine aggregate projection methods with the static microsimulation methods.

Table 6.6 depicts the steps that need to be followed in the procedure for allocating *employment status* and *industry* and modelling survival and migration. It should be noted however that the example depicted is simplified, in order to illustrate the process.

Table 6.6 A simple example of the microsimulation procedure for the modelling of migration and survival

Steps	1st	2nd	. . .	Last
Age, sex and marital status and location (e.g. neighbourhood or small-area level)(given)	Age: 25	Age: 76	. . .	Age: 30
	Sex: Male	Sex: Female		Sex: Male
	Marital Status: single	Marital Status: married		Marital Status: married
	GeoCode: Neighbourhood 1	GeoCode: Neighbourhood 2		GeoCode: Neighbourhood 3
Probability (conditional upon age, sex, location) of person to migrate	0.30	0.05	. . .	0.26
Random number	0.2	0.4	. . .	0.6
Migration status assigned on the basis of random sampling	Migrant	Non-migrant	. . .	Non-migrant
Probability (conditional upon age, sex, location) of person to survive	0.9	0.8		0.9
Random number	0.5	0.9	. . .	0.4
Survival status	Survived	Deceased	. . .	Survived

One of the inherent difficulties of such a task is to determine the interdependencies between individual attributes and events. For instance, the probabilities of an individual participating in the labour force might be conditional upon family status (e.g. having children). However, it could also be argued that family status depends on labour market status.

An additional difficulty associated with dynamic spatial microsimulation models is the lack of sufficient geographical data that would enable the simulation of interactions such as migration flows between areas (e.g. there are no microdata on migration that would enable a reasonably accurate simulation of migration into the future). Due to the lack of suitable data there have been very few examples of spatial microsimulation of dynamic events such as migration (e.g. Ballas *et al.*, 2005b; Rossiter *et al.*, 2009; Kavroudakis *et al.*, 2013). Nevertheless there are ongoing efforts to develop dynamic spatial microsimulation methodologies building on previous work (e.g. the work of Harding *et al.* (2010) and Holm and Mäkilä (2013)) in order to project small-area populations in all study

regions under different scenarios. The models to date include the so-called probabilistic dynamic modelling (e.g. see Ballas *et al.*, 2005b), implicitly dynamic macro approaches (Ballas *et al.*, 2005a) and econometric approaches. There is also great potential to link dynamic microsimulation with another conceptually similar type of individual-level modelling: agent-based models (ABM). As noted above, ABMs are normally associated with the behaviour of multiple agents in a social or economic system. These agents usually interact constantly with each other and the environment they 'live' or move within, and their actions are driven by certain rules. Although this methodology is conceptually very similar to microsimulation (where agents could be the individuals within the households), it has long been argued that ABM might offer a better framework for including behavioural rules into the actions of agents (including an element of random behaviour) and for allowing interactions between agents (Davidsson, 2000). There are a number of good illustrations in a geographical setting (Heppenstall *et al.*, 2005, 2006, 2007; Malleson *et al.*, 2009; Wu *et al.*,

2008) and there is a research agenda to link these two complementary approaches more effectively. Spatial microsimulation could be used to give the agents in ABM their initial characteristics and locations, while ABM could then provide the capacity to model individual adaptive behaviours and the emergence of new behaviours (also see Boman and Holm, 2004).

Using GIS and spatial microsimulation for public policy analysis

Spatial microsimulation models are very powerful tools for the analysis of urban, regional or national government policies. There are numerous examples of applied spatial microsimulation models for national policy analysis by geographers and regional scientists in a wide range of fields including social policy (e.g. Ballas and Clarke, 2000; Ballas *et al.*, 2007a; Chin *et al.*, 2005), poverty small-area estimation and analysis (Ballas, 2004; Tanton, 2011), health (Edwards and Clarke, 2009; Morrissey *et al.*, 2008; Tomintz *et al.*, 2008), agricultural policy (Ballas *et al.*, 2006b; Hynes *et al.*, 2009), international migration (Rephann and

Holm, 2004), educational policy (Kavroudakis *et al.*, 2013) and crime analysis (Kongmuang *et al.*, 2006). In this section we give just a small flavour of how GIS and spatial microsimulation applications can be used to estimate and analyse policy-relevant information.

One of the key application areas of spatial microsimulation involves the estimation of small-area income distributions and the use of these estimates for policy analysis. Here we give a couple of examples of spatial microsimulation models that have been used to that end. We begin with the SimBritain model (Ballas *et al.*, 2005a, 2007b) which combined small-area data from the UK census of population with the BHPS (described in more detail above) in order to produce a small-area survey microdata set for the geographical analysis of poverty and to explore the possible impacts of alternative social policy scenarios. The SimBritain model also involved the use of previous census data and appropriate population projection methods to estimate small-area microdata into future years. Table 6.7 shows an example of the analysis that was conducted at the small-area level (British electoral wards, parliamentary constituencies and metropolitan districts) with this model. The table shows households classified as very poor (all households with income below or equal to

Table 6.7 Living standards of very poor households

Very poor households	1991	2001	2011	2021
Households (% of all households in York)	17.2	17.3	17.8	21.3
Individuals (% of all individuals in York)	14.7	13.3	13.7	20.5
Children (% of all children in York)	21.8	17.7	18.6	38.5
Limiting long-term illness (as a % of all individuals in group)	9.0	7.3	5.4	7.9
Elderly (over 64 years as a % of all individuals in group)	30.1	32.0	33.3	44.2
Individuals in group with father's occupation: unskilled (%)	10.5	6.8	3.3	15.1
Reporting anxiety and depression (% of all individuals in group)	10.6	10.3	7.4	3.1
Individuals who reported that they have no one to talk to (%)	19.9	23.8	31.1	31.5
Promotion opportunities in current job (as % of individuals with a job)	33.7	36.9	51.9	79.7
Feeling unhappy or depressed (%)	19.9	19.0	18.2	12.1
Home computer in accommodation (%)	1.4	1.0	0.5	0.4
House without central heating (%)	26.1	21.4	21.4	31.1
Single-person households (%)	61.6	76.0	77.9	64.4
Cars/households ratio	0.23	0.32	0.38	0.40

Source: SimBritain model after Ballas *et al.* (2005c)

half of the median household income) in the city of York, UK, in different simulation years. As can be seen, the model outputs include variables that are not typically measured by the census of population (such as income, but also 'loneliness' and 'happiness' indicators) and cross-tabulations of as many variables as are deemed useful. For instance, Table 6.8 uses the outputs to draw pictures of the life of households in a similar way that Charles Booth and Seebohm Rowntree did in their original studies of poverty in London and York (as discussed in Chapter 5). It is possible to use the output of the method presented here to make some notes on the life of simulated households, and in this case of typical *very poor households*. This kind of information could also be used for the development of ABMs, which is a potential new development as discussed above.

A key feature of spatial microsimulation models is the ability to perform *what-if policy analysis*. Again, we can use an application of the SimBritain model to illustrate how this is possible. Having estimated small-

area microdata containing policy-relevant information (e.g. individual and household attributes and circumstances, levels of income) it is possible to estimate how many individuals and households are affected by particular policy changes. SimBritain was used to consider and analyse the geographical and socio-economic impact of a set of policies that were introduced by the British government in the late 1990s and which are described in detail in Appendix 6.3. The information provided in Appendix 6.3 was used to identify eligible individuals and households in the spatially microsimulated data set that was created by the SimBritain model. The next step was to estimate the total amounts by which each simulated eligible individual would be better off as a result of these policies and to then aggregate these to any geographical level deemed appropriate for mapping and analysis of any spatial patterns.

Figures 6.8 and 6.9 show examples of such mapping at the parliamentary constituency level in Wales and at the level of electoral ward in the city of York. In particular

Table 6.8 Simulating the lives of individuals and households at small-area levels

Age in years	Status
56 and 52	Married couple, male aged 56, economically active but unemployed. Formerly employed as motor mechanic/auto engineer. Female aged 52, economically inactive. Food expenditure £25 per week. No car. House owned with mortgage. Highest educational qualification of male: GCE O levels. Female has no formal qualifications.
46	Divorced, female, full-time personal services worker (hairdresser on seasonal/temporary job). Finding it quite difficult financially. No dependent children. Believes that all health care should be free, feeling unhappy or depressed. Has one car. Weekly household food expenditure £35.
78	Divorced, female, retired sales assistant. Feels that there is no one who she could count on to listen if she needs to talk. Feels that she is just about getting by financially. No children, no car. Weekly food and grocery expenditure £20.
34 and 21	Married couple, two children (aged eight and five). Two cars. Male has full-time job. He is self-employed (craft and related occupations – construction). Female is economically inactive (family care). Just about getting by financially. Household weekly food and grocery expenditure £40.
34 and 33	Married couple, four children (aged 14, 13, 12 and 9). Male unemployed (previous job: food, drink and tobacco process operative). Female economically inactive (family care). Household weekly expenditure on food £80.
30	Divorced, female, mother of three children (aged 11, 9 and 2), economically inactive (family care), weekly expenditure on food £20, no qualifications, formerly employed as a sales assistant. No car. House rented from local authority.

Source: SimBritain model after Ballas *et al.* (2005c)

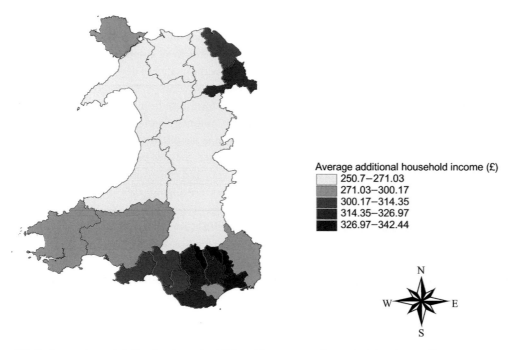

Average additional household income (£)
- 250.7–271.03
- 271.03–300.17
- 300.17–314.35
- 314.35–326.97
- 326.97–342.44

Figure 6.8 Estimated spatial distribution of additional income per household at the parliamentary constitu-ency level for Wales

Source: SimBritain model after Ballas *et al.* (2005c)

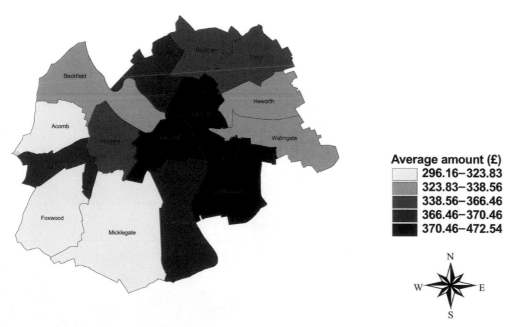

Average amount (£)
- 296.16–323.83
- 323.83–338.56
- 338.56–366.46
- 366.46–370.46
- 370.46–472.54

Figure 6.9 Estimated spatial distribution of additional income per household by electoral ward for the city of York, UK

Source: SimBritain model after Ballas *et al.* (2005c)

they depict the spatial distribution of the additional income per household by area which would result from implementing all the policy reforms shown in Appendix 6.3.

These examples demonstrate how spatial microsimulation and GIS together can be used to highlight the importance of geography and to estimate the *geographical* as well as the *social, temporal* and *economic* impacts of policies. In particular, spatial microsimulation methods and GIS can be used to estimate the geographical impacts of social policies. These can then be compared with the respective impacts of area-based policies, as *social policies* can be seen as alternatives to *area-based policies*. Further, spatial microsimulation methods can be used to analyse social policy in a geographically oriented *proactive* fashion. For instance, spatial microsimulation can be employed to identify deprived localities in which poor individuals and

households are over-represented, and then used to answer questions such as: 'What social policy could be applied, which, all else being equal would most likely improve the quality of life of residents in the inner-city localities of a city?' In other words, new social policies can be formulated on the basis of spatial microsimulation outputs. Spatially oriented social policies can be seen as a substitute or an alternative to traditional area-based policies and direct comparisons of their efficiency and effectiveness can be made.

There are ongoing developments in GIS and spatial microsimulation for policy analysis. A recent example is the SIMALBA[13] model of the Scottish population and a recent application aimed at modelling the geographical impacts of proposed British national fiscal policies, focusing on the capital of Scotland, Edinburgh, and the largest city, Glasgow (Campbell and Ballas, 2013). Using

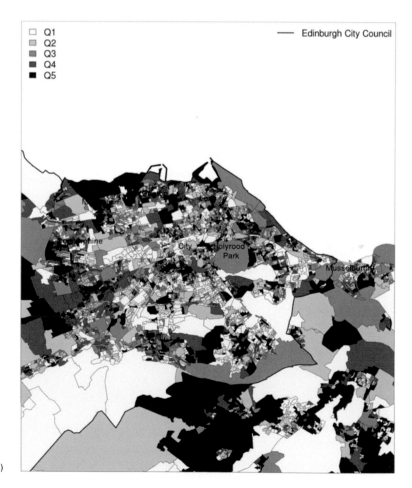

Figure 6.10 Simulated percentage earning over £150,000, Edinburgh, UK: quintiles

Source: Campbell and Ballas (2013)

a similar approach to the one adopted in SimBritain, SIMALBA combines data from the Scottish Health Survey with small-area census data to generate policy-relevant small-area microdata. These microdata can then be used to explore alternative proposed national government policies. One of these proposed policies that was considered and analysed is the introduction of a tax rate of 50% on personal income in the UK for higher incomes. SIMALBA was used to estimate the spatial distribution of people who would be affected by this change and mapped them at the OA level (neighbourhood level). Figure 6.10 shows the estimated distribution of people (at the neighbourhood level) in the city of Edinburgh who earn over £150,000 per year and who would be liable for a 50% rate of income tax if it were to be introduced in Scotland.

Another application area that relates to adding a geographical dimension to debates around the importance of measuring quality of life is where spatial microsimulation has been used to estimate personal happiness and quantify and estimate its value for different types of individuals, living in different areas. As Nobel Laureate Amartya Sen points out:

A person who has had a life of misfortune, with very little opportunities, and rather little hope, may be more easily reconciled to deprivations than others reared in more fortunate and affluent circumstances. The metric of happiness may, therefore, distort the extent of deprivation in a specific and biased way.

(Sen, 1987: 45)

Table 6.9 Measuring subjective well-being in the BHPS: the General Health Questionnaire (GHQ) set of questions as they appear on the BHPS questionnaire

GHQ questions/responses	1	2	3	4
1. Been able to concentrate on whatever you are doing?	Better than usual	Same as usual	Less than usual	Much less than usual
2. Lost much sleep over worry?	Not at all	No more than usual	Rather more than usual	Much more than usual
3. Felt that you are playing a useful part in things?	More than usual	Same as usual	Less so than usual	Much less than usual
4. Felt capable of making decisions about things?	More so than usual	Same as usual	Less so than usual	Much less capable
5. Felt constantly under strain?	Not at all	No more than usual	Rather more than usual	Much more than usual
6. Felt you could not overcome your difficulties?	Not at all	No more than usual	Rather more than usual	Much more than usual
7. Been able to enjoy your normal day-to-day activities?	Much more than usual	Same as usual	Less so than usual	Much less than usual
8. Been able to face up to your problems?	More so than usual	Same as usual	Less able than usual	Much less able
9. Been feeling unhappy and depressed?	Not at all	No more than usual	Rather more than usual	Much more than usual
10. Been losing confidence in yourself?	Not at all	No more than usual	Rather more than usual	Much more than usual
11. Been thinking of yourself as a worthless person?	Not at all	No more than usual	Rather more than usual	Much more than usual
12. Been feeling reasonably happy all things considered?	More so than usual	About same as usual	Less so than usual	Much less than usual

Figure 6.11 Estimated geographical distribution of happiness (% happy more than usual) in Wales, 2001

Source: After Ballas (2010)

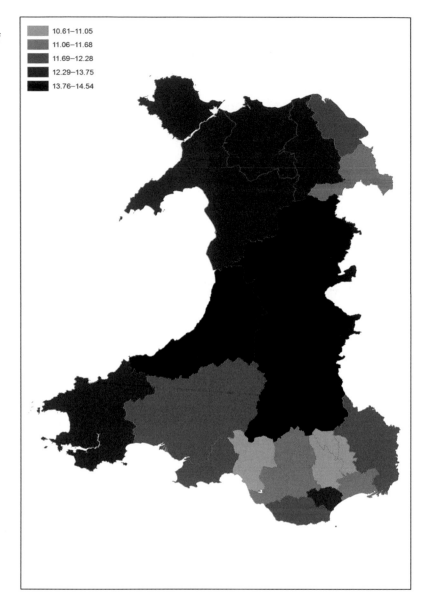

	10.61–11.05
	11.06–11.68
	11.69–12.28
	12.29–13.75
	13.76–14.54

Spatial microsimulation can be ideally suited to estimate happiness, since the degrees of well-being vary significantly between different individuals (different people are made happy by different things, life-courses, etc.). 'SimHappiness' (Ballas, 2010) was used to that end, combining data from the BHPS (and especially the subjective happiness and well-being variables generated from responses to the questions described in Table 6.9) and the UK census of population to add a geographical dimension to happiness research (such as the research on happiness by Clark and Oswald, 2002; Blanchflower and Oswald, 2004; Layard, 2005). SimHappiness was developed to estimate the geographical distribution of individual contentment through the 1990s across Britain. Figure 6.11 shows an output of this model, showing the estimated geographical distribution of happiness in Wales (at the level of Welsh Unitary Authorities).

Concluding comments

This chapter has presented a number of methods for the estimation of policy-relevant geographical information at various spatial scales. A key argument and message that is particularly prominent is that it is increasingly possible to use GIS and related spatial modelling methods to adopt a geographical approach to national policy analysis. The methods presented in this chapter (and especially the spatial microsimulation models) can be used to estimate the geographical distribution of a wide range of extremely policy-relevant social and economic variables. The accompanying practical shows how it is possible to make a start towards estimating and mapping small-area data using GIS. This provides a basis for the use of more advanced methods such as spatial microsimulation which enable the estimation of small-area microdata that can be re-aggregated or disaggregated geographically to provide estimates at any spatial level deemed appropriate for policy analysis as demonstrated in the practical. Although these methods are not available as part of proprietary GIS packages, there are stand-alone open-source GIS-based applications such as Micro-MaPPAS (Ballas *et al.*, 2007b) as well as available software and data (see Further reading and online resources section for more information).

Accompanying practical

The accompanying practical (Practical 2: Combining survey and small-area data) makes use of some of the approaches discussed in this chapter in order to estimate the spatial distribution of consumer expenditure. Specifically, you estimate small-area expenditure on groceries, combining counts of households and surveyed expenditure rates, which vary by geodemographic classification. We also explore how to convert data between different geographies, and how to create high-quality cartographic outputs.

Notes

1 www.ons.gov.uk/ons/guide-method/census/2011/how-our-census-works/how-we-took-the-2011-census/how-we-collected-the-information/questionnaires--delivery--completion-and-return/2011-census-questions/index.html.
2 www.ons.gov.uk/surveys/informationforhouseholdsandindividuals/householdandindividualsurveys/livingcostsandfoodsurveylcf.
3 www.gov.uk/government/collections/family-resources-survey--2.
4 www.understandingsociety.ac.uk/.
5 E.g. see www.arcgis.com/home/item.html?id=71a65d35688a4502b123cbdfc99afdee.
6 http://webhelp.esri.com/arcgisdesktop/9.3/index.cfm?TopicName=Geographically%20Weighted%20Regression%20(Spatial%20Statistics).
7 http://desktop.arcgis.com/en/arcmap/10.3/tools/spatial-statistics-toolbox/h-how-cluster-and-outlier-analysis-anselin-local-m.htm.
8 www.nomisweb.co.uk/.
9 http://census.ac.uk/cdu/experian/household%20income.pdf.
10 Williamson's ESRC report in 2005, reported on a research project carried out between 2000 and 2001. Heady and colleagues are mentioned in this report with regard to the dissemination of the findings.
11 www.neighbourhood.statistics.gov.uk/dissemination/Info.do?m=0&s=1446935569123&enc=1&page=analysisandguidance/analysisarticles/small-area-model-based-income-estimates.htm&nsjs=true&nsck=false&nssvg=false&nswid=1600.
12 Microsimulation was first introduced in Orcutt (1957).
13 Sim referring to simulation, and Alba meaning 'Scotland' in Scottish Gaelic.
14 www.inlandrevenue.gov.uk/.

Further reading and online resources (including free software)

All website URLs accessed 30 May 2017.

■ **Combinatorial Optimisation software (including dummy data set and associated documentation)** by Paul Williamson (University of Liverpool): http://pcwww.liv.ac.uk/~william/microdata/CO%20070615/CO_software.html.

- **Iterative Proportional Fitting and integerisation R code and data:** Lovelace, R., & Ballas, D. (2013) 'Truncate, replicate, sample': A method for creating integer weights for spatial microsimulation, *Computers, Environment and Urban Systems*, www.sciencedirect.com/science/article/pii/S0198971513000240 (open-access article including publicly available R code and data).

- **A recent review of the state of the art and research challenges** by Adam Whitworth (University of Sheffield): Whitworth, A. *et al.* (2013) *Evaluations and Improvements in Small Area Estimation Methodologies*. Discussion Paper. NCRM www.ncrm.ac.uk/research/NMI/2012/smallarea.php and http://eprints.ncrm.ac.uk/3210/. This includes a Spatial Microsimulation R-Library by Dimitris Kavroudakis (University of the Aegean) including R code. Available from: www.shef.ac.uk/polopoly_fs/1.268326!/file/sms_Manual_v9.zip.

- **An introductory text to spatial microsimulation:** Ballas, D., Rossiter, D., Thomas, B., Clarke, G., & Dorling, D. (2005) *Geography Matters: Simulating the Local Impacts of National Social Policies*, Joseph Roundtree Foundation. Available from: www.jrf.org.uk/file/36059/download?token=NkTWwksy&filetype=full-report.

Ballas, D., Clarke, G. P., & Dewhurst, J. (2006) Modelling the socio-economic impacts of major job loss or gain at the local level: a spatial microsimulation framework. *Spatial Economic Analysis*, 1(1), 127–146.

Ballas, D., Kingston, R., Stillwell, J., & Jin, J. (2007) Building a spatial microsimulation-based planning support system for local policy making. *Environment and Planning A*, 39(10), 2482–2499.

Hermes, K., & Poulsen, M. (2012) A review of current methods to generate synthetic spatial microdata using reweighting and future directions. *Computers, Environment and Urban Systems*, 36, 281–290.

Tanton, R., & Edwards, K. (2013) *Spatial Microsimulation: A Reference Guide for Users*, Springer, Berlin.

Appendix 6.1 Details of the core, rotating core and variable component question subject areas from the BHPS Individual Questionnaire (from Taylor *et al.*, 2001)

BHPS Individual Questionnaire

Core	*Neighbourhood and individual*:	*Current employment*:	*Finances*:
	Demographics	Employment status	**Incomes from**:
	Birthplace, residence	Not working/seeking work	Benefits/allowances/pensions/rents/savings/
	Satisfaction with home/neighbourhood	Self-employed	Interest/dividends
	Reasons for moving	Sector private/public	Pension plans
	Ethnicity	SIC/SOC/ISCO	Savings and investments
	Educational background and attainments	Nature of business/duties	Material well-being
	Recent education/training	Workplace/size of firm	Consumer confidence
	Partisan support	Travelling time	Internal transfers
	Changes in marital status	Means of travel	External transfers
	Citizenship	Length of tenure	Personal spending
		Hours worked/overtime	Roles of partners/spouses
		Union membership	Domestic work/childcare/bills/everyday spending
		Prospects/training/ambitions	Car ownership/use/value of car
		Superannuation/pensions	Interview characteristics
		Attitudes to work/incentives	Windfalls
		Wages/salary/deductions	
		Childcare provisions	
		Job search activity	

		Career opportunities	
		Bonuses	
		Performance related pay	

Rotating core	*Health and caring:*	*Employment history:*	*Values and opinions:*
	Personal health condition	Past year	Partisanship/interest in politics
	Employment constraints	Labour force status spells	Religious involvement
	Visits to doctor	Size/sector/nature of business/duties	Parental questionnaire
	Hospital/clinic use	Wages/salary/deductions	
	Use of health/welfare services	Reasons for leaving/taking jobs	
	Social services		
	Specialists		
	Check-ups/tests/screening		
	Smoking		
	Caring for relatives/others		
	Time spent caring for others		
	Private medical insurance		
	Activities in daily living		

Variable components	*Lifetime marital status history (Wave 2):*	*Lifetime fertility and adoption history (Wave 2 and Wave 8 catch-up):*	*Lifetime employment status history (Wave 2):*
	Number of marriages	Birth dates	Start and finish dates
	Marriage dates	Adoption dates	Employment status
	Divorce/widowhood/ Separation dates	Sex of children	
	Cohabitation before marriage	Leaving or mortality dates	*Values and opinions:*
			Aspirations for children
	Lifetime marital status history (Wave 3):	*Lifetime cohabitation history (Wave 2 and Wave 8 catch-up):*	Important events
			Quality of life
	Start and finish dates	Start and finish dates	
	Labour force status	Number of partners	*Credit and debt:*
	Sector/nature of business duties		Investment and savings
		Neighbourhood and demographics:	Commitments
	Health and caring:	Driving licence	
	Children's health	Parents' employment background	*Crime:*
	Other health scales: SF36 (Wave 9)	Family background	Criminal activity on local area
		Difficulties with debt	Perceptions of crime
	Computers and computing (Wave 6/7):	Community and neighbourhood	
	Ownership and usage		
		Employment (Wave 9):	
		National Minimum Wage	
		Work strain	
		Work orientation	

Appendix 6.2 Details of the core question subject areas from the BHPS Household Questionnaire (from Taylor *et al.*, 2001)

BHPS Household Questionnaire: core questions

Size and condition of dwelling	*Household Finances*:
Ownership status	Rent and mortgage, loan and HP details
Length of tenure	Local Authority service charges
Previous ownership	Allowances/rebates
Interview characteristics	Difficulties with rent/mortgage payments
	Household composition
	Consumer durables, cars, telephones, food
	Heating/fuel types, costs, payment methods
	Non-monetary poverty indicators
	Crime

Appendix 6.3 A selection of policies that were evaluated in SimBritain

Working Families' Tax Credit

One of the major policy initiatives that was implemented in the 1990s was the Working Families' Tax Credit (WFTC), which is an allowance paid to low-paid workers with children (Fitzpatrick *et al.*, 2002; Inland Revenue online, 2003[14]). In order to qualify for WFTC individuals would have to fulfil the following criteria:

- They or their partner should work normally full time (16 hours or more a week).
- They have at least one dependent child for whom they are responsible.
- They do not get disabled person's tax credit.
- Their income is sufficiently low.
- Their savings and capital are not worth more than £8,000.
- They are present and ordinarily resident in Great Britain.
- They are not subject to immigration control.

WFTC is calculated by comparing the family income with the applicable amount or threshold figure, which in 2002 was £94.50 per week. If the family income is less than the applicable amount, then the family receives the maximum WTFC. If the family income exceeds the applicable amount, the maximum WFTC is reduced by 55% of the excess (Fitzpatrick *et al.*, 2002). As noted above, in the context of the research reported here all the relative amounts were adjusted to allow for inflation. In the case of WFTC the applicable amount of £94.50 in 2002 was readjusted to its equivalent in 1991 on the basis of the RPI growth of 29.3%. Thus, the adjusted applicable amount that we used was £66.77. Further, all the relevant credits were adjusted before allocating them to eligible households of the simulated database. The table below lists the actual (2002) amounts for the various credits.

Working Families' Tax Credits	Amount per week
Couple or lone parent	£60.00
Child aged:	
under 16	£26.35
16–18	£27.20
30 hours credit	£11.65
Disabled child credit	£35.50

Enhanced disability credit:

Couple or lone parent	£16.25
Child	£46.75

Childcare credit

One child	70% of up to £135
Two or more children	70% of up to £200
Additional partners in a polygamous marriage	£22.70

Minimum wage

Another related major policy development in the 1990s was the introduction of the minimum wage. The minimum wage in October 2002 was £4.50 per hour for individuals at work who are over 21 years old and £3.80 for individuals aged 18–21, which were adjusted to £2.97 and £2.54 for 1991.

Minimum income guarantee

The introduction of the minimum income guarantee was another major policy development that occurred in the late 1990s. This guarantee aimed at topping up the income of elderly individuals or couples to a minimum level (aged 60 or over and with savings less than £12,000). In March 2003 the minimum level was £98.15 for a single person and £149.80 for a couple. These figures were adjusted on the basis of RPI growth to £69.35 and £105.84.

Winter Fuel Payment and free TV licence for the elderly

Another policy initiative aimed at boosting the incomes of the elderly was the Winter Fuel Payment, which was given to individuals aged 60 or over. This amount was £200 in 2003 and was adjusted to £141.31 for 1991. Further, a similar government initiative was the provision of free TV licences to those aged over 75.

In the case of TV licence there is no need to readjust the 2002–3 figure to 1991 as data exist on the TV licence across time. The TV licence was £112 in 2002, whereas in 1991 it was £77.

References

All website URLs accessed 30 May 2017.

Anselin, L. (1995) Local Indicators of Spatial Association – LISA. *Geographical Analysis, 27*, 93–115.

Ballas, D. (2004) Simulating trends in poverty and income inequality on the basis of 1991 and 2001 census data: a tale of two cities. *Area, 36*(2), 146–163.

Ballas, D. (2010) Geographical modelling of happiness and well-being, in J. Stillwell, P. Norman, C. Thomas, & P. Surridge (eds) *Understanding Population Trends and Processes Volume 2: Spatial and Social Disparities*, Springer, Dordrecht, 53–66.

Ballas, D., & Clarke, G. P. (2000) GIS and microsimulation for local labour market policy analysis. *Computers, Environment and Urban Systems, 24*, 305–330.

Ballas, D., Clarke, G. P., Dorling, D., Eyre, H., Rossiter, D., & Thomas, B. (2005a) SimBritain: a spatial microsimulation approach to population dynamics. *Population, Place and Space, 11*, 13–34.

Ballas, D., Clarke, G. P., & Wiemers, E. (2005b) Building a dynamic spatial microsimulation model for Ireland. *Population Place and Space, 11*, 157–172.

Ballas, D., Rossiter, D., Thomas, B., Clarke, G. P., & Dorling, D. (2005c) *Geography Matters: Simulating the Local Impacts of National Social Policies*, Joseph Rowntree Foundation contemporary research issues, Joseph Rowntree Foundation, York.

Ballas, D., Dorling, D., Anderson, B., & Stoneman, P. (2006a) *Assessing the Feasibility of Producing Small Area Income Estimates*, final project report to the UK Office for the Deputy Prime Minister, 31 March 2006.

Ballas, D., Clarke, G. P., & Wiemers, E. (2006b) Spatial microsimulation for rural policy analysis in Ireland: the implications of CAP reforms for the national spatial strategy. *Journal of Rural Studies, 22*, 367–378.

Ballas, D., Clarke, G. P., Dorling, D., & Rossiter, D. (2007a) Using SimBritain to model the geographical impact of national government policies. *Geographical Analysis, 39*(1), 44–77.

Ballas, D., Kingston, R., Stillwell, J., & Jin, J. (2007b) Building a spatial microsimulation-based planning support system for local policy making. *Environment and Planning A, 39*(10), 2482–2499.

Ballas, D., Dorling, D., & Hennig, B. (2014) *The Social Atlas of Europe*, Policy Press, Bristol.

Blanchflower, D. G., & Oswald, A. J. (2004) Well-being over time in Britain and the USA. *Journal of Public Economics*, 88, 1359–1386.

Boman, M., & Holm, E. (2004) Multi-agent systems, time geography, and microsimulations, in M. Olsson & G. Sjöstedt (eds) *Systems Approaches and their Application*, Kluwer Academic, Dordrecht, 95–118.

Bramley, G., & Lancaster, S. (1998) Modelling local and small-area income distributions in Scotland. *Environment and Planning C*, 16, 681–706.

Bramley, G., & Smart, G. (1996) Modelling local income distributions in Britain. *Regional Studies*, 30(3), 239–255.

Brunsdon, C. A., Fotheringham, A. S., & Charlton, M. E. (1996) Geographically weighted regression: a method for exploring spatial non-stationarity. *Geographical Analysis*, 28, 281–298.

CACI Information Solutions (2005) *Wealth of the Nation 2005*, Midnight Communications, London.

Campbell, M., & Ballas, D. (2013) A spatial microsimulation approach to economic policy analysis in Scotland. *Regional Science Policy and Practice*, 5, 263–288.

Census Dissemination Unit (2005) *Experian Data Index*. Available from: http://census.ac.uk/cdu/experian/.

Chin, S. F., Harding, A., Lloyd, R., McNamara, J., Phillips, B., & Vu, Q. N. (2005) Spatial microsimulation using synthetic small area estimates of income, tax and social security benefits. *Australasian Journal of Regional Studies*, 11(3), 303–335.

Clark, A., & Oswald, A. (2002) A simple statistical method for measuring how life events affect happiness. *International Journal of Epidemiology*, 31(6), 1139–1144.

Cloutier, N. R. (1995) Lognormal extrapolation and income estimation for poor black families. *Journal of Regional Science*, 35(l), 165–171.

Cloutier, N. R. (1997) Metropolitan income inequality during the 1980s: the impact of urban development, industrial mix, and family structure. *Journal of Regional Science*, 37(3), 459–478.

Cressie, N. (1995) Bayesian smoothing of rates in small geographic areas. *Journal of Regional Science*, 35(4), 659–673.

Davidsson, P. (2000) Multi agent based simulation: beyond social simulation, in S. Moss & P. Davidsson (eds) *Multi Agent Based Simulations*, Springer, Berlin, 97–107.

Dhanecha, N., Ellerd-Elliott, S., Herring, I., Horsfall, E., Majetic, P., Shome, J., & Snow, J. (2003) *Family Resources Survey: Great Britain 2001–02*. Department for Work and Pensions, London. Available from: www.dwp.gov.uk/asd/frs/2001_02/pdfonly/frs_2001_02_report.pdf.

Edwards, K. L., & Clarke, G. P. (2009) The design and validation of a spatial microsimulation model of obesogenic environments for children in Leeds, UK: SimObesity. *Social Science & Medicine*, 69(7), 1127–1134.

Fay, R. E., & Herriot, R. A. (1979) Estimates of income for small places: an application of James-Stein procedures to census data. *Journal of the American Statistical Association*, 74, 269–277.

Fisher, R. (1997) Methods used for small area poverty and income estimation, US Bureau of the Census, Washington, DC 20233. Available from: www.census.gov/did/www/saipe/publications/files/Fisher97.pdf.

Fitzpatrick, P., *et al.* (2002) *Welfare Benefits Handbook 2002/03*, Child Poverty Action Group, London.

Fotheringham, A. S., Brunsdon, C., & Charlton, M. E. (2002) *Geographically Weighted Regression: The Analysis of Spatially Varying Relationships*, Wiley, Chichester.

Gee, G., & Fisher, R. (2004) Errors-in-variables county poverty and income models, Bureau of the Census, Washington, DC 20233. Available from: www.census.gov/did/www/saipe/publications/files/FisherGee2004asa.pdf.

Hammer, R. B., Blakely, R. M., & Voss, P. R. (2003) The effects of integrating the U.S. Census Bureau's Small Area Income and Poverty Estimates into the Appalachian Regional Commission's designation of economically distressed counties. *Economic Development Quarterly*, 17, 165–174.

Hamnett, C., & Cross, D. (1998) Social polarisation and inequality in London: the earnings evidence 1979–1995. *Environment and Planning C: Government and Policy*, 16, 659–680.

Harding, A., Keegan, M., & Kelly, S. (2010) Validating a dynamic microsimulation model: recent experience in Australia. *International Journal of Microsimulation*, 3(2), 46–64.

Heady, P., Clarke, P., *et al.* (2003) *Model-Based Small Area Estimation Project Report*, Office for National Statistics Model-Based Small Area Estimation Series No. 2, January, Office for National Statistics, London.

Heard, D., Dent, G., Schifeling, T., & Banks, D. (2015) Agent-based models and microsimulation. *Annual Review of Statistics and Its Application*, 2(1), 259–272.

Heppenstall, A. J., Evans, A. J., & Birkin, M. H. (2005) A hybrid multi-agent/spatial interaction model system for petrol price setting. *Transactions in GIS*, 9(1), 35–51.

Heppenstall, A. J., Evans, A. J., & Birkin, M. H. (2006) Using hybrid agent-based systems to model spatially-influenced retail markets. *Journal of Artificial Societies and Social Simulation*, 9(3). Available from: http://jasss.soc.surrey.ac.uk/9/3/2.html.

Heppenstall, A. J., Evans, A. J., & Birkin, M. H. (2007) Genetic algorithm optimisation of a multi-agent system for simulating a retail market. *Environment and Planning B*, 34, 1051–1070.

Hermes, K., & Poulsen, M. (2012) A review of current methods to generate synthetic spatial microdata using reweighting and future directions. *Computers, Environment and Urban Systems*, *36*, 281–290.

Holm, E., & Mäkilä, K. (2013) Design principles for micro models, in R. Tanton & K. Edwards (eds) *Spatial Microsimulation: A Reference Guide for Users*, Springer, New York, 195–207.

Hynes, S., Morrissey, K., O'Donoghue, C., & Clarke, G. (2009) Building a static farm level spatial microsimulation model for rural development and agricultural policy analysis in Ireland. *International Journal of Agricultural Resources, Governance and Ecology*, *8*(3), 282–299.

Kavroudakis, D., Ballas, D., & Birkin, M. (2013) A spatial microsimulation approach to the analysis of social and spatial inequalities in higher education attainment. *Applied Spatial Analysis and Policy*, *6*, 1–23.

Kongmuang, C., Clarke, G. P., Evans, A. J., & Jin, J. (2006) SimCrime: a spatial microsimulation model for the analysing of crime in Leeds, *Working Paper 06/1*, School of Geography, University of Leeds.

Layard, R. (2005) *Happiness: Lessons from a New Science*, Penguin Books, London.

Longley, P., & Tobon, C. (2004) Spatial dependence and heterogeneity in patterns of hardship: an intra-urban analysis. *Annals of the Association of American Geographers*, *94*(3), 503–519.

Lynch, R. G. (2003) Estimates of income and income inequality in the United States and in each of the fifty states: 1988–1999. *Journal of Regional Science*, *43*, 571–588.

Malleson, N., Evans, A., & Jenkins, T. (2009) An agent-based model of burglary. *Environment and Planning B*, *36*, 1103–1123.

Marsh, C. (1993) Privacy, confidentiality and anonymity in the 1991 Census, in A. Dale & C. Marsh (eds) *The 1991 Census User's Guide*, Her Majesty's Stationery Office, London, 111–128.

McLoone, P. (2002) *Commercial Income Data: Associations with Health and Census Measures*, MRC Social & Public Health Sciences Unit, Occasional Paper No. 7, May.

Mitton, L., Sutherland, H., & Weeks, M. (eds) (2000) *Microsimulation Modelling for Policy Analysis: Challenges and Innovations*, Cambridge University Press, Cambridge.

Morrissey, K., Clarke, G., Ballas, D., Hynes, S., & O'Donoghue, C. (2008) Examining access to GP services in rural Ireland using microsimulation analysis. *Area*, *40*, 353–364.

Office for National Statistics (ONS) (2000a) *Family Spending: A Report on the 1998–99 Family Expenditure Survey*, The Stationery Office, Crown Copyright, London.

Office for National Statistics (ONS) (2000b) *Family Resources Survey: Great Britain*, Department of Social Security, London.

Office for National Statistics (ONS) (2015) Information Paper: quality and methodology information – Labour Force Survey. Available from: www.ons.gov.uk/file?uri=/employmentandlabourmarket/peopleinwork/employmentandemployeetypes/qmis/labourforcesurveylfsqmi/qmilfsjan2015finalforpubdocx_tcm77-180685.pdf.

Orcutt, G. H. (1957) A new type of socio-economic system. *The Review of Economics and Statistics*, *39*, 116–123.

Redmond, G., Sutherland, H., & Wilson, M. (1998) *The Arithmetic of Tax and Social Security Reform: A User's Guide to Microsimulation Methods and Analysis*, Cambridge University Press, Cambridge.

Rees, P. (1998) 'What do you want from the 2001 Census?' Results of an ESRC/JISC survey of user views. *Environment and Planning A*, *30*, 1775–1796.

Rephann, T. J., & Holm, E. (2004) Economic-demographic effects of immigration: results from a dynamic spatial microsimulation model. *International Regional Science Review*, *27*(4), 379–410.

Rossiter, D., Ballas, D., Clarke, G. P., & Dorling, D. (2009) Dynamic spatial microsimulation using the concept of GHOSTs. *International Journal of Microsimulation*, *2*(2), 15–26.

Rusanen, J., Muilu, T., Colpaert, A., & Naukkarinen, A. (2001) Local and regional income differences in Finland in 1989–1997: a GIS approach. *Geografiska Annaler*, *83B*, 205–220.

Rusanen, J., Muilu, T., Colpaert, A., & Naukkarinen, A. (2002) Income differences within municipalities in Finland, 1989–1997. *Scottish Geographical Journal*, *188*(2), 69–86.

Sen, A. (1987) *On Ethics and Economics*, Blackwell, Oxford.

Tanton, R. (2011) Spatial microsimulation as a method for estimating different poverty rates in Australia. *Population, Space and Place*, *17*(3), 222–235.

Tanton, R., & Edwards, K. (2013) *Spatial Microsimulation: A Reference Guide for Users*, Springer, Dordrecht.

Taylor, M. F. (ed.) with Brice, J., Buck, N., & Prentice-Lane, E. (2001) *British Household Panel Survey User Manual Volume A: Introduction, Technical Report and Appendices*, University of Essex, Colchester.

Tomintz, M., Clarke, G. P., & Rigby, J. (2008) The geography of smoking in Leeds: estimating individual smoking rates and the implications for the location of stop smoking services. *Area*, *40*, 341–353.

UK Data Service (2017a) *General Lifestyle Survey (General Household Survey)*. Available from: https://discover.ukdataservice.ac.uk/series/?sn=200019.

UK Data Service (2017b) *The Labour Force Survey*. Available from: https://discover.ukdataservice.ac.uk/series/?sn=2000026.

Voas, D., & Williamson, P. (2000) An evaluation of the combinatorial optimisation approach to the creation of synthetic microdata. *International Journal of Population Geography*, *6*, 349–366.

Webber, R. (2004) *The Relative Power of Geodemographics vis à vis Person and Household Level Demographic Variables as Discriminators of Consumer Behaviour*, CASA Working Paper, UCL. Available from: http://discovery.ucl. ac.uk/202/.

Williamson, P. (2000) Income imputation for small areas: interim progress report, paper presented at '*The Census of Population: 2000 and Beyond*', Manchester, 22–23 June.

Williamson, P. (2005) *Income Imputation for Small Areas*, ESRC grant award number H507255166, final report.

Williamson, P., Birkin, M., & Rees, P. (1998) The estimation of population microdata by using data from small area statistics and samples of anonymised records. *Environment and Planning A, 30*, 785–816.

Williamson, P., & Voas, D. (2000) Income estimates for small areas: lessons from the census rehearsal. *BURISA, 146*, 2–10.

Wu, B., Birkin, M., & Rees, P. (2008) A spatial microsimulation model with student agents. *Computers, Environment and Urban Systems, 32*, 440–453.

7 GIS and crime pattern analysis

LEARNING OBJECTIVES

- Becoming familiar with relevant key types of crime data and relevant GIS data sources
- Using GIS for the allocation and management of policing resources
- Using GIS for mapping and analysis of crime patterns
- Using GIS to support crime prevention strategies

Introduction

This chapter explores the potential of GIS to support the work of police and fire authorities around the world. Thus the chapter focuses largely on GIS and crime pattern analysis. The primary uses of GIS technology in law enforcement are in detection and the analysis of crime patterns and in the development and management of preventative measures. This chapter will first discuss the types of data that are potentially available for crime analysis (although this will vary by country) and the issues surrounding usage and sharing. We will then review a series of tools that are available for GIS mapping and the analysis of crime. Key questions we might typically want to address include: when and where should resources (e.g. officers and vehicles) be most effectively deployed? How can future crimes be prevented? Can we find patterns to help us apprehend repeat offenders? What are the displacement effects of crime prevention? These types of questions have an inherently spatial-temporal aspect and GIS is an obvious choice to help in law enforcement.

Crime data

In many countries there are often two different types of crime data available for analysis with GIS. The first are data collected by police forces themselves. In many countries these data have to be made routinely available for inspection and analysis by law. Typical features of basic crime data recorded by police authorities include when and where a crime has been committed and various details of the nature of the offence. These (point) data are collected by the police or law enforcement agencies for various purposes including both strategic and operational planning.

The fundamental issue regarding crime data is that police data can only provide a limited picture, which does not implicitly indicate true patterns of crime. For

Figure 7.1 UK public web-based GIS

Source: http://data.police.uk/

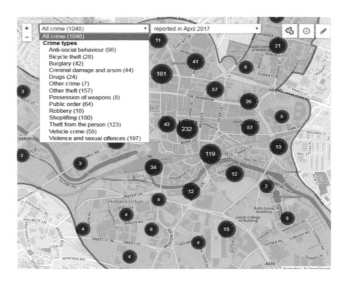

instance, even if a crime takes place it is not guaranteed that the crime will actually be recorded by the police. In fact, Dodd *et al.* (2004) estimated that only 30% of all crime is actually represented by recorded crime data. This is, of course, an average and will vary hugely by offence type and by country. Burglaries, for instance, are more commonly reported for insurance purposes. However, unreported crime is expected to cluster in the same places as reported crime (Chainey and Ratcliffe, 2013), thereby effectively increasing the association between reported crime figures and actual amounts of crime (Malleson, 2010).

Figure 7.1 shows a screenshot of a publicly available GIS site created by the police in the UK. It shows all crimes in a single month in the city centre of Leeds, UK (mainly shoplifting and assaults in bars and night clubs). The user can plot any one of ten crime types and for different months of the year. (In the US the FBI collects crime data from each law enforcement agency across the US and then publishes the 'uniform crime reporting statistics' at various spatial scales.)

The data given by police forces often have limitations. As noted above, only recorded crime can be included. Second, some crime locations might be reported as an approximate location (by victims or eye witnesses) but they are always recorded as a precise point location. This might be more of an issue for certain crime types over others.

Due to the increased developments in mobile phone technology, the Police.UK website shown above is also available through a number of phone applications. While there are a number of different applications, the applications essentially allow users to look at the level and types of crime in a given area based on the phone's internal GPS. The main applications include: Crime Spy UK, Crime Map UK, Crimes Near Me, UK Crime View and Crime Finder. The Crime Finder application is demonstrated in Figure 7.2 and even includes Google Street View capability for the visualisation of the street (see Malleson (2010) for more details).

As an alternative to using data provided by the police, it is also often possible to use survey data. In the UK, for example, it is possible to access the newly named Crime Survey (CS) for England and Wales,

Figure 7.2 Crime Finder mobile phone application

Source: https://data.gov.uk/apps/crime-finder

formerly known as the British Crime Survey (BCS), which measures the amount of crime in England and Wales. The original BCS started in 1982 and initially covered England, Wales and Scotland. In 2011/12 around 67,000 private households across England and Wales were invited to participate in the survey and about 75% of those responded (50,000). The main topics include information about the crimes people (aged 16 and over) have experienced in the last year, attitudes towards the police, attitudes about the criminal justice system and perceptions of crime and anti-social behaviour. From January 2009, 4,000 interviews have also been conducted each year with children 10–15 years old, although the resulting statistics remain experimental. The UK Home Office (the government department responsible for policing and crime reduction) believe that in some respects the CS provides a better reflection of the true level of crime than police statistics, since it includes crimes that have not been reported to, or recorded by, the police. The survey is also a better indicator of long-term trends because it is unaffected by changes in levels of reporting to the police or police recording practices (Malleson, 2010). In the US, the Bureau of Justice is responsible for a similar survey: the 'National Crime Victimisation Survey' which samples from 90,000 households across the US.

The downside to survey data is that they are often poorly referenced in spatial terms. For confidentiality reasons, individuals and their locations are rarely named. Thus it is impossible to plot crimes from the survey with any geographical accuracy – however, it is often possible to reweight such survey data to link the attributes of individuals to census data in order to add small-area geography (see the discussion of microsimulation in Chapter 6).

GIS for the allocation of funding

One of the first major uses of spatial analysis for crime fighting is likely to be to help allocate resources. Indeed, most public service organisations have funding models that are based on geographical variations in socio-economic data. In the UK, the police use such

a formula to allocate £11 billion (2014 figure) between the different police forces shown in Figure 7.3.

The allocation formula includes the following census-based indicators: population total, population density, (youth) unemployment rates, bars per hectare (a proxy for where assaults are more likely to be committed), CACI's geodemographic classification 'Hard Pressed' label (which picks up areas of low income), student housing rates, population inflow (to measure community cohesion) and single-parent households (Home Office, 2013). A GIS containing these layers of information would be a good starting point to be able to not only calculate the allocations on a regular basis, but also to be able to experiment with different weightings in order to see how resources could be allocated differently and perhaps more fairly (especially as the values of these variables change over time).

GIS for mapping and analysing crime patterns

While we explore the use of GIS for crime pattern analysis it is useful to try to answer the questions why is geography important in crime analysis, or why does crime vary spatially? Figure 7.4 shows the spatial variations at a single point in time across London, UK. As the map shows, there are widespread variations across the city. Crime data can be mapped using raw numbers or normalised to produce rates. City centres, for example, can have large actual crime figures (as there are lots of people in town during the day and night) but low rates when expressed as a percentage of the population that resides there. Figure 7.4 shows there is, indeed, a high concentration of crime in the city centre but also in the east of London, with much lower rates to the south-west. These high crime areas tend to correspond with higher levels of poverty. Thus, the first explanation for spatial variations in crime rates is simply socio-economic. Areas with high levels of unemployment, drug use, poor educational attainment, etc. can become breeding grounds for crime.

Figure 7.5 shows the spatial variation in crime classified as homicides (murders) across Washington, DC. Again we can see a clustering effect with much higher

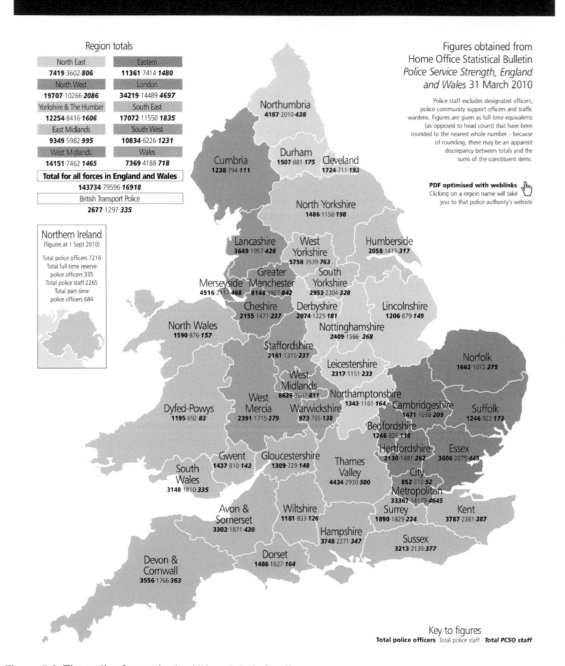

Police service strength in England and Wales at March 2010

*Association of **Police** **Authorities***

Region totals

North East	Eastern
7419·3602·*806*	**11361**·7414·*1480*
North West	London
19707·10266·*2086*	**34219**·14489·*4697*
Yorkshire & The Humber	South East
12254·8416·*1606*	**17072**·11550·*1835*
East Midlands	South West
9349·5982·*995*	**10834**·6226·*1231*
West Midlands	Wales
14151·7462·*1465*	**7369**·4188·*718*

Total for all forces in England and Wales
143734·79596·*16918*

British Transport Police
2677·1297·*335*

Northern Ireland
(figures at 1 Sept 2010)

Total police officers 7216
Total full-time reserve
police officers 335
Total police staff 2265
Total part-time
police officers 684

Figures obtained from
Home Office Statistical Bulletin
*Police Service Strength, England
and Wales* 31 March 2010

Police staff excludes designated officers,
police community support officers and traffic
wardens. Figures are given as full-time equivalents
(as opposed to head count) that have been
rounded to the nearest whole number – because
of rounding, there may be an apparent
discrepancy between totals and the
sums of the constituent items.

PDF optimised with weblinks
Clicking on a region name will take
you to that police authority's website

Northumbria **4187**·2010·*438*

Cumbria **1238**·794·*111*

Durham **1507**·881·*175*

Cleveland **1724**·711·*193*

North Yorkshire **1486**·1158·*198*

Lancashire **3649**·1957·*428*

West Yorkshire **5758**·3539·*763*

Humberside **2058**·1415·*317*

Greater Manchester **8148**·3927·*842*

Merseyside **4516**·2117·*468*

South Yorkshire **2953**·2304·*328*

Lincolnshire **1206**·879·*149*

Cheshire **2155**·1471·*237*

Derbyshire **2074**·1225·*181*

Nottinghamshire **2409**·1566·*268*

North Wales **1590**·876·*157*

Staffordshire **2161**·1315·*237*

Leicestershire **2317**·1151·*233*

Norfolk **1662**·1072·*275*

West Midlands **8626**·3607·*811*

Northamptonshire **1343**·1161·*164*

West Mercia **2391**·1715·*279*

Warwickshire **973**·765·*138*

Cambridgeshire **1471**·1038·*209*

Suffolk **1246**·922·*173*

Dyfed-Powys **1195**·692·*83*

Bedfordshire **1246**·826·*116*

Gwent **1437**·810·*143*

Gloucestershire **1309**·729·*148*

Thames Valley **4434**·2930·*500*

Hertfordshire **2130**·1481·*262*

Essex **3606**·2075·*445*

City **852**·310·*52*

South Wales **3148**·1810·*335*

Metropolitan **33367**·14179·*4645*

Avon & Somerset **3302**·1871·*430*

Wiltshire **1181**·833·*126*

Surrey **1890**·1829·*224*

Kent **3787**·2381·*387*

Hampshire **3748**·2271·*347*

Sussex **3213**·2139·*377*

Dorset **1486**·1027·*164*

Devon & Cornwall **3556**·1766·*363*

Key to figures
Total police officers · Total police staff · ***Total PCSO staff***

Figure 7.3 The police forces in the UK and their funding

Source: Association of Police Authorities: www.webarchive.org.uk/wayback/archive/20121213152137/http://www.apa.
police.uk/publications

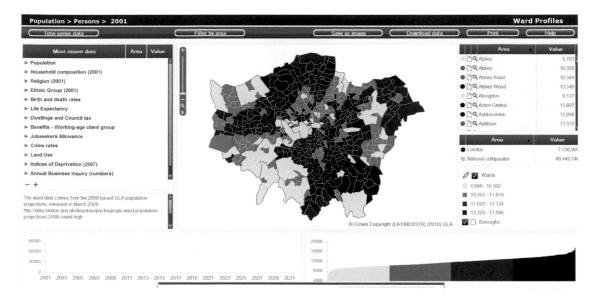

Figure 7.4 Crime across London, 2009

Source: http://londondatastore-archive.s3.amazonaws.com/visualisations/atlas/ward-profiles-2010/atlas.html?detect flash=false (reproduced with thanks to Greater London Authority and London Datastore)

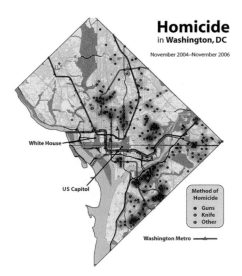

Figure 7.5 Crime rates vary spatially in Washington, DC

Source: https://commons.wikimedia.org/w/index.php?curid= 1414546

rates in certain notorious inner-city or downtown districts, north and east of the US Capitol. You will be able to explore recorded crime rates in Washington, DC, further in the practical activity linked to this chapter.

There have been many studies exploring this correlation between crime and deprivation. A GIS application for exploratory analysis of crime patterns has been widely used by researchers at the University of Liverpool in the UK, who have concentrated particularly on the mapping of crime patterns in relation to socio-economic structure. An early application used ArcGIS to plot maps of crime incidence and examined spatial and temporal trends in these patterns (Hirschfield *et al.*, 1995; Bowers, 1999; Hirschfield and Bowers, 2001). The researchers also used GIS to study the hypotheses that:

■ crime is greater where affluent areas border deprived areas;

■ more crime occurs in areas of high population turnover;

■ crime can be associated with low socio-economic status and social heterogeneity.

The studies plotted the location of disorder incidents in relation to deprived areas, police beats, main roads, pubs and night clubs. The STAC software (Spatial and Temporal Analysis of Crime) was developed to identify clusters or hotspots of criminal activity, which can be displayed in relation to the distribution of affluent and disadvantaged areas from various geodemographic classifications. One of the primary discoveries was a link between crime such as burglaries and changes in household occupancy (as represented by changes in council tax registrations), which conforms to the second of the above hypotheses.

Marxist interpretations of crime claim that such links between crime and deprivation are inevitable given the capitalist system and its inherent ability to create class divisions around income (Peet, 1975; Lowman, 1986). Some argue that in more deprived areas, individuals (especially youths) are more likely to have had a more difficult if not violent upbringing with few good role models to draw on. If more individuals who are likely to commit crime live in certain areas then we have the first part of Cohen and Felson's (1979) crime triangle – a motivated offender. Add a victim in the neighbourhood of the offender, and the absence of a capable guardian, and we complete the crime triangle shown in Figure 7.6.

Brantingham and Brantingham (1981) show how the geography of crime can be better understood by further considering the 'spatial activity map' of criminals. Criminals tend to operate close to home (but not too close), in and around major destinations they are likely to visit (friend's house, shopping centre, even workplace). Thus individuals build up a spatial awareness map over time around key anchor points in their

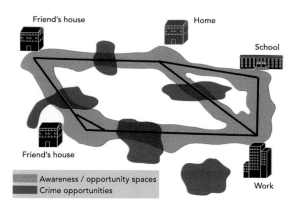

Figure 7.7 Activity space and crime opportunities

Source: Brantingham and Brantingham (1981), redrawn by Malleson (2010)

neighbourhoods. Figure 7.7 demonstrates this idea geographically.

As hypothesised above by Hirschfield and Bowers (2001), we might also expect crime to be higher in areas where affluent areas border lower income areas. Figure 7.8 plots the locations of offenders committing crimes in Roundhay, Leeds, an affluent suburb of the city. Roundhay has above average crime rates compared to other high-income areas in the city. As Figure 7.8 shows, part of the explanation might lie in the fact that Roundhay is bordered to the east and south in particular by low-income areas such as Harehills (shown on the map), Burmantofts and Seacroft. Many other high-income areas are themselves surrounded by high-income or middle-income suburbs. This unique geography provides an opportunity for offenders to quickly escape back into territories they are familiar with in order to escape police attention.

Deprived areas are often associated with more nuisance crimes such as anti-social behaviour, petty theft and vandalism. Wilson and Kelling (1982) referred to this as the 'broken window' theory. This refers to the fact that evidence of minor petty crimes in the landscape tends to encourage others to do the same. Thus a number of police forces around the world have tried to crack down on minor crimes in the belief that this would inevitably prevent more serious crimes. GIS can again help target those locations where petty crimes prevail.

Figure 7.6 The crime triangle

Source: Cohen and Felson (1979), redrawn by Malleson (2010)

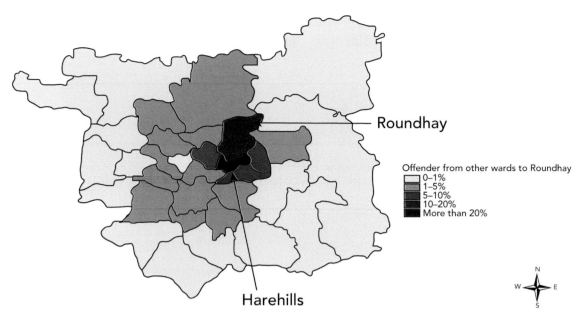

Figure 7.8 Plotting offender locations for crimes committed in Roundhay, Leeds, UK

Another major explanation for spatial variations in crime rates is the physical fabric of an area – especially the location of bars, clubs and shopping centres, etc. – or areas that will simply attract larger populations. For example, in 2013 the newspaper *Yorkshire Post* published the top ten crime hotspot areas in the city of Leeds, UK. Seven of these were major shopping destinations, with a nightclub, a festival site and a hospital interestingly making up the rest (it is largely the accident and emergency unit of the hospital, which has to deal with many drunken revellers in the early hours of the morning causing problems for staff, that often needs police support).

If space is important in crime pattern analysis then so too is time. In fact, crime pattern analysis should be spatial-temporal. Figure 7.9 shows clusters of crime activity by day of the week and time of the day in Leeds, UK – with the location of recorded crimes during a Saturday daytime (cluster A) reflecting a major shopping centre, and the cluster of recorded crimes on a Saturday evening corresponding with a student area (cluster B).

As Figure 7.9 shows, the night pattern of crime is greater, more widespread and generally more serious. Understanding such patterns has enormous impli-

cations for real-time policing, crime prevention and resource allocation.

An extension of crime mapping is the understanding of repeat victimisation. For example, the risk of burglary doubles after a victimisation, but risk does then rapidly decay with time. This is said to be related to offender confidence (she/he knows that the property is able to be burgled). Figure 7.10 is taken from the work of Hirschfield and Bowers (2001) in part of Liverpool, UK. Although a rather crude GIS map by today's standards, plotting the location of the victims of repeat burglary does help to show an important additional feature – that properties most likely to be repeat victims are closer to the main road network (thus again allowing quicker getaways).

This knowledge has been put to good use in Manchester in the UK. Police in the Trafford district of the city worked with GIS analysts from The Bartlett Centre for Advanced Spatial Analysis in London to examine the geography of burglaries and repeat burglaries. They showed that burglars were more likely to return to previously successful 'hit' locations than to choose new locations. This again was for reasons of offender confidence and the fact that the properties of the victim's neighbours are more likely to have similar

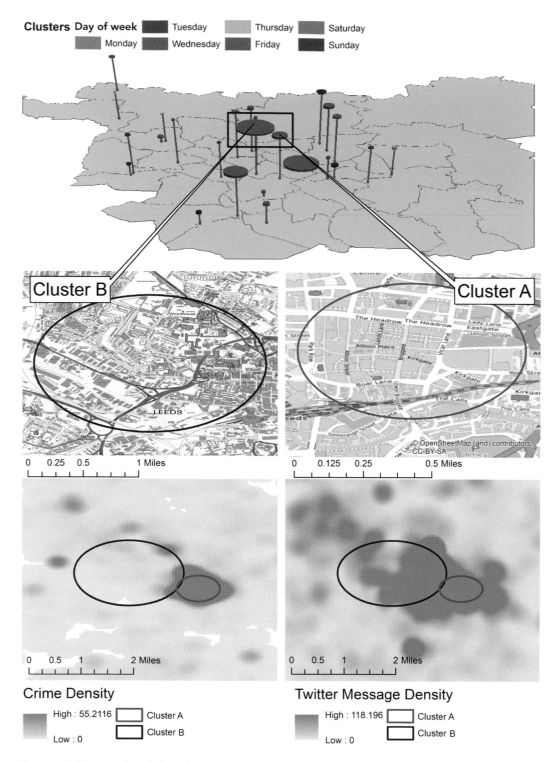

Figure 7.9 Temporal variations in clusters of crime in Leeds, UK

Source: Malleson and Andresen (2015)

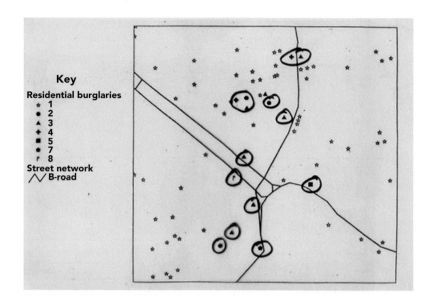

Figure 7.10 A map of repeat victimisation for burglaries in Liverpool, UK

Source: Hirschfield and Bowers (2001)

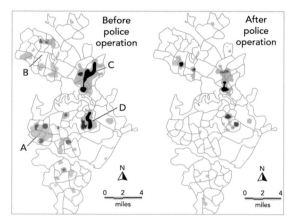

Figure 7.11 Mapping the impact of 'Operation Anchorage' in Canberra, Australia, 2001

Source: Ratcliffe (2002)

designs (hence could be broken into in the same manner as the original property chosen on the street). This allowed Trafford Police to develop a new model for reducing burglary, which has become known in the UK as the Trafford Model. By targeting resources at areas recently burgled Greater Manchester Police were able to reduce burglaries in Trafford by 26% in six months (see Fielding and Jones (2012) for more details).

Following a police initiative to reduce crime (such as the Trafford Model above), GIS can examine the result-ant changes in the spatial patterns of crimes committed. Figure 7.11 shows the reduction in burglaries across Canberra in Australia following 'Operation Anchorage' in 2001, where special 'burglary-reduction teams' were deployed in selected areas with high burglary rates (Ratcliffe, 2002).

Crime profiling

GIS is increasingly being used in conjunction with psychology techniques to analyse spatial crime patterns to, in turn, estimate where offenders are most likely to live. This can help narrow down the search for criminals in the detective exercise following a (major) crime. This is becoming increasingly popular in attempts to capture serial murderers/rapists, etc., but is also more common today for many other applications; e.g. prolific burglary. The idea of geographical profiling is to use GIS to create a 'probability surface' of most likely offender home location based on key facts available (Rossmo, 1999; Canter et al., 2000; Canter, 2010; Canter and Youngs, 2008).

The most famous criminal never to be caught in the UK is probably Jack the Ripper. He lived in Victorian London in the infamous Whitechapel area, an area in 19th-century London that was synonymous with

Figure 7.12 Pinpointing the likely residence of Jack the Ripper

Source: *Daily Mail* (2014) www.dailymail.co.uk/sciencetech/article-2650534/Jack-Rippers-address-revealed-Geographic-profiling-pinpoints-street-serial-killer-lived.html

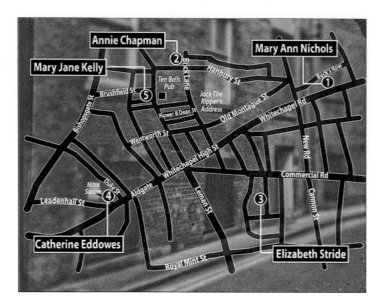

poverty, crime and prostitution. Jack is believed to have killed five prostitutes in and around the area but was never caught. Using modern GIS and geographical profiling new evidence can be brought to bear on the case. His victims were known to live in lodgings close to the Ten Bells public house (see Figure 7.12) although they drank and picked up clients in all the local pubs. If a polygon is drawn around the location of his victims it is argued that he probably lived close to the centre of that polygon. If you then add the fact that the victim lived close to the Ten Bells pub it could be argued that he himself was likely to frequent that pub and live nearby. Expert Canadian criminal profiler Kim Rossmo undertook this analysis and pinpointed 'Flower and Dean Street' as the most likely location of Jack's residence (see Figure 7.12). Given the fact the police had several possible suspects on their files from that street at that time, then it is possible they could have concentrated their enquiries on that much smaller group of possible suspects.

The same type of profiling has recently been undertaken to find a so-called copycat killer operating in Yorkshire, UK in the 1980s. Nicknamed the 'Yorkshire Ripper', Peter Sutcliffe was found and convicted for the murder of 13 victims. However, he gave the police the run around for many years. Again modern GIS and geographical profiling might have helped capture Sutcliffe earlier. Using key evidence that he seemed to know the areas well and victims were killed close to major roads (it turned out he was a truck driver), profiling has shown that the most likely location of Sutcliffe was around Heaton and Manningham – the location, as it turns out, where he did actually live.

Estimating expected versus actual crime rates

In many countries crime rates are often produced as league tables – the top crime areas at the top going down to low crime areas at the bottom. That is all well and good but such a table is often very predictable – most deprived areas at the top of the list, leafy, green suburban areas at the bottom. GIS and spatial analysis can be used to help create alternative league tables that might be more interesting. Harper *et al.* (2001) report the results of this procedure for the whole of the UK in a joint project with the UK Home Office. The study created 'families' of police authority zones that had similar socio-economic profiles – i.e. one family might be the result of clustering areas of lower income, terraced housing, low car ownership, etc. Once the families were created, Harper *et al.* could

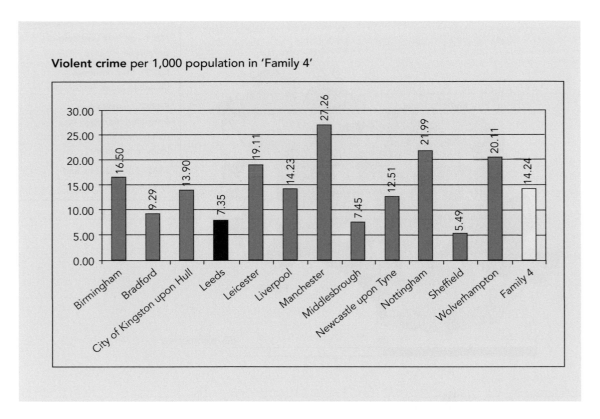

Violent crime per 1,000 population in 'Family 4'

Figure 7.13 Creating 'families' for police comparison purposes using geodemographics

compare crime within these families rather than simply between them (more typical of normal crime lists). Figure 7.13 shows the results for 'Family 4', largely inner-city areas of the UK's major cities. It shows how violent crime varied across the family – the Home Office could then ask why violent crime was much lower in inner cities of Leeds and Sheffield compared to Manchester and Nottingham. Hence, good police practice could, in theory, be exchanged between good performing and poorer performing police forces.

GIS and arson mapping

Alongside the police, the fire service provides a major emergency service provision and, once again, GIS and spatial analysis can provide very useful data support roles. Again, an interesting question is why would the distribution of fire incidents have a spatial dimension? Of course there are obvious urban and rural differences. Bush fires in countries such as Australia and the US have been very much in the news recently especially as hot summers provide ideal conditions for such fires to spread rapidly. Chen *et al.* (2003) used GIS to map the distribution and impacts of bushfires.

But even within urban areas geography seems important. Most importantly, fires are more likely to start in households within lower income areas. This is for a number of reasons. First those households might have older electrical appliances, with older wiring. Second, lower income households are less likely to be able to afford smoke detectors. Third, such households lie in areas more prone to arson. The number of malicious calls and incidents of arson are more common than many people suspect.

Corcoran *et al.* (2007a, 2007b, 2011) have analysed the spatial variations in incident types for fires

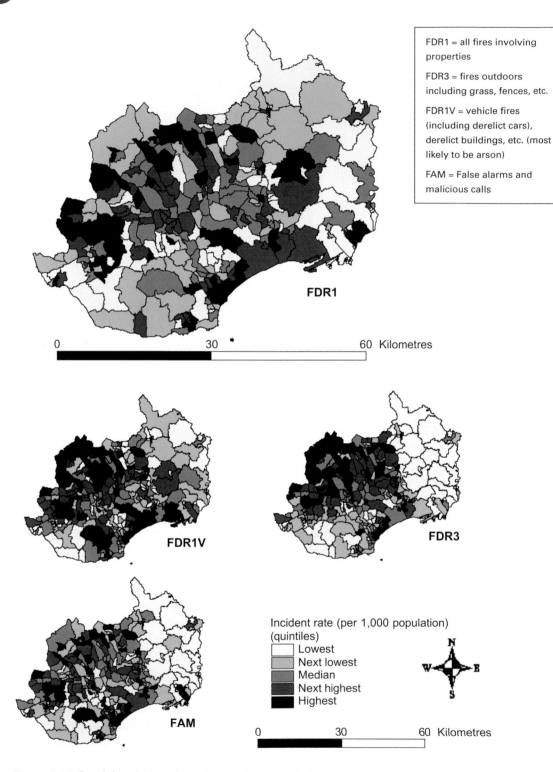

FDR1 = all fires involving properties

FDR3 = fires outdoors including grass, fences, etc.

FDR1V = vehicle fires (including derelict cars), derelict buildings, etc. (most likely to be arson)

FAM = False alarms and malicious calls

FDR1

0   30   60 Kilometres

FDR1V

FDR3

Incident rate (per 1,000 population) (quintiles)

Lowest
Next lowest
Median
Next highest
Highest

FAM

0   30   60 Kilometres

Figure 7.14 Spatial variations in malicious fires, South Wales

Source: Corcoran *et al.* (2007a)

across South Wales in the UK. Figure 7.14 shows these patterns, with marked concentrations visible in the coastal belt (which links the main cities of Cardiff and Swansea) and in the valleys (the old coalmining areas which are now suffering from high levels of poverty). Thus, as with many types of crime, there is a striking correlation between areas of deprivation and high fire incidence, especially for arson and malicious fires.

Figure 7.15 Various types of arson, Leeds, UK

Source: Clarke *et al.* (2006)

GIS is commonly used in crime analysis to identify 'hot spots'. Hot spot analysis can use various techniques to identify clusters of crime activity that are statistically significant and hence unlikely to occur randomly. Eftelioglu *et al.* (2016) provide a good introduction to techniques for identifying hot spots. Our own analysis has revealed clusters of arson events in Leeds. Hot spots were located in low-income areas to the south and, particularly, to the (inner) east of the region, in deprived council estates built in the 1960s as slum clearance projects. Figure 7.15 shows how these different types of environment foster arson – Figure 7.15a shows the build up of rubbish dumped by local residents in the driveway of an abandoned house: this is easily turned into 'fuel' by local youths and set alight. The solution to this problem is to constantly clear rubbish as soon as it starts to accumulate. Figure 7.15b shows the results of drug dealers abandoning a drug den in an old tower block close to the city centre. Once the police become aware of such a den the drug dealers leave and set fire to the inside to remove fingerprints, etc.

Concluding comments

This chapter has provided an introduction to the application of GIS in the crime and fire service sectors. We have also demonstrated that GIS can be applied to crime issues at a variety of levels from illustrative mapping to more complex spatial analysis. The fundamental processing options for GIS in crime analysis include:

- Basic spatial mapping of crime geography, which permits a visual impression of crime distribution, offence types and victim and offender locations.
- Pattern detection analysis of point level data, which offers a statistical interpretation of crime hotspots. This might confirm or denounce hypotheses of distribution suggested by simple map reviewing.
- Location analysis to study crime in relation to other static spatial features such as schools, pubs, cash machines, etc.

- Spatial aggregation of crime data for comparison to contextual information on demography, land use, socio-economic conditions, etc.
- Temporal series analysis to indicate trends in criminal activity and to monitor and manage crime policy.
- Application of these results and analyses to target crime prevention strategies and link efforts with other governmental organisations and authority departments.
- Analysis of crime strategy results and their success or failure.

The Trafford example above shows how crime can be reduced by adding a GIS and spatial analysis capability. There are a number of other illustrations of police forces around the world advocating the benefits of GIS, usually on their websites. The ESRI website (2007), for example, quotes Police Chief Crisp of Columbia Police Force in the US as saying 'the crime rate for Columbia has fallen dramatically with the implementation of GIS mapping' and 'using GIS mapping helps us concentrate efforts to maximize resources in the most effective manner possible. It has helped produce the lowest crime rate that Columbia has seen within the past 15 years'.

The final words can be given to Joe Kezon, GIS manager for the Chicago Police Force in the 2000s (quoted in Chen, 2004). He commented that using GIS technologies allowed officers to make better-informed decisions about which areas of the city need additional police power: 'The commander of the Deployment Operations Center saw the importance of having the ability to do some mapping and analysis that would allow them to make key judgments of where they should create police deployment areas.' The net effect, according to Kezon, was an 18% drop in murders compared with the same period the year before. That alone must be a good reason to employ GIS!

Accompanying practical

This chapter is accompanied by Practical 9: Crime analysis. The practical gives you an opportunity to explore data related to actual recorded crimes in the US city of Washington, DC. We walk you through the process of obtaining, downloading and importing crime data from an external source. You practise handling this data in the GIS, including aggregation to larger spatial units and hot spot analysis to identify clusters of neighbourhoods with high crime rates.

Further reading

Wortley, R., & Mazerolle, L. (2008) *Environmental Criminology and Crime Analysis*, Willan Publishing, London.

References

All website URLs accessed 30 May 2017.

Bowers, K. (1999) Exploring links between crime and disadvantage in north-west England: an analysis using geographical information systems. *International Journal of Geographical Information Science*, 13(2), 159–184.

Brantingham, P. J., & Brantingham, P. L. (1981) Notes on the geometry of crime, in P. J. Brantingham & P. L. Brantingham (eds) *Environmental Criminology*, Prospect Heights, New York, 27–54.

Canter, D. (2010) Offender profiling, in J. M. Brown & E. A. Campbell (eds) *The Cambridge Handbook of Forensic Psychology*, Cambridge University Press, Cambridge, 236–241.

Canter, D., & Youngs, D. (eds) (2008) *Applications of Geographical Offender Profiling*, Ashgate Publishing, London.

Canter, D., Coffey, T., Huntley, M., & Missen, C. (2000) Predicting serial killers' home base using a decision support system. *Journal of Quantitative Criminology*, 16(4), 457–478.

Chainey, S., & Ratcliffe, J. (2013) *GIS and Crime Mapping*, Wiley, Chichester.

Chen, A. (2004) GIS Fights Crime in Chicago, e-week, posted 31 May 2004.

Chen, K., Blong, R., & Jacobson, C. (2003) Towards an integrated approach to natural hazards risk assessment using GIS: with reference to bushfires. *Environmental Management*, 31(4), 0546–0560.

Clarke. G. P., Vickers, D., *et al.* (2006) Evaluation of the Leeds Arson Task Force: Final report to West Yorkshire Fire Service (unpublished) (copies available from the first author).

Cohen, L., & Felson, M. (1979) Social change and crime rate trends: a routine activity approach. *American Sociological Review*, 44, 588–608.

Corcoran, J., Higgs, G., Brunsdon, C., Ware, A., & Norman, P. (2007a) The use of spatial analytical techniques to explore patterns of fire incidence: a South Wales case study. *Computers, Environment and Urban Systems, 31*(6), 623–647.

Corcoran, J., Higgs, G., Brunsdon, C., & Ware, A. (2007b) The use of comaps to explore the spatial and temporal dynamics of fire incidents: a case study in South Wales, United Kingdom. *The Professional Geographer, 59*(4), 521–536.

Corcoran, J., Higgs, G., & Higginson, A. (2011) Fire incidence in metropolitan areas: a comparative study of Brisbane (Australia) and Cardiff (United Kingdom). *Applied Geography, 31*(1), 65–75.

Dodd, T., Nicholas, S., Povey, D., & Walker, A. (2004) Crime in England and Wales 2003/2004. Home Office Statistical Bulletin. Home Office, London.

Eftelioglu, E., Shekhar, S., & Tang, X. (2016) Crime hotspot detection: a computational perspective, in O. E. Isafiade & A. B. Bagula (eds) *Data Mining Trends and Applications in Criminal Science and Investigations*. IGI Global, Hershey, PA, 82–111.

ESRI (2007) Columbia, South Carolina, Police Department Uses GIS for Improved Policing, ESRI Summer Watch, July. Available from: www.esri.com/news/arcwatch/0707/feature.html.

Fielding, M., & Jones, V. (2012) Disrupting the optimal forager: predictive risk mapping and domestic burglary reduction in Trafford, Greater Manchester. *International Journal of Police Science & Management, 14*(1), 30–41.

Harper, G., Williamson, I., See, L., & Clarke, G. P. (2001) *Family Ties: Developing Basic Command Unit Families for Comparative Purposes*, Briefing Note, 4/01. Available from: http://tna.europarchive.org/20071206133532/homeoffice.gov.uk/rds/prgpdfs/brf401.pdf.

Hirschfield, A., & Bowers, K. (2001) *Mapping and Analysing Crime Data: Lessons from Research and Practice*, Taylor & Francis, London.

Hirschfield, A., Brown, P., & Todd, P. (1995) GIS and the analysis of spatially-referenced crime data: experiences in Merseyside, UK. *International Journal of Geographical Information Systems, 9*(2), 191–210.

Home Office (2013) *Home Office National Crime Atlas*. Available from: http://webarchive.nationalarchives.gov.uk/20110218142900/http://rds.homeoffice.gov.uk/rds/ia/atlas.html?detectflash=false.

Lowman, J. (1986) Conceptual issues in the geography of crime: toward a geography of social control. *Annals of the Association of American Geographers, 76*(1), 81–94.

Malleson, N. (2010) *Agent-Based Modelling of Burglary*. PhD thesis, School of Geography, University of Leeds.

Malleson, N., & Andresen, M. A. (2015) Spatio-temporal crime hotspots and the ambient population. *Crime Science, 4* (10). DOI: https//doi.org/10.1186/s40163-015-0023-8.

Peet, R. (1975) The geography of crime: a political critique. *The Professional Geographer, 27*(3), 277–280.

Ratcliffe, J. (2002) Burglary reduction and the myth of displacement. *Trends & Issues in Crime and Criminal Justice, 232*. Australian Institute of Criminology, Canberra. Available from: http://www.jratcliffe.net/wp-content/uploads/Ratcliffe-2002-Burglary-reduction-and-the-displacement-myt.pdf.

Rossmo, D. K. (1999) *Geographic Profiling*, CRC Press, Boca Raton, FL.

Wilson, J. Q., & Kelling, G. L. (1982) Broken windows. *Critical Issues in Policing: Contemporary Readings, 7*, 395–407.

8 GIS for retail network planning and analysis

LEARNING OBJECTIVES

■ Introduce the fundamental principles underpinning store location research
■ Using GIS to identify demand and supply-side factors for retail
■ GIS and localised marketing campaigns
■ Using GIS to identify store catchment areas and market shares
■ GIS and spatial interaction models for retail planning
■ The role of GIS for public sector retail planning
■ GIS for the identification, mapping and analysis of food deserts

Introduction

The retail sector is a major commercial user of GIS software and spatial data. In this chapter we outline a series of examples of GIS for retail network planning and analysis. Many of these examples demonstrate that GIS is often coupled with spatial modelling tools in order to derive powerful insight into retail systems. Location-based decision making undoubtedly represents one of the most important functions within any retail organisation. Despite the rapid growth in internet retailing across the world (Clarke *et al.*, 2015) it is through their network of stores that most contemporary retailers still interact most closely with their customers. In the highly competitive UK grocery sector, for example, one of the key battlegrounds during the 'store-wars' era of the 1970s and 1980s (Wrigley,

1987) involved the rapid acquisition of sites suitable for new store development and the diffusion of brands spatially from a core headquarter location. It was during this 'store-wars' era that we saw the development of site location teams within major grocery retailers and, embedded within them, a highly competitive strategy towards new store development. Today, most of the major UK grocery chains have specialised teams of in-house location analysts, who carry out sophisticated spatial analysis to identify new sites, estimate market areas associated with new and existing stores and forecast revenue in advance of new store investment (Birkin *et al.*, 2002, 2017). Reynolds and Wood (2010) note that UK grocery retailers tend to carry out the most sophisticated site location research in the retail industry and are more likely to have their own specialised in-house teams than retailers in any other sector, managing some of the largest store portfolios. These retailers also benefit from some of the most powerful consumer insights driven by loyalty schemes, Electronic Point of Sale (EPOS) data and geodemographics (see the section on demand estimation, below).

A recent set of surveys of UK location planning teams identified their primary role as being to support the financial business case for new stores (Wood and Reynolds, 2011, 2012; Reynolds and Wood, 2010). An important component of their work thus involves an assessment of the trading potential of a proposed site and the prediction of store revenue in advance of investment. Site location teams are thus fundamental to many areas of a retailer's operations and operate at both a strategic level (e.g. evaluating sites and generating revenue predictions in advance of major investment decisions) and an operational level (e.g. assisting marketing teams with store-based demographic information or monitoring store performance against forecast revenue predictions in day-to-day operations).

There are now many useful introductions to store location research spanning the different techniques as they have been fashionable over time (Davies, 1977; Fenwick, 1978; Davies and Rogers, 1984; Bowlby *et al.*, 1984a, 1984b; Clarke, 1998; Birkin *et al.*, 2002, 2017). Although we wish to concentrate here on the role of GIS it is necessary to also consider alternative methods, albeit more briefly. Early store location decisions were inevitably made by senior executives on gut feeling or retail nous accumulated over many years (Davies, 1977). But in the 1970s and 1980s the analogue technique became more widespread. Analogue techniques were (and still are) very common procedures for site location in the UK and the US especially. The basic approach involves attempting to forecast the potential sales of a new (or existing) store by drawing comparisons (or analogies) with other stores in the corporate chain that are alike in physical, locational and trade area circumstances. This can be done 'manually' or through regression techniques. Hence, if you are evaluating a new store site in, say, Cambridge, UK, can you find an existing store location around the UK that has the same (or similar) population and trading characteristics as Cambridge? Alternatively, the procedure might

work by trying to find sites that are analogous with the top performing stores within the company. That is, if a store in, for example, Oxford, UK, is performing very strongly, can the analyst find sites elsewhere in the country that match the characteristics of the Oxford site? Additionally, a similar approach has been to follow the behaviour of other (larger) retailers and base store location decisions on whatever decisions they make. This has been labelled the 'parasitic' approach and is more common for smaller retailers.

The success of the analogue approach depends on whether or not you can find similar sites across the country and whether you believe you can successfully transfer the trading characteristics across geographical locations. In reality, a wide variation in performance is frequently found between 'similar' outlets in a retail chain. Moreover, even if a similar geographical catchment is found to the new store, what happens if the analogous store is currently over- or underperforming? Laing *et al.* (2003) provide an interesting discussion on the analogue procedure. If key drivers of success can be suggested by analogue, revenue predictions have increasingly utilised regression analysis across a broad set of variables (which can be compared across stores), such as total sales area, catchment demographics and the degree of competition (see Birkin *et al.* (2002, 2017) and below for more detail on the use of analogues and regression analysis for store revenue estimation).

During the 1980s, census data became available in electronic form and the introduction of GIS and desktop PCs allowed retail location planning teams to develop computerised spatial forecasting models to estimate store revenue. At the time, Wrigley (2014: 30) asserted that 'never before have the skills of locational analysts, developed and practised by geographers and planners been so closely identified with the commercial imperatives of retailers'. Retailers were able to carry out analysis and visualisation, utilising the wealth of spatially referenced data at their disposal following the introduction of loyalty cards and the widespread availability of census and geodemographic data, allowing location analysis to grow in sophistication. GIS provided retailers with tools to undertake drive time analysis, allowing them to identify the size and characteristics of the population that live within thresholds of individual stores. Coupled with knowledge of competitor presence, store catchment population characteristics could now be used to predict sales and revenue.

Before we examine the use of GIS for retail analysis in more detail, it is useful to examine the fundamental principles which underpin store location research.

The building blocks for analysis

Whether developing new retail sites or ensuring current sites are making the most of their corporate brand, or planning what retail networks should look like in the future, the data used in the analysis are vitally important. As such, the following section looks at the various data needed for network planning and the issues impacting on the quality of those data.

Demand estimation

Fundamental to retail network planning is the use of demand data (e.g. population counts, understanding of different household types, expenditure estimates, etc.), for it is crucial to understand where people live and how much potential expenditure is available at a small-area level. However, for many retailers this information is difficult and expensive to obtain. For instance, some retailers have their own data collected through store cards (e.g. Tesco Clubcard) or store credit cards (e.g. Topshop, M&S and Debenhams). These data are used as a way to understand the location and characteristics of their consumers and the catchment area associated with a given store. Those retailers that do not have access to this sort of information have to collate information from other sources. These include information generated from official government surveys, market research or consumer data purchased from companies such as Experian and Acxiom Ltd. This might include geodemographic classifications – designed to segment the entire population of an area (e.g. the UK) according to the type of area in which they live. Such classifications are usually based on a

broad range of indicator variables that cover housing tenure, socio-economic information and demographic characteristics. Examples in the UK include CAMEO by Callcredit, ACORN by CACI, Mosaic by Experian and OAC by the ONS (see Chapter 5 for a recap on geodemographics). However, these data can often be expensive to acquire and the methodologies behind their production are usually kept hidden. In addition, the journey to work can also be especially important to consider in certain geographical locations. Using straightforward population data, retail demand in Canary Wharf in London, for example, would be small whereas we know that with a large, predominately young professional daytime work population such demand is likely to be considerable (see Berry *et al.*, 2016). So to reproduce market performance effectively we need to represent expenditure originating from the workplace as well as in relation to residential locations.

Alternatively, retailers also have the option of using data from the various censuses undertaken around the world. The census forms the most comprehensive and detailed data available in most countries and has been collected by the ONS in England and Wales every decade since 1801 (with the exception of 1941). The attributes collected form a comprehensive view of the population, comprising numerous demographic, social and economic characteristics. The data are geographically referenced and the finest area at which they are released is the Output Area (OA) level. However, the UK census data have no information on income or expenditure (as we discussed at length in Chapters 5 and 6) which is problematic when trying to estimate demand (although note that this information is present in other censuses around the world – e.g. in the US). If these data are missing, a combination of both small-area census data and up-to-date consumer data can be used to create a more up-to-date and robust data source. Mindful of the inadequacies of both commercially available data and available census variables, Birkin *et al.* (2002) discuss examples for the estimation of financial services expenditures in relation to diverse factors including age, gender, household size, dependent children, occupation, income and ethnicity.

Another issue are the boundary data in which the demand data are produced. In commercial applications,

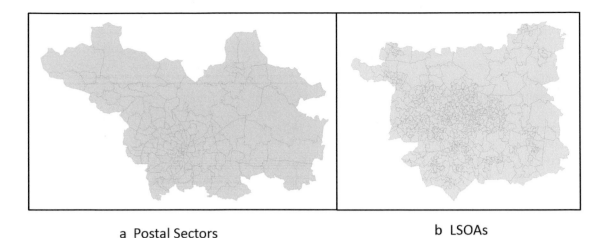

a Postal Sectors

b LSOAs

Figure 8.1 Differences between postal geographies and administrative boundaries in Leeds, UK

postal geographies have long been a popular boundary data set for retail analysis due to the range of contextual geodemographic and behavioural indicators that were available using these geographies (which are also important for marketing). The standard hierarchical format from the lowest level is postcodes, postal sectors, postal districts and postal areas. Postal geographies can be problematic, however, as postcodes are continually moving and changing (Raper *et al.*, 1992). Census output geographies used in England and Wales (coverage varies in Scotland and Northern Ireland) include OAs, Lower Layer Super Output Areas (LSOAs), Middle Layer Super Output Areas (MSOAs), alongside electoral wards and Local Authority Districts (LADs) or unitary authorities (UAs). Similarly to postal geographies, these geographies are hierarchical so that the smaller areas can be aggregated into larger ones. More recently, the release of specific geographies such as workplace zones (WZs), designed for the release of data related to workplace populations, recognises the importance of small-area geographies that are fit-for-purpose for the range of small-area analysis carried out by commercial users such as the retail sector. However, postal geographies and administrative boundaries do not match exactly. Figure 8.1 highlights the differences between the two for Leeds.

In addition, organising any data into discrete areal units presents a set of theoretical problems such as the MAUP and the ecological fallacy. First of all, the MAUP can create difficulties when analysing aggregate data for discrete geographical areas (Openshaw, 1984). The problem is dual faceted: the first relates to scale, the second to zoning. Essentially, the issue is that patterns identified in data at one scale of aggregation might not present themselves at a different level of aggregation. The ecological fallacy is a special case of the MAUP, whereby individual-level relationships inferred from relationships observed at the aggregate level might not be valid (see also the discussion in Chapter 2 on these issues).

Supply-side factors

In addition to demand data, a number of important choices regarding the representation of supply points have to be made in network planning. Whether retail facilities are identified as stand-alone outlets or they are grouped into centres is the first crucial decision. For many retail activities – automotive, petrol and supermarkets are all obvious examples – it looks natural to represent flows to individual outlets. In addition, the distinction between outlet-based systems and centre-based systems is becoming increasingly blurred. The incorporation of 'stand-alone' retail facilities into retail parks often means that such locations are, in fact, more akin to centres than stand-alone outlets (see Thomas *et al.*, 2004): the presence of both competitors and complementary retailers adding to the overall 'attractiveness' of the outlet.

Table 8.1 Drivers of retail attractiveness in different retail markets

	Supermarket	*Retail*	*Auto*	*Finance*	*Petrol*
Spatial aggregation	Fine zones	Medium zones	Coarse zones	Medium zones	Medium zones + network
Brand loyalty	Moderate/ strong	Moderate	Moderate	Strong (transactions)	Moderate
Attraction components					
Space	5	3	4	1	4
Parking	5	1	3	1	5
Accessibility	5	4	2	3	5
Product range	4	4	4	3	1
Prices	5	5	5	5	5
Adjacencies	1	5	1	5	2
Opening hours	4	3	1	3	5
	Complexity of distribution	Simple/ moderate	Simple/ moderate	Complex	Simple
	Trip type	Moderate	Long	Moderate	Long
	Segmentation	Multi-purpose	Single-purpose	Multi-purpose	Distress
		Fundamental	Important	Moderate	Marginal

Note: 1 = Relatively unimportant; 5 = Very important.

Source: Birkin *et al.* (2010)

Most retailers will, of course, have information on their own stores (size, turnover, location, workforce, etc.). However, in order to gain information on the location of competitors, they might have to purchase data from private sector sources (if available). If detailed information can be obtained on store attributes, this information can then be used to measure the attractiveness of a given shopping destination to be used in a GIS or spatial modelling framework. Size and turnover have traditionally been used to measure the attractiveness of shopping centres; however, we know that other elements affect the attractiveness of a shopping centre, such as brand, parking facilities, price and consumers' perception (Birkin *et al.*, 2010). Table 8.1, taken from Birkin *et al.* (2010), highlights the various attributes required and their importance to each retail sector.

GIS for retail analysis and planning

The retail industry makes extensive and sophisticated use of GIS and the analysis of spatial data in order to understand consumers, evaluate store performance and develop store networks. Planning authorities and local government also make use of GIS to evaluate access to retail services and to assess proposals for new stores or retail centres. We shall discuss public sector GIS-based retail planning issues later in the chapter.

Major retail corporations employ large teams of analysts who routinely use GIS to support both operational and strategic decision making. Insight derived from GIS analysis of customer and store trading data includes:

- understanding the characteristics and spatial distribution of consumers;
- planning localised marketing campaigns;

- identifying locations to launch new products or store formats;
- using consumer characteristics to estimate likely retail spending;
- identifying store catchment areas;
- estimating market shares across small-area neighbourhoods;
- forecasting new/proposed store revenue.

The first major advantage of GIS, as we have seen throughout the book so far, is visualisation. Data in spreadsheet or database management packages are impossible to analyse for spatial patterns. Once mapped the data are often said to be 'brought to life'. GIS visualisation enables retailers to map any aspect of demand or supply, but also to identify the links between demand and supply by considering consumer flows (derived from loyalty card or survey data) and providing an indication of how and where people

shop. Let us take each in turn. First, GIS can be useful for mapping all kinds of data relating to demand-side issues.

Figure 8.2 shows an example of mapping core census data for Montreal, Canada, as part of a wider retail GIS study of accessibility (see further discussion below). Here, Apparacio *et al.* (2007) plot income levels against a social deprivation index, to highlight a cluster of low-income areas in central Montreal. They later test the hypothesis that retail access is worse for residents of these areas compared to those in more affluent suburbs.

Second, we can use GIS to plot store location networks, to understand the spatial patterns of different types of stores and perhaps identify market gaps. Figure 8.3 shows the location of supermarkets in London, Ontario, Canada, in 1961 and 2005 (Larsen and Gilliland, 2008).

First, Figure 8.3 shows the expansion of stores into suburban neighbourhoods and the closure of many

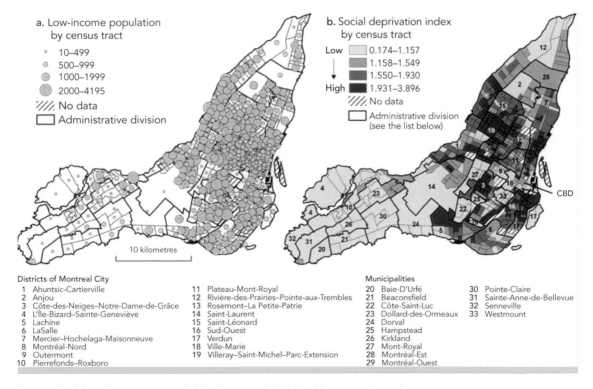

Figure 8.2 Mapping census variables in a retail GIS for Montreal, Canada

Source: Apparacio *et al.* (2007)

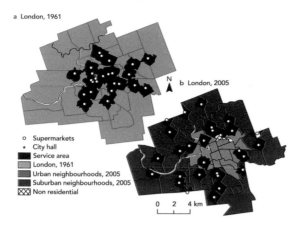

a London, 1961

N

b London, 2005

o Supermarkets
* City hall
■ Service area
▨ London, 1961
▨ Urban neighbourhoods, 2005
■ Suburban neighbourhoods, 2005
▨ Non residential

0   2   4 km

Figure 8.3 Mapping immediate retail catchments in London, Ontario, Canada, in 1961 and 2005

Source: Larsen and Gilliland (2008)

central stores since 1961. Second, buffers of one kilometre have been placed around each store to see which areas of the city might be poorly served and therefore possible commercial opportunities in the future. As we shall discuss later in the chapter, such maps are also useful for identifying potential 'food deserts' for public sector planners.

For retailers the most common use of the buffer procedure in GIS is to estimate the catchment area of existing or potential new stores. The first step might be to estimate the typical catchment area for their existing stores in the chain. Then a buffer of the average size could be placed around a potential new store location. Figure 8.4 shows the example presented by Russell and Heidkamp (2011) in their study of retail accessibility in New Haven, Connecticut, US (see also the section on GIS for public sector retail planning, below).

Figure 8.4 shows that buffers can be drawn either based on straight-line distance (Figure 8.4a) or on travel distance along a road network (Figure 8.4b). Suppose now we wish to overlay the population that lives in these buffers in order to work out some measure of potential demand or catchment size. The areas demarcated by the buffers in Figure 8.4 are likely to be made up of a number of small census zones or tracts or perhaps postal zones (or whatever spatial scale is being used). So a key question is how can the GIS be used to estimate the demand that lies within these buffers?

Census tracts that fall entirely within a buffer are easy to deal with – we can simply take the total population in each zone and sum to give an overall total. However, towards the edge of a buffer the census zones or postal areas might only be partly included. The task of estimating demand within these divided census tracts is called interpolation. Figure 8.5 helps to explain and understand interpolation better.

In Figure 8.5, the buffer cuts the census zone at its eastern end. Let us assume that the census zone contains 10,000 people and that the buffer contains around 10% of the area of the census zone. The standard interpolation procedure would suggest we need to add 1,000 people into the buffer for this part of the intersection calculation. The problem with this procedure is that the population might not be evenly distributed across the census tract being split in this way, thus making the estimation of the population contained with the buffer problematic. If, for example, the population is actually clustered in the western area of the zone, the population living within the area cut by the buffer might be very small or even zero. Alternatively, if the population is concentrated in the eastern area of the census zone then perhaps most of the 10,000 people should realistically be allocated to the buffer.

The overlay procedure in GIS is useful in retail catchment analysis. Following Benoit and Clarke (1997) we can use GIS overlay procedures to search for 'ideal zones'. Their example related to the search for a new site in Leeds, UK, for a discount retailer. GIS could be used as a type of sieve, removing layers of information not required. Thus, to find a new site for a discounter we could remove all postal sectors with a high social class background, all postal sectors containing young families (on the assumption that the more elderly might use discounters more) and all postal sectors containing an existing (competitor) discount retailer. By overlaying these characteristics one at a time we are left with zones that only contain older, low-income residents with no discount retailers within their locality. A population size threshold could also be added. Figure 8.6 shows the outcome of this type of analysis in Leeds where a number of ideal zones emerge, of which postal sector LS10 3 seems to be the most promising. The end product shows the location

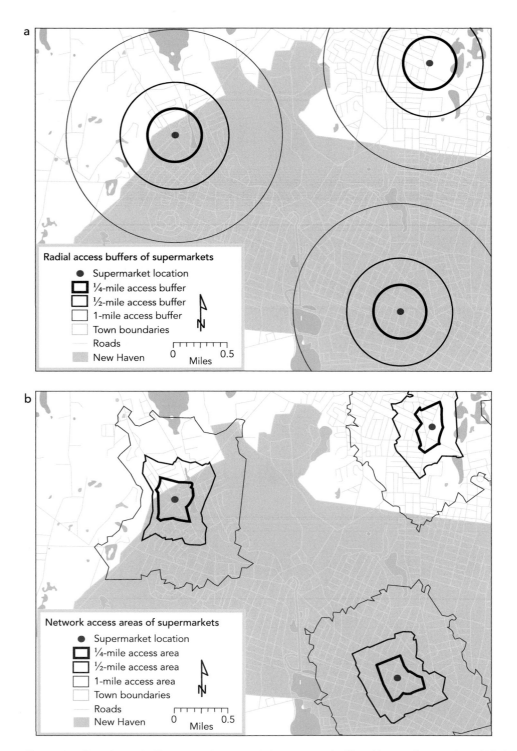

Figure 8.4 Drawing a buffer around a potential new store in New Haven, Connecticut, US: (a) straight line; (b) drive times

Source: Russell and Heidkamp (2011)

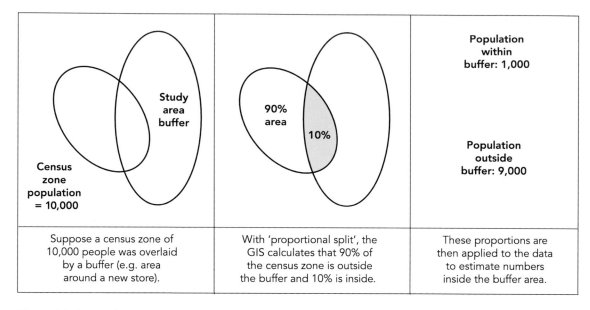

Figure 8.5 Interpolation procedure within GIS

Source: Birkin *et al.* (2017)

Figure 8.6 The end result of 'sieving' data to find optimal or ideal zones

Source: Benoit and Clarke (1997), redrawn in Birkin *et al.* (2017)

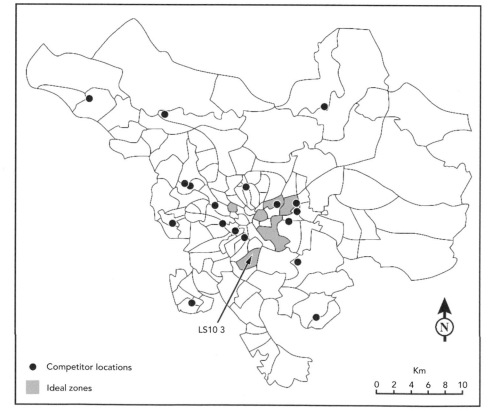

of large numbers of elderly and low-income households with little or no competitor locations.

Another way of overlaying or combining data is to turn variables into a standard indicator score based on a type of scoring system. The following example is taken from Spatial Insights' (undated) study of potential new locations for additional pawnbrokers or pawn shops in the city of Houston, Texas, in the US. Pawn shops are retail outlets where typically low-income residents can sell goods and possibly buy them back later (usually after payday). Figure 8.7 plots the current distribution of pawn shops in Houston along with a supply/demand index which, in turn, is portrayed as a GIS 'hot spot' style indicator of demand. This index is formed by combining (or overlaying) five key census variables. The first is total population. The distribution

of population across Houston is divided into five equal categories. Then, an individual census tract scores one point for a low population and (up to) 5 points if it falls within the top band. The second variable is household income. This time low = 5 points, high = 1 point, as low-income residents are more likely to use pawn shops. The third variable is the number of rented households (a proxy again for low social class: again low = 1 point, high = 5 points). The fourth variable is household size: large families again tend to come from the poorer backgrounds in US cities. So, again low = 1 point, high = 5 points. Finally, population density is also measured. A high population density is also associated with lower income groups, so low = 1 point, high = 5 points. Thus by overlaying the scores on all five variables a very low-income area can be identified

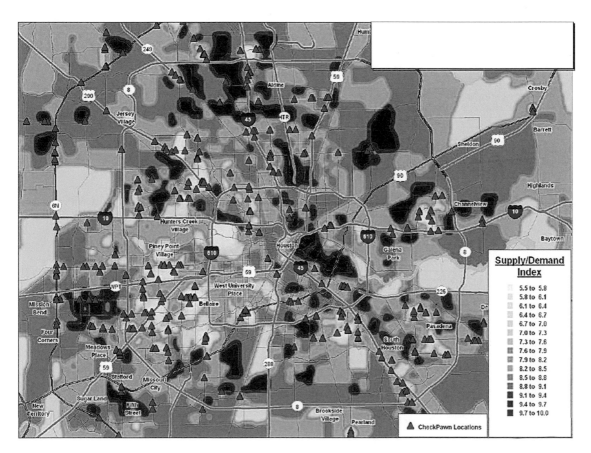

Figure 8.7 Deriving hot spots of demand for potential new pawnbrokers in Houston, Texas, US

Source: Spatial Insights (undated)

by a very high score (maximum 25). This score was then factored for simplicity to a mark of zero to ten. These areas are plotted on Figure 8.7 as high hot spot areas and the retailers can quickly evaluate the current location of pawn shops against these high indicator scores (giving a type of index of potential).

So the buffer and overlay procedure seems to have a number of potentially useful applications in retail analysis. However, there are problems when this technique is used to try to estimate potential store revenues. This is a vital component of the store location planning exercise – yes, we can look for ideal zones or areas of high potential, but how much revenue will a new store actually attract? Let us introduce an example relating to a retailer operating in the US. So far in the methodology we have shown how such retailers estimate a buffer size which can then be drawn onto a potential new site location. Figure 8.8 illustrates an

example where a US retailer has earmarked a new site in a typical US city. They have then estimated the catchment area as a buffer of one mile around this potential new store location. The key next question is how much of the demand within the buffer, which can be estimated by overlay and interpolation in total, will actually end up at the new store? If there are no competitors within the one-mile boundary then maybe the new store will be able to capture much of the demand that presently has to leave the catchment area. However, it is rare that one-mile buffers in large city regions contain no existing competitors (for most retail goods and services). If we suppose that in fact there are three existing competitor stores in the one-mile buffer then the picture is more complicated, but probably more realistic. According to a number of studies (e.g. Beaumont, 1991) the most likely allocation procedure is the fair share method. That is, the

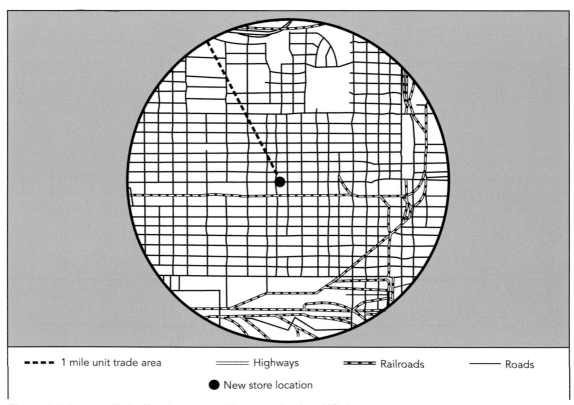

Figure 8.8 A one-mile buffer demarcated for a retailer in a US city

Source: Birkin *et al.* (2017)

four retailers will each get approximately 25% of the business. This can be factored by brand or size so that the split is not quite so even. However, it remains a fairly crude solution methodology. One alternative is to assume the consumer will travel to the nearest store within the catchment area (*dominant store analysis*).

To compound the problems it is possible to think of other flaws in this technique for estimating individual store revenues. The first is the rather arbitrary nature of the one-mile buffer. Although this might work generally across the store chain, trade might be more likely to be skewed in certain directions because of the location of the competitors, not only within the buffer but also outside it. For example if there were no competitors for two miles to the east of the new store location then in reality the store could get considerable trade from households outside the buffer boundary to the east (buffer inflow). Similarly, if there is a large com-petitor outlet just outside the buffer to the north, then perhaps the new store will get much less trade from the northern parts of the buffer as trade crosses the artificial boundary (buffer outflow). A second problem relates to the fact that there is no account taken of distance decay within the buffer. For example, the new store is actually far more likely to receive trade from nearby streets (which might or might not be very populous) and little from streets near the edge of the buffer. Hence the assumption of fair share allocation from within the whole buffer is likely to be problematic. This problem can be reduced by identifying a primary, secondary and tertiary catchment area within the one-mile buffer, but there will still be problems of assigning proportions of demand from each of these to the new store. Finally, in very densely populated areas with lots of competitors, the individual catchment areas will become so blurred as to make any attempts of allocating the demand almost impossible (the so-called overlapping catchment area problem).

A potential solution to this problem is to try to identify a unique buffer for every store. This will mean some having very small buffer sizes and others very large (depending on the amount of competition nearby). An effective way to do this in GIS is to use Thiessen polygons, also known as Voronoi polygons. Thiessen polygons can be constructed so that each polygon contains exactly one of the stores so that any location within the polygon is closer to that store than to a store within any other polygon. Figure 8.9 shows the plotting of Thiessen polygons for two rival grocery stores in a region of the US (Pearson, 2007). Using Thiessen polygons as trade areas, analysts can evaluate the retail outlets in terms of socio-economic and/or demographic attributes of each outlet's trade area. However, the issue of trade flowing across

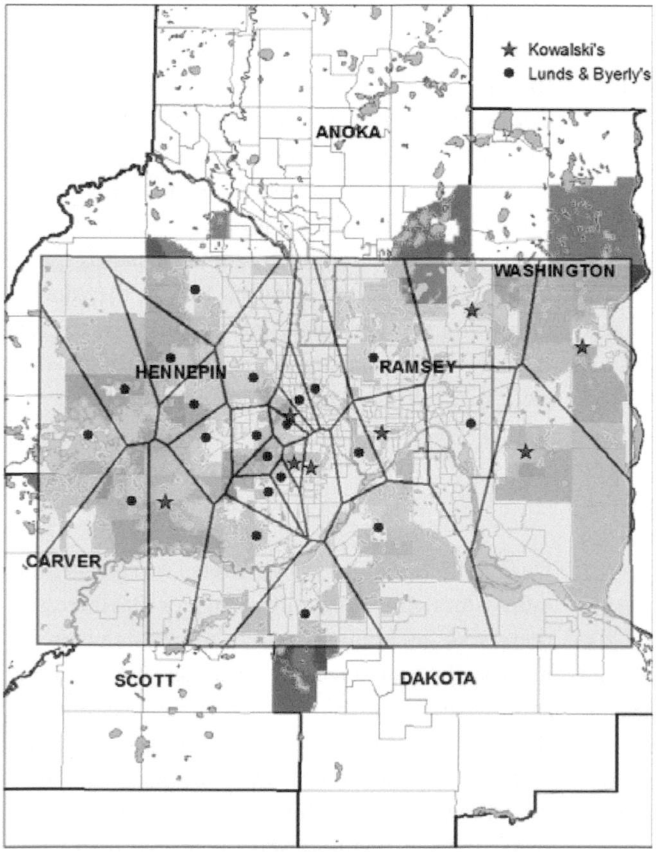

Figure 8.9 Use of Thiessen polygons for trade area demarcation

Source: Pearson (2007)

Thiessen polygon boundaries is huge and again grave doubts might be given for using this technique to accurately forecast retail sales.

Adding a modelling capability to the GIS

As noted above, GIS is excellent for mapping data and exploring spatial patterns using buffer and overlay, sieve mapping, etc. However, it is less accurate and therefore less useful for new store revenue estimation. Thus in this section we look briefly at modelling techniques more commonly used for this purpose. Some modelling techniques are available directly in some proprietary GIS packages, such as regression modelling (as noted in Chapter 6). Even the spatial interaction models discussed below are available now in ArcGIS, although certainly in the case of these models it is hard to disaggregate the models within a GIS package and even harder to calibrate using statistical routines. Thus, more often than not, these models need to be run using specialist software and loose coupled back to the proprietary GIS for mapping and any further spatial analysis.

Statistical models

Regression modelling represents one of the more scientific methods of network planning and builds on the philosophy of the analogue procedure discussed in the first section of this chapter. Regression analysis works by defining a dependent variable such as store turnover and attempting to correlate this with a set of independent or explanatory variables (e.g. size, catchment population and brand). If there is a degree of correlation between these dependent and independent variables then the regression model can be used in a predictive context by using the independent variables as predictors of the dependent variable. Coefficients are also calculated to weight the importance of each independent variable in explaining the variation in the set of dependent variables. The model can be written as:

$$Y_i = a + b_1 X_{1i} + b_2 X_{2i} + b_3 X_{3i} + ... + b_m X_{mi} \quad (8.1)$$

where Y_i is turnover (the dependent variable) of store i, X_{mi} are independent variables, b_m are regression coefficients estimated by calibrating against existing stores and a is the intercept term.

Although multiple regression allows for greater sophistication and objectivity than the more manual analogue techniques, there are still a number of problems. The primary weakness is that they evaluate sites in isolation, without considering the full impacts of the competition or the company's own global network. As with the analogue method, a second major weakness is the problem of 'heterogeneity of sample stores'. That is, how easy is it to find a sample of stores which have similar trading characteristics and catchment areas (see Ghosh and McLafferty, 1987)? Additionally, regression models assume that the explanatory variables in the models be independent of each other and uncorrelated. In many retail applications this is not the case. More specifically, independent variables such as floor-space and car parking spaces might be strongly correlated. This can lead to unreliable parameter estimates and severe problems of interpretation (multicollinearity). However, through careful analysis and interpretation many of these problems can be overcome. Perhaps the most important limitation is that regression models fail to handle adequately spatial interactions or customer flows. That is, they do not model the processes (spatial interactions) that generate the flows of expenditure between residential or workplace areas (demand zones) and retail outlets.

Mathematical modelling

During the 1980s and 1990s, as retail markets became more saturated and the competition for sites increased, there was an important shift towards the use of more sophisticated techniques in network planning. Improvements in information technology and the development of GIS software led to major improvements in spatial modelling techniques. One such modelling technique which became a core feature of retail network planning was the spatial interaction model (SIM). SIMs were originally developed from the gravity-based principles of Newton's scientific theory of Universal Gravitation (hence they are most commonly called gravity models in the retail industry) and over time these principles have been refined to produce much more complex models (Birkin *et al.*, 2010; Newing *et al.*, 2015). By definition, SIMs are used to simulate or predict the interactions or flows (e.g. people, households, expenditure) between origins and destinations. Wilson (1974) defines four different variations of the SIM in his 'family of SIMs'. With regard to the retail sector, the most commonly used SIM is the production constrained model, shown below:

$$S_{ij} = A_i O_i W_j \, exp^{(-\beta d_{ij})} \quad (8.2)$$

where S_{ij} is the flow of people (or expenditure) from residential area i to retail store j, O_i is a measure of demand (expenditure) in area i, W_j is a measure of attractiveness of retail store j, d_{ij} is a measure of the distance (e.g. time, miles) between i and j, β is the distance decay parameter and A_i is a balancing factor to ensure that all demand is allocated to grocery stores within the modelled region, written as:

$$A_i = \frac{1}{\Sigma_j W_j \, exp^{(-\beta d_{ij})}} \quad (8.3)$$

The model allocates flows of expenditure between origin and destination zones on the basis of two main hypotheses: (i) Flows between an origin and destination will be proportional to the relative attractiveness of that destination vis-à-vis all other competing destinations. (ii) Flows between an origin and destination will be proportional to the relative accessibility of that destination vis-à-vis all other competing destinations. Furthermore, the model works on the assumption that in general, when choosing between centres that are equally accessible, shoppers will show a preference for the more attractive centre. When centres are equally attractive, shoppers will show a preference for the more accessible centre. Note, however, that these preferences are not deterministic. Thus when choosing between equally accessible centres, shoppers will not always choose the most attractive. The models are therefore able to represent well the stochastic nature

of consumer behaviour. By summing the flows coming into each store or shopping centre the models can estimate the revenue of those outlets. This is a crucial output of the modelling procedure.

In the retail industry, SIMs can be calibrated to reproduce existing interaction patterns between populations and shopping centres from observed data (e.g. Tesco Clubcard data) to facilitate the estimation of store turnovers. Having allocated expenditures between all retailers, the models can be used to predict local market shares and compare actual turnovers to model predictions – that is, what would the model expect a certain outlet to be achieving in sales terms? For example, is a store that turns over £1 million per annum doing well or badly in relative rather than absolute terms? Having identified the variations in market share (or penetration) the retailer might then be keen to improve its performance by opening new outlets in the areas that currently have a low market share. The models can then be used to test the impact of new store openings through a variety of 'what-if?' scenarios. This powerful capability has meant many of the UK retailers (Tesco, Asda, Sainsbury's, etc.) and UK consultancy firms (Callcredit, CACI, Experian and Javelin) have invested heavily in producing SIMs for market analysis. Figure 8.10 demonstrates how the individual market share figures for an example 70,000-square foot Asda store in Leeds can be calculated by a SIM similar to that in Equation 8.2.

Over time, a number of derivations of the SIM have been developed to ensure the technique is applicable in different sectors of the retail market. For example, Fotheringham (1983) presented the 'competing destinations' model, arguing that the standard SIM needed to be adapted to include a competing destinations term that recognised that outlets or centres in very close proximity to each other were really a single destination in the eyes of consumers. Ottensmann (1997) and Birkin *et al.* (2010) also introduced the concept of elastic demand into SIMs, a concept that replicates the increase in demand for a specific service in an area, if that service is introduced or increased. The theory is that increased access to a service will increase demand for that service. For example, the opening of a new cinema, bingo hall, fast food out-

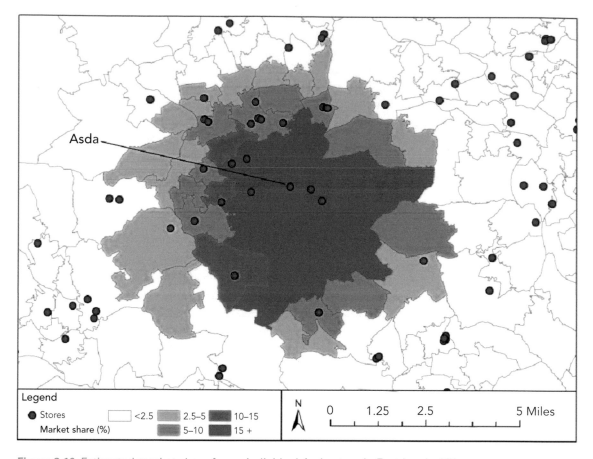

Figure 8.10 Estimated market share for an individual Asda store in East Leeds, UK

let, etc. will stimulate local demand from people who might not have bothered to travel longer distances to enjoy such services beforehand. In addition to these theoretical developments, it has become commonplace for modern SIMs to be highly disaggregated, to account for more complex behaviour in a given system. Wilson (1974) initially made attempts to more accurately represent different behaviour in commuting to work through the disaggregation of the SIM to represent different modes of travel in a transport model.

Nevertheless, despite the benefits associated with SIMs, care must still be taken when using them. For instance, SIMs are data intensive which means they require good quality data (demand and supply). Poor quality data cause issues with calibration and ultimately accuracy. In the retail sector, it is likely that firms would be very reluctant to divulge their own customer flow data. Conversely, even if one were to gain access to, say, Tesco's Clubcard data, calibrating a model exclusively to one retailer's customers rather than the whole market could create a bias that might manifest itself in a false value of a parameter for other competitor types (Birkin *et al.*, 2010). Equally troublesome are missing real-world flows in the data and the way in which these are dealt with, as calibrating the model to a set of averages does not necessarily result in the correct patterns. In addition, boundary issues come through modelling a system that is enclosed within a boundary but, in real life, that boundary does not form an effective divide. In a given retail system, the outcome is that any stores near the boundary edge will have unrealistic and normally very high levels of market share. To deal with this problem, Birkin *et al.* (2010) suggest a boundary-free approach to spatial interaction modelling (i.e. entire regions or nations modelled concurrently), which is certainly more possible today with high performance PCs and super-computers.

GIS for public sector retail planning

Guy (2006) provides an excellent account of retail planning developments since the 1950s and 1960s, especially from a UK perspective. In the first chapter of his book, Guy discusses the purpose of retail planning, in terms of different economic, social and environmental issues. Interestingly, he reminds the reader that retail planners have an economic duty to promote retailing – so they should 'allow the retail sector to grow and change whilst maintaining a significant level of profitability' (Guy, 2006: 3). That said, there are a number of social issues which in turn raise spatial issues and which are therefore more significant perhaps for a GIS approach. In particular, Guy (2006) notes the importance of planners maintaining a good balance or mix of shop types, protecting the vitality of small shopkeepers and of the town centre (especially in Europe). Indeed, most European countries have seen a tightening of planning guidelines towards out-of-town developments, effectively making it more difficult to open such centres without proving a real need in the local communities affected, and providing equity in shopping provision such that no particular socio-economic group should be disadvantaged.

In this section we shall explore the use of GIS to help this final goal in particular, although we will look at impact assessment as well as this also impinges on these social issues of protecting smaller retail centres and independents. GIS can be seen to be a key instrument in examining variations in retail provision and also accessibility by local residents. All such studies can also be couched in terms of a growing literature on 'food deserts' – the latter term referring to parts of our cities or regions where access to larger supermarkets in particular is poor.

The simplest measure of access is the distance to the nearest store, usually measured from the centre of an existing demand zone (census block or post code). This can be made more sophisticated by taking a population-weighted mean minimum distance along a street or road network. Smoyer-Tomic *et al.* (2006) provide a neat illustration of this for supermarket accessibility in Edmonton, Canada (see Figure 8.11).

There are many other methods for estimating access. Figure 8.12 shows a very good illustration of how simple nearest distances (in this case 1,000 metres) can be made more sophisticated by the overlay of different types of stores and some census data (in this case related to household income). Apparacio *et al.* (2007) show clearly that access to branded supermarket chains was even worse in the low-income central areas of Montreal, Canada, than access to supermarkets in general.

A number of other accessibility studies have examined the difficulties faced by certain types of consumers in accessing the larger, high-quality supermarket outlets. Building on Figure 8.11, Smoyer-Tomic *et al.* (2006) were able to look at distances needed to be travelled (estimated in a variety of ways) against population size, low-income households, households with no car and the population over 65. The resulting analysis revealed nine areas of concern within Edmonton (see Figure 8.13), which they could further disaggregate into areas of concern for 'elderly residents', 'low-income residents', 'seniors and low-income residents', 'low-income residents and no car owners'.

While sieve mapping in GIS can be used to find ideal sites for making profits (see above) it can also be used to find ideal sites for new stores based on low existing access for certain groups of consumers. In addition to the above examples, Eckert and Shetty (2011) illustrate the combination of a large number of key variables to plot census blocks 'of concern' in Toledo, Ohio, US, in terms of poor accessibility. They overlay data to find census block groups with the following characteristics (a sieve map in effect as all zones that do not have these qualities can be eliminated):

■ Average accessibility to nearest food retailer is over one mile.

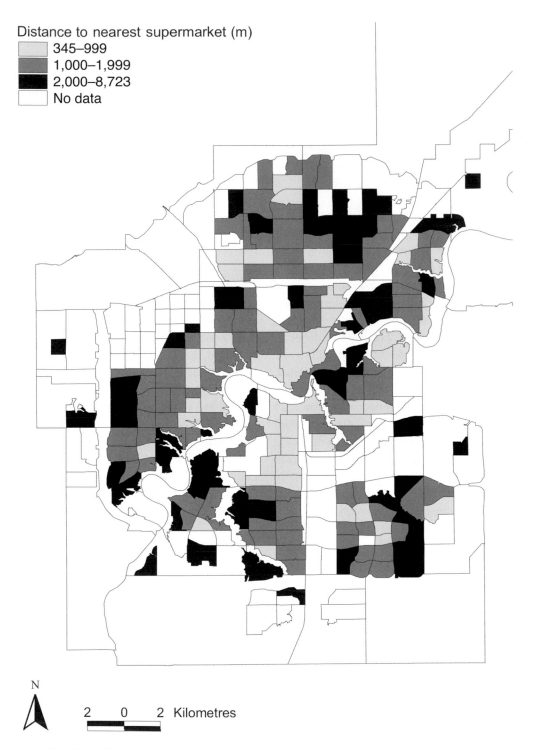

Distance to nearest supermarket (m)
- 345–999
- 1,000–1,999
- 2,000–8,723
- No data

N

2   0   2 Kilometres

Figure 8.11 Neighbourhood-level population-weighted mean minimum street network travel distance to the nearest supermarket in Edmonton, Canada

Source: Smoyer-Tomic *et al.* (2006)

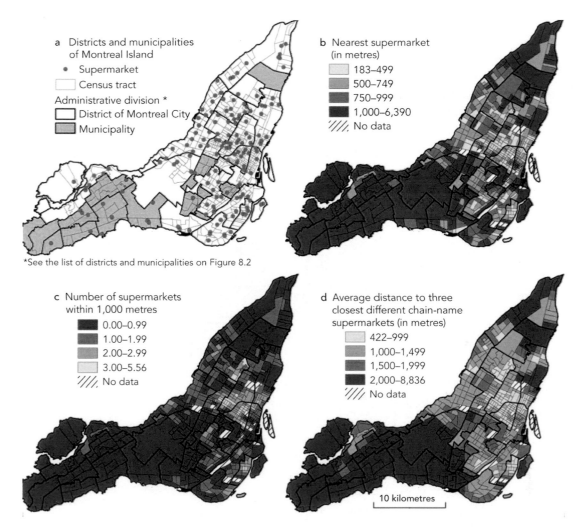

Figure 8.12 Spatial distribution of supermarket accessibility in Montreal, Canada, 2001

Source: Apparacio *et al.* (2007)

- The percentage of the population below poverty level is above the city average.
- The percentage of households without a car is above the city average.
- The percentage of population receiving public assistance is above the city average.
- Median household income of the block group is below the city average.

They found 28 blocks met those criteria (Eckert and Shetty, 2011). Figure 8.14 shows those 28 zones 'of concern'.

Many of the studies in the literature on food deserts use some kind of distance measure based on straight-line or distance on a road/street network. Others have favoured an accessibility score based on an index, such as that derived by Hansen (1959). Clarke *et al.* (2002) developed this indicator approach by estimating accessibility based on the outcomes of a SIM as defined in the preceding section of this chapter. This is arguably more powerful than the Hansen style indicator as it takes into account the actual interactions that people are most likely to make rather than simply the set of opportunities in the area (Bertuglia *et al.*, 1994). Figure

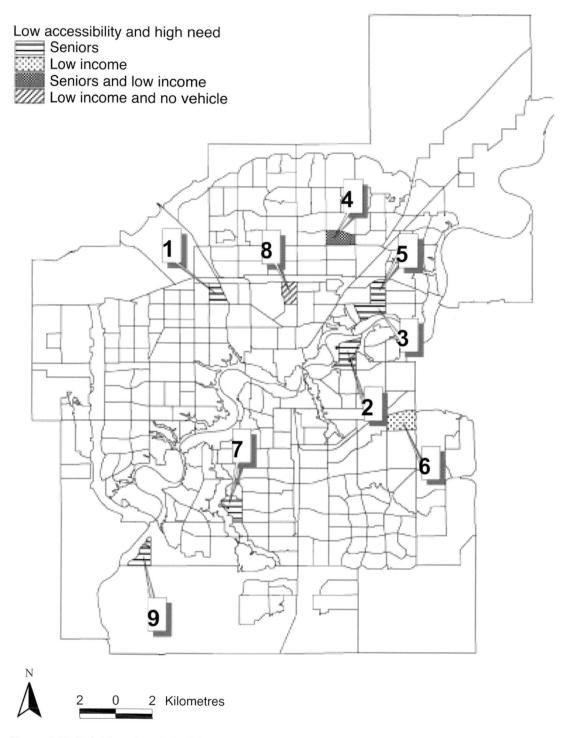

Figure 8.13 Neighbourhoods in Edmonton, Canada, with high population need and low supermarket accessibility, disaggregated by consumer type

Source: Smoyer-Tomic et al. (2006)

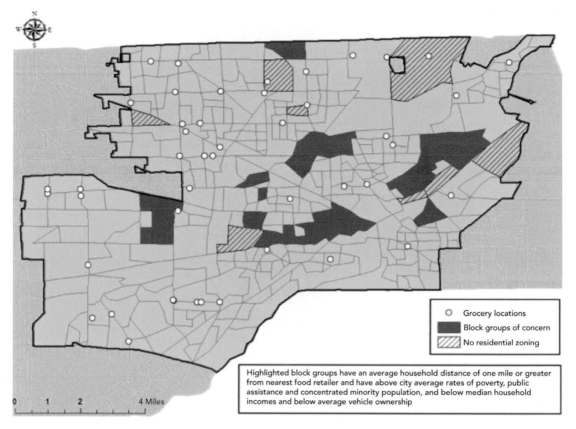

Highlighted block groups have an average household distance of one mile or greater from nearest food retailer and have above city average rates of poverty, public assistance and concentrated minority population, and below median household incomes and below average vehicle ownership

- ○ Grocery locations
- ■ Block groups of concern
- ▨ No residential zoning

0   1   2   4 Miles

Figure 8.14 Areas of concern regarding poor access to grocery retailing in Toledo, Ohio, US

Source: Eckert and Shetty (2011)

8.15 shows the resulting access scores for Cardiff in Wales in 2002. The low-scoring areas typically coincide with low-income housing estates such as Rhiwbina, Fairwater and Llanrumney.

The identification of poor access or food deserts is of obvious interest to retail planners (see also Raja *et al.* (2008) for an interesting discussion of food deserts and potential ethnic disparities in access, and McEntee and Agyeman (2010) for the use of GIS to identify potential food deserts in rural areas). The question then arises – what can be done about the problem? Many of the papers cited above discuss policy alternatives. Retail planners could investigate the possibility of 'spatial infill' by encouraging small supermarkets (or convenience stores) to help plug the gaps in access. For retailers concerned about low revenue potential in low-income

food deserts, planners could consider financial incentives such as tax breaks or help with costs of purchase and refurbishment. In the UK we have even seen the encouragement of large superstores in low-income food deserts – especially the Tesco Extra stores opened in collaboration with local authorities under a 'social regeneration' banner in the UK (Wrigley *et al.*, 2003).

Finally in this chapter we briefly examine how GIS could be useful for planners to estimate the impacts of large-scale retail developments such as out-of-town retail parks, shopping centres or very large superstores. As we argued above in relation to the private sector, GIS is typically not enough on its own to handle estimating the impacts of such major developments, especially in relation to revenue estimation. Ironically in the 1960s and 1970s many local

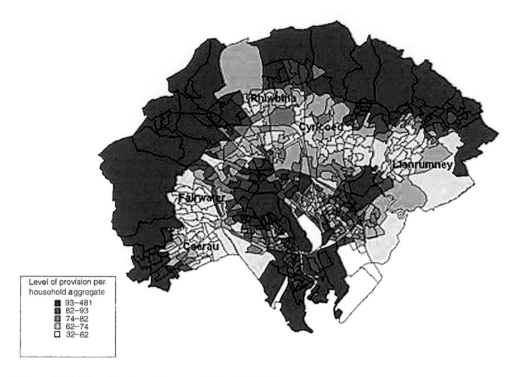

Figure 8.15 Mapping food deserts in Cardiff, UK

Source: Clarke *et al.* (2002)

authorities in the UK used SIMs to help in this type of analysis. However, after a number of conflicting model results in a number of planning enquiries these models were effectively banned. Khawaldah *et al.* (2012) and Birkin *et al.* (2015) review the alternative methods adopted since the 1980s and compare the results against using SIMs. In the following discussion, we briefly discuss the likely stages in a UK planning enquiry (the analysis for which is normally undertaken by retail consultants acting on behalf of commercial developers). Here we follow the steps outlined by Drivers Jonas (1992; also see England, 2000).

1   Identify the catchment area of the proposed retail development, the area from which the centre draws the majority of its revenue. The information about that ideally should come from a household survey to show the existing shopping patterns in the area, although often this information is not available. The catchment area might be subdivided into drive time isochrones, and these isochrones might be further subdivided into primary, secondary and tertiary zones for greater spatial accuracy in assessing impact.

2   Estimate the expenditure within the catchment area, derived from the population and its per capita expenditure.

3   Estimate the turnover of existing shopping centres. This information can be estimated by using national average turnover/floor-space ratios for individual firms (or the centre as a whole), or by using survey data and proportioning centre turnover on the basis of the customer patronage seen in the survey (i.e. as a percentage of the total spend).

4   Estimate the turnover of the new shopping proposal. This is normally estimated on the basis of company average turnover/floor-space ratios for retailers known to be entering the centre (or trading performance of similar centres where such data are not available).

5   Estimate the amount of spending in each existing centre which will be diverted to make up the new centre's turnover, and the locational source of that spending.

GIS might be typically used at all stages of this process. The main difficulties with this procedure are, first, that many smaller centres get omitted from the analysis. When consumers record their preferences for shopping destinations in surveys, they commonly overstate the importance of the larger centres in the area and underplay (or omit) the importance of smaller centres. Thus, when the analysis comes to impact assessment the smaller centres (probably most affected by new developments) can be completely ignored. Second, the estimation of potential revenue using national turnover/floor-space ratios is fraught with danger – smaller developments will clearly be estimated to take too much revenue based on national figures while very large new centres will be under-predicted. The conclusion of Khawaldah *et al.* (2012) and Birkin *et al.* (2015) is that SIMs would do a much better job more often than not.

Concluding comments

In this chapter we have explored the use of GIS in retail location planning from the perspective of both the private and public sectors. It has been argued that GIS is a powerful tool for visualisation and catchment area analysis. Key data sets can be mapped and analysed using key analytical routines such as buffer and overlay. The illustrations throughout the chapter show the widespread use of GIS in many retail organisations and consultancies. Despite its usefulness, however, we have also argued that for certain operations GIS might not be currently powerful enough – especially, for example, for the estimation of store revenues. For this important task, we argue that spatial models (statistical and/or mathematical) are more powerful and have a better track record of success (see Birkin *et al.*, 2016). If these models can be incorporated directly into a GIS package then there are advantages in terms of data linkage and visualisation. However, mathematical models in

particular are harder to incorporate into black box systems where the models cannot be effectively disaggregated and calibrated to fit the system of interest. Thus loose coupling is probably the best solution for linking such models to the other spatial analysis routines in a GIS.

Accompanying practicals

This chapter is accompanied by two practical activities.

Practical 4a: Retail site location analysis, provides hands-on experience of using GIS to evaluate outputs from a retail modelling exercise. Using a fictional (but realistic) retail network in the UK, we give you experience handling data related to both the demand and supply sides. We give you the opportunity to assess retail provision, retail demand and store performance. You also examine outputs from a SIM in order to assess impacts of a new store opening.

Practical 4b: Public sector retail planning, gives you the opportunity to consider concepts related to access to retail stores. Using an example from Washington, DC, you identify potential food deserts by considering access to food stores.

References

All website URLs accessed 30 May 2017.

Apparacio, P., Cloutier, M. S., & Shearmur, R. (2007) The case of Montreal's missing food deserts: evaluation of accessibility to food supermarkets. *International Journal of Health Geographics*, 6(4). DOI: https://doi.org/10.1186/1476-072X-6-4.

Beaumont, J. R. (1991) GIS and market analysis, in D. J. Maguire, M. Goodchild, & D. W. Rhind (eds) *Geographical Information Systems: Principles and Applications*, Longman, London, vol. 2, 139–151.

Benoit, D., & Clarke, G. P. (1997) Assessing GIS for retail location analysis. *Journal of Retailing and Consumer Services*, 4(4), 239–258.

Berry, T., Newing, A., Davies, D., & Branch, K. (2016) Using workplace population statistics to understand retail store performance. *The International Review of Retail, Distribution and Consumer Research*, 26(4), 375–395.

Bertuglia, C. S., Clarke, G. P., & Wilson, A. G. (1994) *Modelling* the City: Planning, Performance and Policy, Routledge, London.

Birkin, M., Clarke, G. P., & Clarke, M. (2002) *Retail Geography and Intelligent Network Planning*, Wiley, Chichester.

Birkin, M., Clarke, G. P., & Clarke, M. (2010) Refining and operationalizing entropy-maximizing models for business applications. *Geographical Analysis*, 42(4), 422–445.

Birkin, M., Khawaldah, H., Clarke, M., & Clarke, G. P. (2015) Applied spatial interaction modelling in economic geography: an example of the use of models for public sector planning, in C. Karlsson, M. Andersson, & T. Norman (eds) *Handbook of Research Methods and Applications in Economic Geography*, Edward Elgar, Cheltenham, 491–512.

Birkin, M., Clarke, G. P., & Clarke, M. (2017) *Retail Location Planning in an Era of Multi-Channel Growth*, Routledge, London.

Bowlby, S., Breheny, M., & Foot, D. (1984a) Store location: problems and methods 1: Is locating a viable store becoming more difficult? *Retail and Distribution Management*, 12(5), 31–33.

Bowlby, S., Breheny, M., & Foot, D. (1984b) Store location: problems and methods 2: Expanding into new geographical areas. *Retail and Distribution Management*, 12(6), 41–46.

Clarke, G. P. (1998) Changing methods of location planning for retail companies. *GeoJournal*, 45, 289–298.

Clarke, G. P., Eyre, H., & Guy, C. (2002) Deriving indicators of access to food retail provision in British cities: studies of Cardiff, Leeds and Bradford. *Urban Studies*, 39(11), 2041–2060.

Clarke, G. P., Thompson, C., & Birkin, M. (2015) Exploring the geography of e-commerce in UK retailing. *Regional Studies, Regional Science*, 2(1), 370–390.

Davies, R. L. (1977) Store location and store assessment research: the integration of some new and traditional techniques. *Transactions of the Institute of British Geographers*, 141–157.

Davies, R. L., & Rogers, D. S. (1984) *Store Location and Store Assessment Research*, Wiley, Chichester.

Drivers Jonas (1992) *Retail Impact Assessment Methodologies: Research Study for the Scottish Office*, Scottish Office, Edinburgh.

Eckert, J., & Shetty, S. (2011) Food systems, planning and quantifying access: using GIS to plan for food retail. *Applied Geography*, 31(4), 1216–1223.

England, J. (2000) *Retail Impact Assessment: A Guide to Best Practice*. Routledge, London.

Fenwick, I. (1978) *Techniques in Store Location Research: A Review and Applications*, Retailing and Planning Associates, Corbridge.

Fotheringham, A. S. (1983) A new set of spatial-interaction models: the theory of competing destinations. *Environment and Planning A*, 15(1), 15–36.

Ghosh, A., & MacLafferty, S. (1987) *Locational Strategies for Retail and Service Firms*, Lexington Books, Lexington, MA.

Guy, C. (2006) *Planning for Retail Development: A Critical View of the British Experience*, Routledge, London.

Hansen, W. G. (1959) How accessibility shapes land use. *Journal of the American Institute of Planners*, 25(2), 73–76.

Khawaldah, H., Birkin, M., & Clarke, G. (2012) A review of two alternative retail impact assessment techniques: the case of Silverburn in Scotland. *Town Planning Review*, 83(2), 233–260.

Laing, R. D., Clarke, I., Mackaness, W., Ball, B., & Horita, M. (2003) The devil is in the detail: visualising analogical thought in retail location decisionmaking. *Environment and Planning B: Planning and Design*, 30, 15–36.

Larsen, K., & Gilliland, J. (2008) Mapping the evolution of food deserts in a Canadian city: supermarket accessibility in London, Ontario, 1961–2005. *International Journal of Health Geographics*, 7(16). DOI: https://doi.org/10.1186/1476-072X-7-16.

McEntee, J., & Agyeman, J. (2010) Towards the development of a GIS method for identifying rural food deserts: geographic access in Vermont, USA. *Applied Geography*, 30(1), 165–176.

Newing, A., Clarke, G. P., & Clarke, M. (2015) Developing and applying a disaggregated retail location model with extended retail demand estimations. *Geographical Analysis*, 47, 219–239.

Openshaw, S. (1984) *The Modifiable Areal Unit Problem*. Geo Books, Norwich.

Ottensmann, J. R. (1997) Partially constrained gravity models for predicting spatial interactions with elastic demand. *Environment and Planning A*, 29(6), 975–988.

Pearson, J. (2007) A comparative business site-location feasibility analysis using geographic information systems and the gravity model. Volume 9, *Papers in Resource Analysis*. Saint Mary's University of Minnesota Central Services Press, Winona, MN. Available from: www.gis.smumn.edu.

Raja, S., Ma, C., & Yadav, P. (2008) Beyond food deserts: measuring and mapping racial disparities in neighborhood food environments. *Journal of Planning Education and Research*, 27(4), 469–482.

Raper, J., Rhind, D., & Shepherd, J. (1992) *Postcodes: The New Geography*, Longman, Harlow.

Reynolds, J., & Wood, S. (2010) Location decision making in retail firms: evolution and challenge. *International Journal of Retail & Distribution Management*, 38(11/12), 828–845.

Russell, S. E., & Heidkamp, C. P. (2011) 'Food desertification': the loss of a major supermarket in New Haven, Connecticut. *Applied Geography*, 31(4), 1197–1209.

Smoyer-Tomic, K. E., Spence, J. C., & Amrhein, C. (2006) Food deserts in the prairies? Supermarket accessibility and neighborhood need in Edmonton, Canada. *The Professional Geographer, 58*(3), 307–326.

Spatial Insights (undated) *Market Potential GIS Case Study: Market Potential Analysis using ReCAP Retail Location Data: Check Cashing and Pawn Brokers.* Available from: www. directionsmag.com/entry/a-market-potential-gis-case-study-market-potential-analysis-for-check-cashi/122646.

Thomas, C. J., Bromley, R. D. F., & Tallon, R. (2004) Retail parks revisited: a growing competitive threat to traditional shopping centres? *Environment and Planning A, 36*, 647–666.

Wilson, A. G. (1974) *Urban and Regional Models in Geography and Planning*, Pion, London.

Wood, S., & Reynolds, J. (2011) The intra-firm context of retail expansion planning. *Environment and Planning A, 43*(10), 2468–2491.

Wood, S., & Reynolds, J. (2012) Managing communities and managing knowledge: strategic decision making and store network investment within retail multinationals. *Journal of Economic Geography, 12*(2), 539–565.

Wrigley, N. (1987) The concentration of capital in UK grocery retailing. *Environment & Planning A, 19*(10), 1283–1288.

Wrigley, N. (2014) *Store Choice, Store Location and Market Analysis (Routledge Revivals)*, Routledge, London.

Wrigley, N., Warm, D., & Margetts, B. (2003) Deprivation, diet, and food-retail access: findings from the Leeds 'food deserts' study. *Environment and Planning A, 35*, 151–188.

9 GIS and health care planning and analysis

LEARNING OBJECTIVES

- Understanding the geographical aspects of health and health care provision
- Using GIS to map and analyse spatial health patterns and their socio-economic and geographic determinants
- Using GIS to identify health-related hot spots
- Geographical access to health care provision and location planning/spatial decision support
- Using GIS for the management and targeting of health resources

Introduction

Health itself is difficult to define, but is generally considered as a state of well-being, with the World Health Organization (WHO) claiming that health is 'a state of complete physical, mental and social well-being and not merely the absence of disease and infirmity' (WHO, 1946). Although we often consider health (or more specifically ill-health) to be an individual factor, health and health care are important concepts from a geographical perspective. It is clear that both between countries and within countries there can be substantial health inequalities, whether those are differences in mortality rates or in access to primary care (doctors or GPs) or secondary care (hospitals). However, within the broader social sciences, there is an interesting debate around the relationship between health

and place, and the causes of such health inequalities. Health is experienced at the individual level, but health issues are inherently geographical. Your own health is determined by pre-disposing genetic factors, but also by personal geographies including where you were brought up and where you live, work and socialise now, all of which affect your risk of ill-health, your access to health care and the quality of treatment you receive.

This fascinating debate is explored in more detail elsewhere (Kearns and Joseph, 1993; Curtis and Rees Jones, 1998; Curtis, 2010; Gatrell and Elliott, 2014). Although some researchers argue that processes operating at the level of the individual are far more important than the significance of different places, others argue that these processes that influence an individual's health experience can operate differently in different places (Curtis and Rees Jones, 1998).

Certainly, many health geographers would argue that health inequalities are a combination of the characteristics of individuals (age, gender, social class, ethnicity, etc.: often labelled compositional effects) and the place in which they live or were brought up (often labelled contextual effects). GIS allows policy makers to explore these health variations in more detail and perhaps try to ascertain, for each case study, whether compositional or contextual effects are more important. For certain, GIS allows more local interventions to be considered as policy options perhaps alongside national policy interventions that might impact on entire populations.

This chapter thus focuses on the use of GIS to evaluate health and health care from a geographical perspective. GIS has a major role to play in evaluating small-area health patterns, identifying where populations and health care need are located and in the targeting of resources. As we have seen with income, deprivation and geodemographics, GIS is most widely used as a visualisation tool to explore inequalities in small-area health patterns, but also as a more powerful tool for location planning/spatial decision support. Such uses include identification of optimal sites for health care facilities and evaluation of access to and provision of care. GIS is a fundamental tool for this purpose.

The chapter begins with a brief review of why geography is important in health care planning and analysis. Then we shall consider the use of GIS and spatial analysis to help manage health care provision, access and utilisation.

Geographical components of health and health care policy

Health care is increasingly focusing on small-area variations in health in order to provide patient-centred health care services. There are ongoing academic debates about the origins of these small-area variations, but commonly they are considered to be driven by differences in individual composition and other contextual factors. Shaw *et al.* (1999, 2005) note that gender and wealth are key variables that affect individual health and there is a longstanding relationship between small-area deprivation and health outcomes,

with more deprived areas being associated with higher mortality rates than relatively less deprived areas. In spite of this realisation, inequalities in health between relatively more deprived and affluent areas continue to increase in many countries.

In the UK, the 1980 Black Report presented the findings of a working group set up in 1977 to examine health inequalities (Black *et al.*, 1980). The report suggested that the UK National Health Service (NHS), set up in 1948, had effectively failed in its drive towards equality and identified that major health inequalities by social class still exist. Similar findings were reported in the 1998 Acheson Report, an independent inquiry into health inequalities, which found that overall health outcomes had improved, but that the class gap had widened, thus suggesting inequality in health outcomes by social class at a

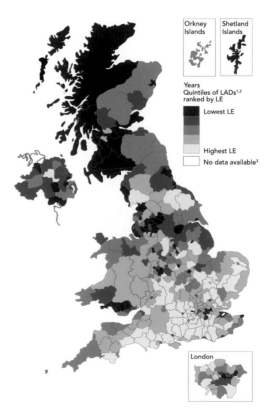

Figure 9.1 Variations in life expectancy (LE) for males at birth by local authority district, UK, 2010–12

Source: ONS (2014)

local level. A 1999 publication titled *The Widening Gap: Health Inequalities and Policy in Britain* (Shaw *et al.*, 1999) outlined some of the more extreme local-level health inequalities in Britain, and demonstrated the clear impacts of social class on life expectancy, rates of limiting long-term illness and infant mortality.

Shaw *et al.* (2005) and Dorling (2013) provide excellent overviews of health inequalities in Britain (especially since the 2000s) exploring the link with deprivation and poverty. Figure 9.1 shows more recent UK data and clearly highlights a high degree of spatial clustering of low life expectancy around locations such as Scotland, South Wales and northern England, and a broader distribution of long life expectancy across much of the south-east of England. Even within deprived areas there are pronounced variations in mortality rates. Tunstall *et al.* (2011) outline that broader socio-demographic factors, including

ethnicity, have a pronounced impact on small-area health variations. Since there is a tendency for many of these groups to cluster, especially within urban areas, there are clear issues here for health care planning.

Figure 9.2 shows similar spatial variations for predicted heart disease (CVD) in the US between 2014 and 2024 (Yang *et al.*, 2015). In this case, the east–west variations are as striking as the north–south variations in the UK (showing a clear link again with US variations in poverty which show a similar pattern), although the deep south also shows high risk.

There are also likely to be substantial spatial variations in health within a city or part of a city. Figure 9.3 shows an example of a recent London atlas produced by the geographer James Cheshire. Using the London underground map is a novel way of showing how (in this case) infant mortality rates vary across London, in comparison to child poverty rates mapped as an overlay.

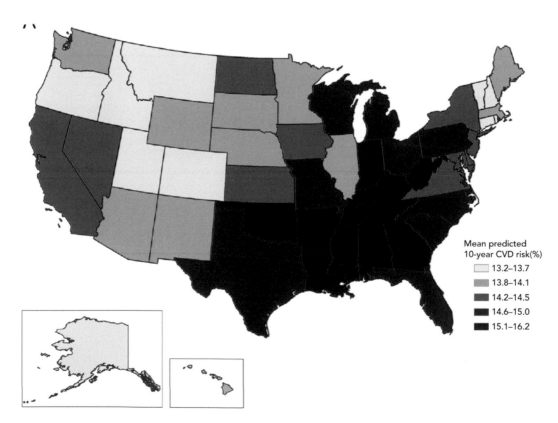

Figure 9.2 Variations in likely future patterns of male heart disease in the US

Source: Yang *et al.* (2015)

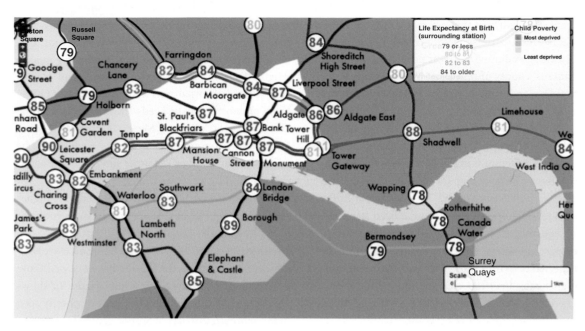

Figure 9.3 Spatial variations in life expectancy in London plotted against child poverty

Source: Cheshire (2012)

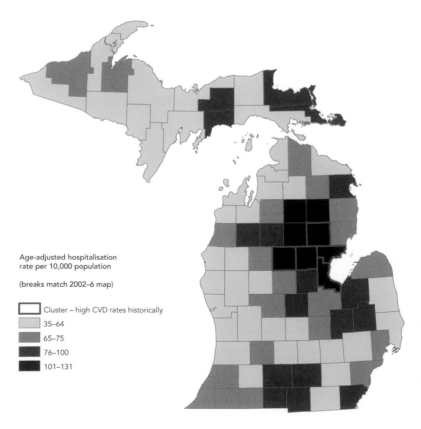

Age-adjusted hospitalisation
rate per 10,000 population

(breaks match 2002–6 map)

☐ Cluster – high CVD rates historically
☐ 35–64
☐ 65–75
☐ 76–100
☐ 101–131

Figure 9.4 Variations
in hospitalisation rates
for heart disease across
Michigan, US

Source: CDC (undated),
www.cdc.gov/dhdsp/maps/gisx/
mapgallery/maps/pdf/mi-chd-
hosprates.pdf

193

Similarly, Figure 9.4 shows a map of Michigan in the US, by county, displaying four levels of coronary heart disease age-adjusted five-year hospitalisation rates, by patients' county of residence. The map clearly shows higher rates of heart disease in the downtown, less affluent area of the city.

In Chapters 7 and 8 we noted the use of GIS to create hot spot maps. There are many examples of these in the GIS literature in relation to health care. Figure 9.5 shows a hot spot map for colorectal cancer between 1998 and 2003 across Iowa, US (Beyer and Rushton, 2009). The proportion of cases has been standardised by age and sex and smoothed using adaptive filter density estimation. Each rate is based on the closest 50 expected cases on a three-mile grid. Red areas indicate higher rates than expected and blue rates lower than expected (given the statewide rate). Many of these hot spots are in areas of lower income.

Residents of lower income areas might also have poorer lifestyles in terms of higher propensities to smoke and consume alcohol and lower propensities to consume fresh fruit and vegetables (and generally have a poorer diet than more affluent persons). Morris *et al.* (2017) profiled women consumers in a large UK survey by geodemographics (see Chapter 5 for a review of geodemographics). They used data-driven dietary patterns in order to show how these varied by geodemographic group nationally. Then they mapped estimated dietary patterns across Leeds, UK, based on local geodemographics. The better quality 'Higher Diversity Traditional Omnivore' and 'Health Conscious' dietary patterns were seen to be most likely to be present in North Leeds (in the higher income areas) while 'Traditional Meat, Chips and Pudding Eaters' and 'Monotonous Omnivores' were seen to be more likely in the deprived inner suburbs to the south and east of the city centre (Figure 9.6).

Poor diet has been linked in many studies to higher rates of obesity. GIS has also been used in recent years to explore spatial variations in patterns of obesity. Drewnowski *et al.* (2007, 2014) have rare access to obesity data for Seattle in the US and have mapped those distributions with interesting results. Figure 9.7 shows an example of this, with higher rates of obesity seen in the southern districts of Seattle, generally in poorer areas, non-white locations with lower property values (which they have shown to be an important predictor of obesity patterns in Seattle).

But obesity is not simply associated with poor diet and low-income areas. Edwards and Clarke (2009) undertook a geodemographic profile of children in Leeds who were categorised as overweight or obese. They showed that there was a broad correlation with deprivation (the usual correlation found in the

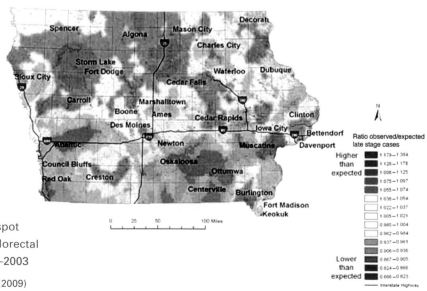

Figure 9.5 Hot and cold spot mapping of late-stage colorectal cancer in Iowa, US, 1998–2003

Source: Beyer and Rushton (2009)

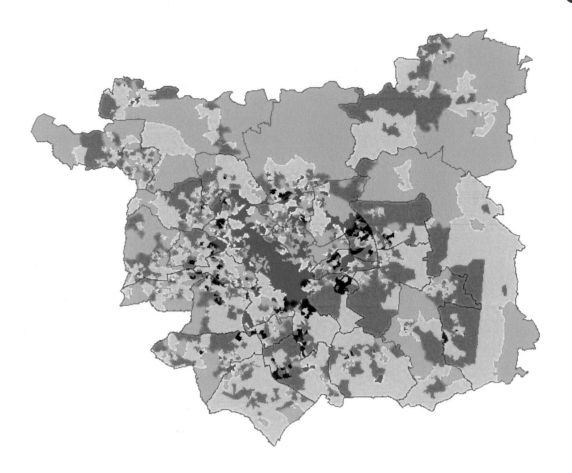

Percentage consuming a Traditional Meat, Chips and Pudding Eater dietary pattern

11.45–13.94 13.95–16.93 16.94–19.34 19.35–22.67 22.68–25.49

Figure 9.6 Locating the geography of poorer diets, Leeds, UK: estimated 'Traditional Meat, Chips and Pudding Eater' dietary pattern in Leeds at the OA level

Source: Morris *et al.* (2017)

literature as noted above) but that some more affluent areas did have obese children (perhaps to do with more sedentary lifestyles). Based on this analysis they then ranked schools as hot or cold spots for obesity based on what we might expect rates to be given the socio-demographics of the catchments of each school. Thus, schools in high-income areas with high obesity rates were hot spots, while schools in low-income areas with low obesity rates were cold spots.

In addition to low income and poverty, such hot spots could be caused by the physical environments in which people live, work and socialise. By its very nature, exposure to environmental factors tends to be spatially clustered, but behavioural or environmental factors are also inherently linked to small geographic areas, such as neighbourhoods, since individuals exhibiting these behaviours tend to spatially cluster. In Chapter 1 we showed the famous map in spatial epidemiology of

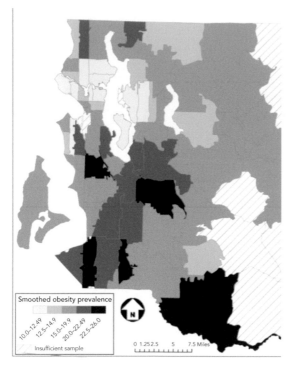

Figure 9.7 Obesity in Seattle, Washington, US

Source: Drewnowski *et al.* (2007)

cholera patients in Soho London in the 19th century. John Snow was the GP who located the source of a cholera outbreak in Soho by mapping the locations of patients with cholera. By focusing on the centre of the distribution map he created, the contaminated drinking tap was able to be identified as the most likely source of the outbreak. Since then many other spatial mapping exercises have been undertaken. Figure 9.8 shows the work of Stan Openshaw and colleagues in 1988, mapping each incidence of childhood leukaemia in the North of England and then plotting the hot spots (or blobs as he called them). The research had anticipated finding greater numbers of childhood leukaemia around Sellafield nuclear power station (Cluster b to the west of the map). However, Openshaw *et al.* were able to find equally alarming clusters in other locations not predicted in advance. Cluster a was, for example, the largest. Subsequent investigation suggested that a waste incinerator plant in Gateshead (near Newcastle) was the most likely source of the contamination.

Figure 9.9 shows another example of potential small-area variations in environmental contamination, this time in New York. Corburn *et al.* (2006) plotted

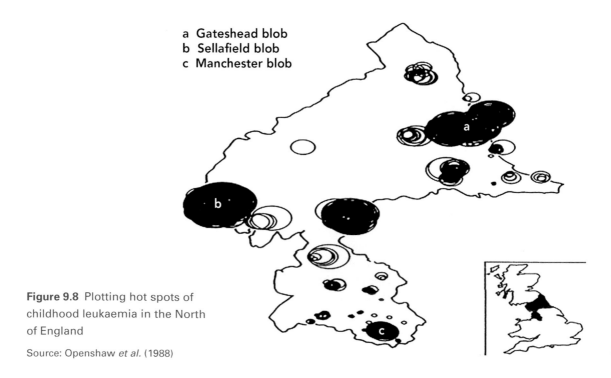

a **Gateshead blob**
b **Sellafield blob**
c **Manchester blob**

Figure 9.8 Plotting hot spots of childhood leukaemia in the North of England

Source: Openshaw *et al.* (1988)

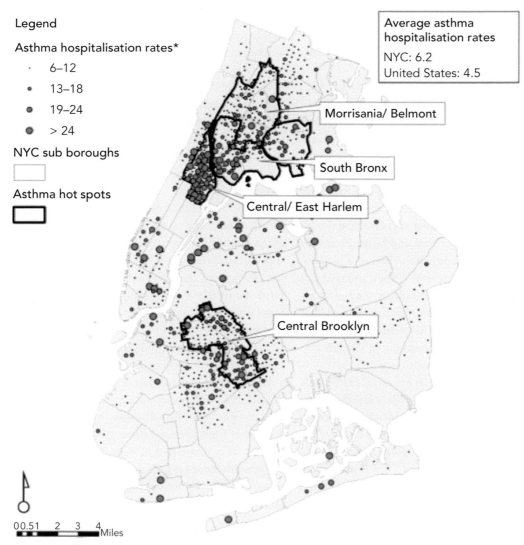

Legend

Asthma hospitalisation rates*

· 6–12

• 13–18

● 19–24

● > 24

NYC sub boroughs

Asthma hot spots

Average asthma
hospitalisation rates
NYC: 6.2
United States: 4.5

Morrisania/ Belmont

South Bronx

Central/ East Harlem

Central Brooklyn

00.51 2 3 4 Miles

* Hospitalisations per 1,000 persons under the age of 14

Figure 9.9 Variations in hospitalisation rates for asthma in New York, 1996–2000

Source: Corburn *et al.* (2006)

asthma hospitalisation rates by small area across Belmont, Bronx, Harlem and Brooklyn. They showed the higher rates of hospital cases around the poorest neighbourhoods, a mixture of contamination from poor housing, high inner-city traffic counts and pollution from inner-city noxious industries.

Using GIS to analyse health care provision

We have established that demand for health services is spread unevenly across space, broadly in line with population distribution, but also driven by a complex range of individualised and area-based factors. Health services, however, have to be provided at discrete locations, such as hospitals, GP surgeries, pharmacies and walk-in centres. Due to this, it is inevitable that there will always be inequalities in the provision of health care, as it is impossible to provide a universal coverage of services at the point of demand.

There is a hierarchy of health care provision, with pharmacies and GP surgeries being widely distributed at a neighbourhood level, while more specialised services such as hospitals tend to be more centralised. Access to primary care is important as a 'gateway' to the whole health system. One major issue for health care planners is the need for specialisation. Expensive facilities and specialised staff cannot be deployed uniformly and must be centralised at key hospital sites to achieve the economies of scale and high patient volumes required to make provision viable. For patients, this is likely to result in higher travel cost and time and therefore it is important to consider notions of access to health care. GIS is a fundamental tool here, and has been used widely to consider the physical separation of supply and demand – taking account of distance and travel times to appropriate health services.

In one such example, Haynes et al. (1999) calculated access to hospitals and GP surgeries using straight-line distance from UK census ward centres (to hospitals), or census enumeration districts (to GPs). In this case, small-area population distribution was used as a proxy for demand for services, taking account of underlying conditions such as deprivation. They found that many inner-city deprived communities, where health needs might be greater, enjoy good access to health care facilities, which are often located nearby. However, many of the examples that follow identify that straight-line distance is a very crude measure of access, with factors such as cost, time and inconvenience being important considerations, alongside more practical measures of access which include availability of public transport, opening hours, availability of appointments, etc.

Martin et al. (2002) used GIS-based analysis to investigate access measures between general district hospitals and the underlying distribution and health needs of the population in Cornwall, UK. Cornwall is a rural area in south-west England, characterised by poor geographic access. They incorporated drive time data to measure access to district hospitals, using rates

of limiting long-term illness as a proxy for small-area health need. A digital road network was used to create a series of raster-based population layers and travel time 'cost' surfaces – these are continuous surfaces made up of a series of cells, such that from any given point, the estimated travel time to the nearest hospital can be calculated, taking account of the underlying road network and road speeds (Figure 9.10). They demonstrate the improvements in evaluating health care access that can be achieved by using road travel time in place of straight-line distance.

Lovett et al. (2002) carried out a similar investigation in East Anglia, another rural area in the UK characterised by poor access. They specifically considered travel times to GP practices. With access to patient registers, they were able to calculate sophisticated cost surface access measures based on both road travel time and public transport (public bus and other community transport services). Figure 9.11 provides an example of the travel time surfaces they produced. They identified that while only 10% of residents faced a car journey of over ten minutes to reach a GP practice, for those without access to private transport, 13% were not able to reach a GP surgery by bus after taking account of timetables and the availability of a return journey. They found that those residents with higher health needs tended to be located in remote rural areas, where low personal mobility and lack of any form of public transport made access to primary care almost impossible.

This form of analysis reveals important concerns regarding access to primary care (with implications for referrals to secondary care), and implies the existence of an 'inverse care law', whereby those most in need of care (in this case elderly or more deprived rural residents with poor health) lack effective health care provision that is accessible to their needs. Lovett et al. (2002) also demonstrated that measures of access need to take account of public transport, since many of the groups with low personal mobility and lacking access to private transport have greater health needs. In addressing this need, Martin et al. (2008) present a software tool created to analyse public transport timetable data in order to analyse bus travel times. They explored bus travel times to Derriford Hospital in Devon, England,

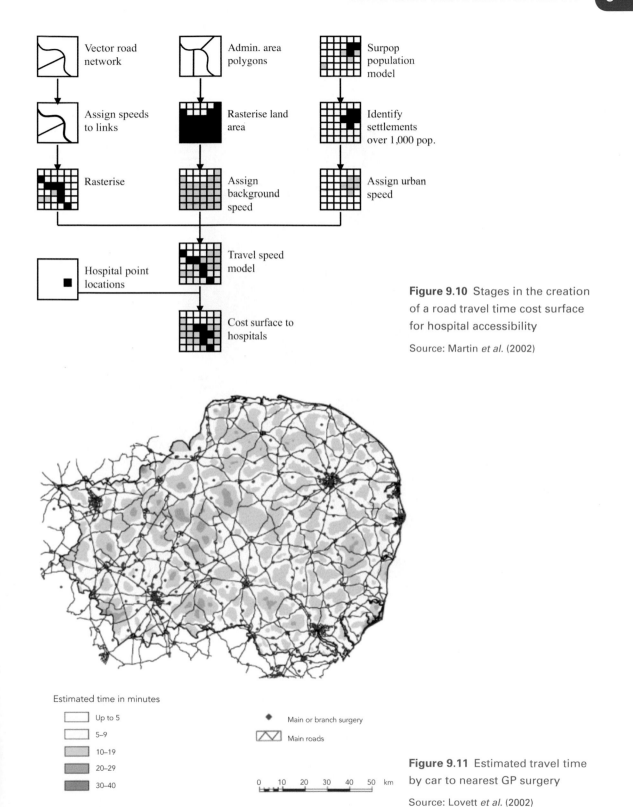

Figure 9.10 Stages in the creation of a road travel time cost surface for hospital accessibility

Source: Martin *et al.* (2002)

Estimated time in minutes

	Up to 5
	5–9
	10–19
	20–29
	30–40

◆ Main or branch surgery

Main roads

0 10 20 30 40 50 km

Figure 9.11 Estimated travel time by car to nearest GP surgery

Source: Lovett *et al.* (2002)

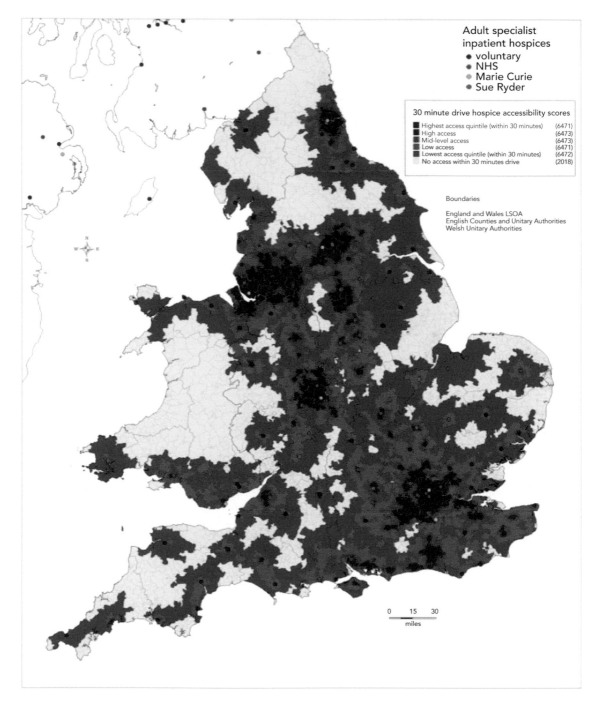

Figure 9.12 Access to adult specialist inpatient hospices in England and Wales based on travel time

Source: Gatrell and Wood (2012)

taking account of the underlying demographic and socio-economic characteristics of the population and, in common with Lovett *et al.* (2002), identified high concentrations of less mobile elderly populations and those with self-reported limiting long-term illness suffering poor access to health care services.

There is also increasing concern surrounding access to palliative or end-of-life care for patients that are close to death and who would prefer to die at home or within a specialist hospice, rather than within a hospital or other health facility. Gatrell and Wood (2012) examine variations in small-area geographic access to specialist hospices relative to estimated need. In this case, need (or demand) is determined by cancer mortality rates, as a major cause of premature death in the UK. Gatrell and Wood (2012) measure access relative

to demand using travel time at the LSOA level, each containing around 1,500 residents. They also account for the size/capacity of each facility, and consider their accessibility scores relative to area-based deprivation. Figure 9.12 shows an extract from their findings and suggests that pronounced geographic variations in access to these services exist.

Estimates of accessibility often go hand in hand with more sophisticated measures of access. Spatial interaction models are a well-established technique for calculating accessibility or provision indicators based on how far people travel (or would have to travel) to reach facilities (the size or quality of which can also be taken into account). Good examples appear in Clarke and Wilson (1994) and Clarke *et al.* (2002). Morrissey *et al.* (2010, 2015) first estimated the number

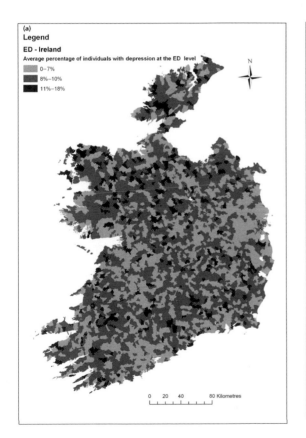

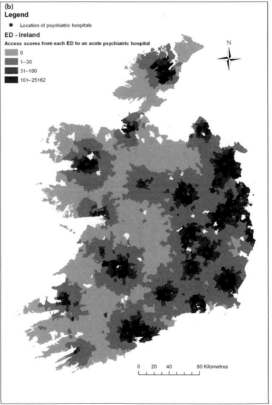

Figure 9.13 Measuring both small-area variations in depression in Ireland and access to mental health care facilities

Source: Morrissey *et al.* (2010)

of persons with depression in Ireland using spatial microsimulation (see again Chapter 6 and good overviews provided by Tanton and Edwards (2012) and O'Donoghue *et al.* (2013)). Figure 9.13a shows the spatial pattern of depression as estimated. Greater concentrations can be seen in the more rural but less affluent census tracts to the north and west of Ireland. Figure 9.13b then shows an accessibility score based on spatial interaction models. There is clearly a very worrying mismatch here – areas with high rates of estimated depression have the lowest access to mental health care facilities (so many of which are centralised around the capital of Dublin).

Health care access has improved at a local level in a number of countries with the introduction of walk-in centres and smaller GP practices. These provide dispersed, localised provision of services (although this might be at the expense of facilities, staffing levels and opening hours). Tomintz *et al.* (2008) investigate the demand for and distribution of one form of localised health care provision, neighbourhood-level stop smoking clinics in Leeds, UK. Stop smoking services tend to be colocated with other health facilities, but might offer restricted service hours or appointment times. For example, Tomintz *et al.* (2008) note that there were a total of 51 smoking cessation service points in Leeds, but that only nine offer services on a Friday. Consequently, provision of (and access to)

these services varies on different days of the week, and more complex modelling is required. You will be able to consider the provision of these services in Leeds in the associated practical activity.

Tomintz *et al.* (2008) also used a spatial microsimulation approach to estimate the distribution of smokers across Leeds, and then used location-allocation modelling (introduced in Chapter 4) to investigate the extent to which the current provision of these services matches the optimum spatial distribution and locations for these services based on the inferred demand, at any given time. Figures 9.14 and 9.15 show extracts from the paper, and identify that access to these services for the target population varies considerably by day of the week, with service provision and access being worse on a Friday and Saturday, when fewer clinics are open. Using location-allocation modelling, they identify that a re-distribution of service locations, while maintaining the overall daily levels of provision, could result in considerable improvements in overall patient access (even though some individuals may have to travel further), especially on a Friday.

The type of analysis presented by Tomintz *et al.* (2008) assists health care planners in identifying hypothetically ideal locations for services such as smoking cessation, but it must be acknowledged that due to long-term investment in resources and infrastructure in particular locations, health service provision

	Longest distance to travel to the nearest centre (km)	Average distance (km)	% of smokers who need to travel further than 5 km
Tuesday			
19 existing centres	11.0	2.1	13.8
19 'optimal' centres	10.0	2.6	0.4
Friday			
9 existing centres	11.0	2.5	39.3
9 'optimal' centres	10.5	3.1	9.4

Figure 9.14 Location-allocation results for smoking cessation services in Leeds, UK, on selected days of the week

Source: Tomintz *et al.* (2008)

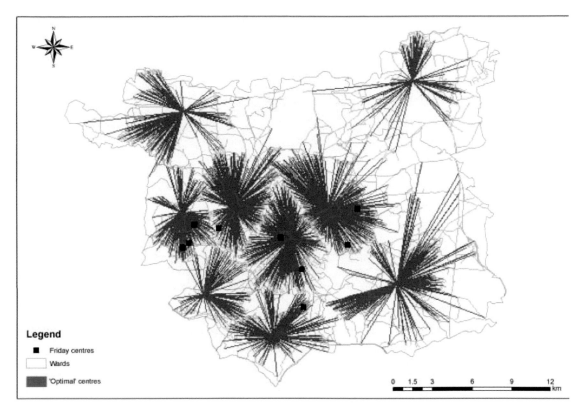

Figure 9.15 Actual smoking cessation centres compared with optimal centres for Leeds, UK, on a typical Friday

Source: Tomintz *et al.* (2008)

cannot constantly shift locations and service delivery points in order to meet the ever-changing distribution of demand. Nonetheless, increasingly, colocation of services such as stop smoking clinics within other facilities such as supermarkets could enable short-term provision of services in an area where existing infrastructure does not exist but a need has been identified.

Tomintz *et al.* (2008) used a location-allocation model for finding optimal locations. Other forms of optimisation models are increasingly seen in the literature to provide similar solutions. Figure 9.16 shows an example based on the optimal location of ambulance centres in Hong Kong (Sasaki *et al.*, 2010).

They used a different form of optimisation model, a genetic algorithm, to optimally locate ambulance stations within a GIS environment. Figure 9.16a shows the existing distribution of emergency cases. Figure

9.16b shows the current location of 27 ambulance stations serving Hong Kong, while Figure 9.16c shows the optimal location of 27 ambulance stations based on emergency cases in 2007. Finally, Figure 9.16d shows the optimal location of 27 ambulance stations based on predicted emergency cases for 2030. The solid circles show optimal locations that are the same as current locations, while the ringed circles show new locations; hollow circles with a cross show current sites not selected during optimisation. The modified GGA was run to select 27 ambulance site locations evaluated on the network distance between each census centroid, weighted by the count of emergency cases. For each census area, the nearest site location was also calculated to indicate the catchment or service area for each ambulance location. Of the 27 locations, 23 were existing stations but four were not, indicating that some improvement in ambulance accessibility could

Figure 9.16 Emergency cases and current and optimal ambulance locations with their catchment areas in Hong Kong

Source: Sasaki *et al.* (2010)

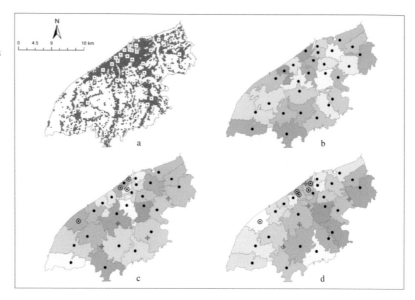

be achieved with some re-allocation of resources (see also Comber *et al.* (2011) for another example of finding optimal locations for ambulance stations).

Access to health care has also been increased through the introduction of non-geographic services such as NHS direct – a nurse-led advice line and web-based service where health advice can be sought at any location and time with no need for direct doctor contact. These services have an important role in providing patient reassurance in the out-of-hours period and might appeal to certain sectors of the community, such as young males, who might be less likely to visit their GP for face-to-face consultations. Turnbull *et al.* (2011) investigate the use of telephone-based advice from primary care centres (PCCs) to manage out-of-hours calls to GPs. Much of the out-of-hours services are managed by telephone, with some patients being asked to travel to the PCC or receiving a follow-up home visit. These services should improve health access for those unable to reach the PCC. Turnbull *et al.* (2011) identify that geography – and specifically distance – play an important role in determining the advice given, with patients living further than six kilometres from the primary care centre more likely to be given telephone advice only, and less likely to be seen face-to-face. Consequently, in spite of efforts to improve access through the use of telephone-based

services; it appears that geographical inequalities in terms of equity access exist.

Inequalities in service provision might not always result from poor access or availability. Even where access and the availability of services offered are considered to be good, there might be inequalities between different areas in terms of the quality of services provided, or the resulting health outcomes. For example, after receiving primary care, small-area variations may exist in referral or admission rates, or in expenditure on certain health conditions. In order to improve patient outcomes and make effective use of the health care resources provided, the UK NHS is taking steps to identify and reduce these variations. Visualisation and spatial data analysis are essential tools in identifying variations in activity rates (e.g. referral), expenditure, quality and outcomes.

Concluding comments

This chapter has identified that health and the management of health care provision and service delivery are inherently geographical concepts. Individual health is driven by personal geographies which can affect risk of ill-health, access to health care and health outcomes. In particular, there is a clear relationship between small-area deprivation and health needs, although health policy seeks to reduce these inequalities and variations in health, striving for a consistent level of care relative to need.

We have identified that GIS is an important tool for visualising health inequalities, identifying small-area health needs and evaluating access to health care facilities. We have explored a range of examples in which GIS and spatial data analysis have been applied in order to evaluate the need for and provision of care to meet the needs of small-area populations. Tools to identify area-based health, driven by underlying demographic and socio-economic indicators have been identified, and a series of studies have been presented which apply increasingly sophisticated measures of health access. We have seen how modelling techniques (techniques such as microsimulation and location-allocation models), when coupled with GIS, can be applied to evaluate the provision of services, and we have explored indicators of health outcomes.

Accompanying practical

The accompanying practical (Practical 5: Health care analysis) explores the interrelationships between health needs and service provision. You consider access to smoking cessation clinics in the UK city of Leeds, drawing on the discussion above. You briefly visualise estimated 'need' for smoking cessation services and use buffer and overlay techniques to consider provision of smoking cessation clinics. We introduce a new and powerful technique – location-allocation modelling – in order to optimise the provision of these services.

References

All website URLs accessed 30 May 2017.

Beyer, K. M. M., & Rushton, G. (2009) Mapping cancer for community engagement. *Preventing Chronic Disease*, 6(1), A03.

Black, D., Morris, J., Smith, C., & Townsend, P. (1980) *Inequalities in Health: Report of a Research Working Group*, Department of Health and Social Security, London.

Centers for Disease Control and Prevention (CDC) (undated) *Michigan hospitalisation rates for coronary heart disease by county*. Centers for Disease Control and Prevention, Atlanta, GA. Available from: www.cdc.gov/dhdsp/maps/gisx/mapgallery/mi_chd_hosprates.html.

Cheshire, J. (2012) *Lives on the line: life expectancy and child poverty as a Tube map*. Available from: http://spatial.ly/2012/07/lives-on-the-line/.

Clarke, G. P., & Wilson, A. G. (1994) A new geography of performance indicators for urban planning, in C. S. Bertuglia, G. P. Clarke, & A. G. Wilson (eds) *Modelling the City*, Routledge, London, 55–81.

Clarke, G. P., Eyre, H., & Guy, C. (2002) Deriving indicators of access to food retail provision in British cities: studies of Leeds, Bradford and Cardiff. *Urban Studies*, 11, 2041–2060.

Comber, A. J., Sasaki, S., Suzuki, H., & Brunsdon, C. (2011) A modified grouping genetic algorithm to select ambulance site locations. *International Journal of Geographical Information Science*, 25(5), 807–823.

Corburn, J., Osleeb, J., & Porter, M. (2006) Urban asthma and the neighbourhood environment in New York City. *Health & Place*, 12(2), 167–179.

Curtis, S. (2010) *Space, Place and Mental Health*, Ashgate Publishing Ltd, London.

Curtis, S., & Rees Jones, I. (1998) Is there a place for geography in the analysis of health inequality? *Sociology of Health and Illness*, 20(5), 645–672.

Dorling, D. (2013) *Unequal Health: The Scandal of Our Times*, Policy Press, Bristol.

Drewnowski, A., Rehm, C. D., & Solet, D. (2007) Disparities in obesity rates: analysis by ZIP code area. *Social Science & Medicine*, 65(12), 2458–2463.

Drewnowski, A., Moudon, A. V., Jiao, J., Aggarwal, A., Charreire, H., & Chaix, B. (2014) Food environment and socioeconomic status influence obesity rates in Seattle and in Paris. *International Journal of Obesity*, 38(2), 306–314.

Edwards, K. L., & Clarke, G. P. (2009) The design and validation of a spatial microsimulation model of obesogenic environments for children in Leeds: SimObesity. *Social Science and Medicine*, 69(7), 1127–1134.

Gatrell, A. C., & Elliott, S. J. (2014) *Geographies of Health: An Introduction*, Wiley, Chichester.

Gatrell, A. C., & Wood, D. J. (2012) Variation in geographic access to specialist inpatient hospices in England and Wales. *Health & Place*, 18(4), 832–840.

Haynes, R., Bentham, G., Lovett, A., & Gale, S. (1999) Effects of distances to hospital and GP surgery on hospital inpatient episodes, controlling for needs and provision. *Social Science and Medicine*, 49, 425–433.

Kearns, R. A., & Joseph, A. E. (1993) Space in its place: developing the link in medical geography. *Social Science & Medicine*, 37(6), 711–717.

Lovett, A., Haynes, R., Sunnenberg, G., & Gale, S. (2002) Car travel time and accessibility by bus to general practitioner services: a study using patient registers and GIS. *Social Science and Medicine*, *55*, 97–111.

Martin, D., Wrigley, H., Barnett, S., & Roderick, P. (2002) Increasing the sophistication of access measurement in a rural healthcare study. *Health & Place*, *8*, 3–13.

Martin, D., Jordan, H., & Roderick, P. (2008) Taking the bus: incorporating public transport timetable data into health care accessibility modelling. *Environment and Planning A*, *40*, 2510–2525.

Morris, M., Clarke, G. P., Edwards, K. L., Hulme, C., & Cade, J. E. (2017) Exploring small area geographies of obesity in the UK: evidence from the UK Women's Cohort Study, in J. Lombard, G. P. Clarke, & E. Stern (eds) *Applied Spatial Modelling and Planning*. Routledge, London, 280–299.

Morrissey, K., Clarke, G. P., Hynes, S., & O'Donoghue, C. (2010) Examining the factors associated with depression at the small area level in Ireland using spatial microsimulation techniques. *Irish Geography*, *43*(1), 1–22.

Morrissey, K., Clarke, G. P., Williamson, P., Daly, A., & O'Donoghue, C. (2015) Mental illness in Ireland: simulating its geographical prevalence and the role of access to services. *Environment and Planning B: Planning and Design*, *42*, 338–353.

O'Donoghue, C., Ballas, D., Clarke, G. P., Hynes, S., & Morrissey, K. (2013) *Spatial Microsimulation for Rural Policy Analysis*, Springer, Berlin.

Office for National Statistics (ONS) (2014) *Life Expectancy at Birth and at Age 65 by Local Areas in the United Kingdom: 2006–08 to 2010–12*, ONS, London.

Openshaw, S., Charlton, M., Craft, A. W., & Birch, J. M. (1988). Investigation of leukaemia clusters by use of a geographical analysis machine. *The Lancet*, *331*(8580), 272–273.

Sasaki, S., Comber, A. J., Suzuki, H., & Brunsdon, C. (2010) Using genetic algorithms to optimise current and future health planning – the example of ambulance locations. *International Journal of Health Geographics*, *9*(4).

Shaw, M., Dorling, D., Gordon, D., & Davey Smith, G. (1999) *The Widening Gap: Health Inequalities and Policy in Britain*, Policy Press, Bristol.

Shaw, M., Davey Smith, G., & Dorling, D. (2005) Health inequalities and New Labour: how the promises compare with real progress. *British Medical Journal*, *330*, 1016–1021.

Tanton, R., & Edwards, K. (eds) (2012) *Spatial Microsimulation: A Reference Guide for Users* (Vol. 6). Springer Science & Business Media, Berlin.

Tomintz, M., Clarke, G., & Rigby, J. (2008) The geography of smoking in Leeds: estimating individual smoking rates and the implications for the location of stop smoking services. *Area*, *40*(3), 341–353.

Tunstall, H., Mitchell, R., Gibbs, J., Platt, S., & Dorling, D. (2011) Socio-demographic diversity and unexplained variation in death rates among the most deprived parliamentary constituencies in Britain. *Journal of Public Health*, *34*(2), 296–304.

Turnbull, J., Pope, C., Martin, D., & Lattimer, V. (2011) Management of out-of-hours calls by a general practice cooperative: a geographical analysis of telephone access and consultation. *Family Practice*, *28*(6), 677–682.

World Health Organization (WHO) (1946) Preamble to the Constitution of the World Health Organization as adopted by the International Health Conference. *International Health Conference*, New York, 19–22 June.

Yang, Q., Zhong, Y., Ritchey, M., Loustalot, F., Hong, Y., Merritt, R., & Bowman, B. A. (2015) Predicted 10-year risk of developing cardiovascular disease at the state level in the US. *American Journal of Preventive Medicine*, *48*(1), 58–69.

10 GIS for emergency planning

LEARNING OBJECTIVES

- How emergency planning relates to GIS
- GIS tools for understanding vulnerability, hazards and exposure
- Ways in which GIS and spatial analysis can be used in different aspects of emergency planning

Introduction

Readers might ask, why devote an entire chapter of a social sciences GIS textbook to emergency planning? At first glance this appears to be a topic falling solidly within the area of expertise of practitioners and policy makers, not necessarily students or researchers in the social sciences. Upon further consideration, however, emergency planning emerges as virtually the quintessential social science and GIS combination – risks and impacts of emergencies are spatial, but they are also related to population characteristics, information transmission, government structures and a host of other societal factors. As the chapter title suggests, a raft of possibilities exists for actual GIS applications in emergency planning. More than that, though, emergency planning provides an exemplar for those working in other areas: many of the methods, concepts and spatial perspectives employed here could easily be retooled for other subjects. Finally, emergency planning provides a blueprint for thinking more generally about how the social sciences and GIS can combine to provide information and understanding about 'real-world' issues.

The phrase 'emergency planning' is deceptively simple. It conjures images of disaster response teams moving into a location struck by earthquake, flooding, hurricane or tornado, providing evacuation assistance, health care or food and water. In fact, each word, 'emergency' and 'planning', represents a range of activities, all informed by spatial data and GIS research. Emergencies, covered in more depth below, might be environmental in origin but this is not necessarily the case: terrorist attacks or disease outbreaks are also emergencies. Emergency planning also refers to the provision of emergency services such as police and fire services in order to respond to more frequent emergencies such as domestic fires (see again the discussion in Chapter 7). Planning suggests an element of forethought and prearrangement. Although the emergency might come as a surprise, all levels of government aim to be prepared for the unexpected. This preparation indeed includes consideration of evacuation routes and knowledge about population centres and areas of potential demand during an emergency. However, it also involves providing up-to-date coverage of disaster impacts and eventual recovery efforts. In fact, planning activities can be thought of as an umbrella, under which proactive, active or real-time and reactive actions fall.

Anticipation of disasters or emergencies means understanding (and even imagining) the types of emergencies that might occur, as well as their likelihood, location and areal impact. Simulations can also help expose unintended consequences and spill-over effects, and assist with response planning. For example, simulating fires along different locations of a subway line can help estimate access to the site and response times depending on where the fire occurs. Should an emergency occur, often multiple immediate responses will be possible. Advance planning helps consider alternatives and identify best options under various circumstances. Emergency planning is equally important during and after a disaster, however. As a crisis is occurring, responses must be recalibrated or redirected according to experiences on the ground or unexpected challenges. Finally, in the aftermath, coordinating clean-up and recovery measures necessitates not only advance planning but also in-the-moment planning that incorporates new information, stakeholder needs and other factors.

Underpinning a good part of before, during and after emergency planning is access to appropriate data and effective location and organisation of emergency services. Locations of fire stations, for example, must meet the day-to-day needs of the local population but must also be equipped to respond quickly to the unexpected emergency. High-quality and recent population and infrastructure data are also key. It is not sufficient to know where an emergency has occurred: effective planning means also knowing how many are likely to be impacted and what the toll on infrastructure might be. Where will be the heaviest hit locations and how many people will be involved? Will they be able to leave the area unassisted or will assistance be required? Will

evacuation via road be possible? Will communication via mobile phones be possible?

At every stage of preparation or response and in every aspect of management, emergency planning is inherently spatial. Emergencies, of course, occur *somewhere* and the extent of the event interacts with the people and things within that area – this is core GIS territory. Research and planning around movement in and out of an afflicted area (evacuation for instance) also lends itself naturally to a GIS environment. Even visualisation in 'real time' of an emergency almost demands the use of a GIS. GIS and a spatial approach provide the necessary tools and techniques to engage in emergency planning. They also enhance understanding of the data inputs required for such planning – GIS fundamentals such as spatial scale, accessibility and interaction help determine the efficacy of any emergency planning. The necessary integration of information and viewpoints is also well suited to a GIS environment. Regardless of the methods used in disaster planning, an enormous assortment of data must be assembled and integrated in such a way that it can be analysed and interpreted efficiently. A GIS offers this capability.

Within applied geography and GIS, the field of emergency planning or management is healthy and even growing. Accessing current research and practical guides on the use of GIS in this area is easy, requiring the simplest of internet searches (a few recommendations for further reading are also provided at the end of this chapter). Because research in this area encompasses so many applications and includes research in 'preparedness, response, recovery, and mitigation' (Emrich *et al.*, 2011: 321), the conceivable uses of GIS or spatial analysis in emergency planning are myriad. Armed with basic concepts and understanding of common GIS applications, students and researchers should be equipped to engage with the academic literature and also contribute to planning in the world of practice.

This chapter has two purposes. The first is to provide the building blocks of good emergency planning research and investigation using GIS, from key concepts to data considerations to common techniques and methods. The second purpose is to illustrate the utility of GIS for emergency planning through examples that emphasise the application, rather than individual techniques or data inputs. As with much research in the social sciences, little is accomplished with just one method. Rather, the goal is to answer questions and provide solutions or recommendations. Thus this section of the chapter emphasises application and shows how methods introduced in Part I of the book can be combined in novel ways to create new information.

What is an emergency?

A simple definition of an emergency is that it is an unexpected and harmful or dangerous event or occurrence. Using this definition, everything from an auto accident to a house fire to a hurricane is considered an emergency. Within the realm of emergency planning, then, some additional criteria must be met. These criteria are related to scale, impact and, perhaps, some aspect of predictability. Scale conveys an impression of the magnitude of the event. Events qualifying as emergencies or disasters are typically of a larger scale, measured in terms of area affected, the response required, length of time the area experiences and recovers from the event, or the number of people or properties affected. Magnitude can also be measured or thought of in terms of impact. An auto accident might impact a household, a hurricane an entire city or region. Impact can be assessed in terms of human lives affected or lost, property damage, range of resources to be mobilised or, again, some time component.

Defining emergencies in terms of predictability is a bit more challenging. As science and documentation of past experience improves, our ability to anticipate when and where a disaster might occur (and the quickest and most effective ways of responding) should also improve. That said, an element of unexpectedness or suddenness is often attributed to emergencies. We might know that an area is prone to flooding or earthquakes, but we are unable to predict exactly when they might occur. This means that resources must be allocated and plans readied, without knowing precisely when or where they will need to be mobilised.

Table 10.1 Some key concepts and terminology

Below are several terms that, although also used in common language, take on specific meaning within the fields of emergency planning or hazards research. These concepts form the backbone of much published research in this area.

Hazard: A potential source of danger or incapacitation. Hazards with spatial implications can be natural or biological in origin, but also human-induced. A hazardous situation, such as a tropical storm or earthquake, becomes a *disaster* or *emergency* when its occurrence has societal impacts – i.e. on people, property, or some aspect of the environment that society depends upon. Hazards can be characterised by their location of course, but also their likely area of impact, duration and the frequency with which they occur (van Westen, 2013).

Exposure: A measure of who and what might be affected by a hazard. At a very basic spatial level, this might be thought of as the area (and the people and things within that area) likely to be impacted by a particular hazard.

Vulnerability: The characteristics of people and places that make them more likely to be affected by a hazard, should it occur. In many cases, exposure is a good proxy for vulnerability. In addition, however, particular population characteristics (e.g. age or poverty levels) and place characteristics (e.g. type of government or quality of infrastructure) may exacerbate vulnerability.

Risk: The combination of the characteristics of the hazard, exposure and vulnerability provides a sense of the probability that a hazard will have deleterious effects. In many cases, hazards cannot be moved, but through efforts in terms of exposure and vulnerability its eventual impacts can be lessened.

Resilience: The ability of a population or area to recover from or adapt to emergencies or disasters.

This limitation can lead to important data challenges, which are discussed below. One way in which planning is accomplished is by way of assessing exposure, vulnerability and risk (see Table 10.1).

Hazards and eventual emergencies are typically classified as natural, biological or human-made. Natural emergencies may be weather-related, such as blizzards, hurricanes or tornadoes, or geologic in origin, such as earthquakes or volcanoes. Biological

emergencies are typically diseases or illnesses and human-made emergencies could be terrorist attacks but also, for example, oil spills. Many hazards or emergencies are combinations of two or more of the above: mud slides, for example, might start from a storm or earthquake but are also due to human impacts on the landscape from economic development and building. Figure 10.1 shows hazards the city of New York has judged to be important (and relevant) to the city, as

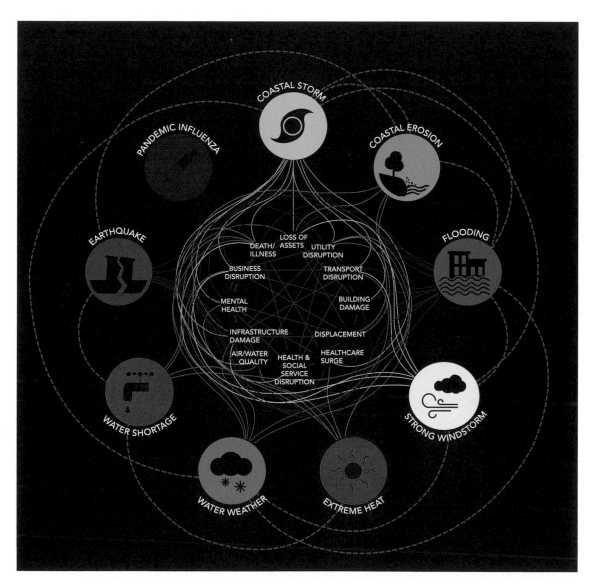

Figure 10.1 A selection of hazards and hazard impacts as identified by the New York City government

Source: www1.nyc.gov/assets/em/downloads/pdf/hazard_mitigation/nycs_risk_landscape_chapter_4_intro.pdf

well as the myriad and complex impacts these hazards can have. Note the wide range of impacts each hazard can have.

Hazards, and subsequent emergencies, can take on varied spatial signatures. Some, like volcanoes or tsunamis, occur in locations that can be very specifically delineated. Others, such as tornadoes or hurricanes, only occur in certain type of areas, but these areas can be quite large and the likelihood of occurrence in any one area might be quite small. Still other types of emergencies, such as disease epidemics, could start almost anywhere; the transmission of the illness will be an important characteristic of the emergency and the subsequent response. Finally, there are emergencies such as terrorist attacks, for which risk at any one particular location is difficult to assess, such that planning is necessarily more abstract and therefore more difficult.

Although this chapter is aimed at those wanting to learn more about GIS and emergencies, it is worth noting that the approaches discussed here – everything from data to techniques to applications – are partially or completely applicable for any sort of impact analysis. That is, there is a *status quo*, some measurable landscape of people or property or firms, and we would like to conduct a 'what-if?' analysis. In the case of emergencies, the question is 'what if this hazard occurs?' The question might also easily be adapted to ask about any sort of shock to the existing landscape (e.g. what if a water main breaks in a certain part of the city or what if a bridge closes due to storm damage?).

Data requirements

The nature of emergency planning is such that data inputs can be quite demanding. Poor data lead to poor results that do not accurately reflect vulnerability, risk or best response. Also, because emergency planning is multi-faceted (involving not only advance preparation but also response and recovery) and the nature of emergencies is multi-layered (requiring information about the hazard but also exposure and vulnerability) the range of data needed can be vast. A complete inventory of data requirements would be impossible

here. Instead, this section highlights some main types and characteristics of data needed for GIS applications in emergency management.

At a minimum, emergency planning within a GIS framework (but also elsewhere) will require information about the *hazard* and about *exposure* to the hazard. Both types of information are spatial; we need locations of hazards, as well as non-spatial attributes of these hazards. Often, additional spatial modelling will take place that estimates the spatial impact of any hazard. This modelling, which combines subject matter and GIS expertise, results in maps and data that indicate the likely severity of any hazard for a given location. The hazard risks becoming an emergency or disaster when it has some societal impact. To judge the potential societal impact, we need information about the people, property, infrastructure and environment in the area of the hazard. Hazards such as possible disease outbreaks require intermediate information about modes of transmission. Combining these layers of information allows us to measure vulnerability. In some cases, the hazard analysis might already have been completed and can be used as a finished GIS layer. The same can be said for vulnerability. On other occasions, raw data will first need to be manipulated and analysed. When visualised, the above data – hazard and vulnerability – give a sense of potential areal, economic and demographic impact of an emergency. Response, management and recovery phases will require additional data about transportation and communication networks, as well as mechanisms for on-the-ground updates.

Table 10.2 lists some of the more common types of data required for GIS use in emergency planning and shows a GIS interface with a website monitoring outbreaks of influenza. In every case, scale – temporal and spatial – is of paramount importance. Temporal scale refers to the time period the data reference. Unfortunately, although some types of hazards might not change a great deal over time, the inputs to vulnerability assessments likely do. Population size and distribution changes and the characteristics of those living in an area might shift over time. Thus, recent, complete data are important. The spatial scale of the data is also important. Hazards can impact very small

Table 10.2 Some of the more common types of data required for GIS use in emergency planning

1 **Demographic and household-level:** These data often come from national or regional statistics offices and can be complemented with local data and knowledge. Important variables are total population counts, but also information related to potential vulnerability, such as age, income and education levels, and possibly other demographic characteristics. The size of spatial unit should match the scale of analysis. So, for instance, if the goal is to identify regions susceptible to hurricanes, data at the province or county level might be sufficient. If the goal is to estimate the number of people likely to be affected by hurricane-related winds or flooding, the unit will need to be much smaller in order for results to be useful. These GIS data are, almost without exception, in vector format.

2 **Economic and civic:** This information – firm locations and characteristics, school locations, tourist sites or land parcel information, for example – could come from national statistics offices, but is more likely to be collected at the appropriate scale at the municipal or regional level. These data are usually in vector format.

3 **Environmental:** Raw data on natural hazard locations and past history of events (e.g. hurricane and tornado paths and locations) are often available via national natural resources agencies or academic institutions charged with aggregating and assimilating the necessary raw information. Regions and cities will also undertake hazard assessments. Increasingly these data are available in spatial format, as GIS gains currency as a primary emergency planning tool. These data may be in vector or raster format.

4 **Transportation and other infrastructure:** The spatial component of these data is generally available at a range of spatial scales (nation to region to city). Information about flows and use of different types of infrastructure can be more difficult to find or access.

5 **Social media:** As increasing numbers of people around the world use social media such as Twitter or Facebook to communicate, interest is growing in using these platforms to both collect information about emergencies and also communicate with areas affected. Twitter feeds on a particular topic, for instance, can be downloaded, often with locational information, either latitude and longitude or city. In the case of epidemics, diffusion of awareness as well as illness can potentially be tracked. For example, Flu Near You (https://flunearyou.org/) uses the internet and smartphone applications to help keep track of flu occurrence and symptoms in the United States. You have the opportunity to make use of social media data in Practical D.

Figure 10.2

An example of user-contributed information: Flu Near You

Source: https://flunearyou.org/

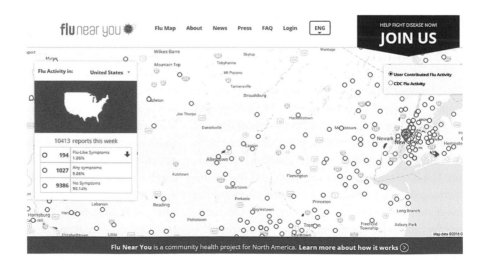

areas; vulnerability data collected at too large a scale will render any locational targeting impossible. The key is to have all data organised and ready to use so that when an emergency occurs, response-related analysis can commence immediately without the need for new or updated information.

Aside from the data issues discussed above, other aspects of emergency planning within a GIS framework also demand attention. Primary among these are challenges related to dynamic and changing situations and those stemming from user characteristics. In spite of advanced analysis and simulation, emergencies when they occur will involve innumerable unexpected elements. In addition, as response and recovery set in, the need for new, updated data grows. In short, the unpredictable and dynamic nature of emergencies and disasters means that facts on the ground are constantly changing and the need for timely data is high.

Another aspect of emergencies is the large number of actors involved at multiple levels of government and geography. GIS offers a unique environment for sharing data and coordinating responses. Challenges can arise, though, when not all are conversant with GIS analysis and when many users are contributing and using information. As both practitioners and the general population become more conversant with spatial data and as use of smartphones and internet increases, new opportunities for data collection and sharing

appear. Social media are increasingly viewed as opportunities for collecting and disseminating information about emergencies (see Figure 10.2) but other innovations are also taking place. Figure 10.3 shows Google's Person Finder, a web application initially launched after the 2010 Haiti earthquake, providing a database of 'missing' persons in the immediate aftermath of a major disaster, enabling friends, relatives, press and other agencies to help people reconnect post-disaster.

One final note about data. Emergency planning is one of the few areas covered in this book that typically requires a combination of vector and raster data (see Chapter 1 for a refresher). Moreover, the combination of data types, scales and sources can mean that a great deal of cleaning and editing are necessary before the various layers will align properly.

Examples: GIS for emergency planning

As discussed in the introductory section of this chapter, the use of GIS in both applied and academic research in emergency management and planning is vast and growing. The remainder of this chapter provides a few selected examples of how GIS informs emergency planning. For particulars on data and techniques, as well as best practices, published literature on a

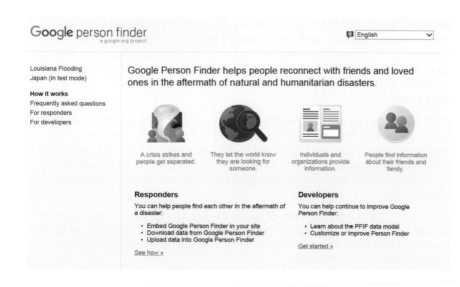

Figure 10.3 Google Person Finder: an example of how the internet and social media allow new ways of sourcing information, both spatial and aspatial

Source: http://google.org/personfinder/global/home.html

specific topic (e.g. determining evacuation routes) is best consulted for state-of-the-art information.

GIS for emergency visualisation

Visualisation – seeing data – is useful at every stage of emergency planning (see Table 10.3). Two different examples are presented below. First, proactive hurricane vulnerability and evacuation information prepared for the general New York City public and, second, estimates of impacts of Hurricane Katrina, which hit the US city of New Orleans and the surrounding area in 2005.

In late October 2011 Hurricane (or Superstorm) Sandy hit New York City, destroying homes and infrastructure and killing dozens. The disaster was a combination of a strong storm system and an enormous urban agglomeration, New York City. Much of the impact of the storm was due to its unexpected strength and effect on city infrastructure and housing. Having learned the lesson that advanced education of the public is valuable, New York City now provides interactive maps (Figure 10.4) that allow users to assess the vulnerability of different areas of the city, as well as locations of evacuation centres. The information provided in Figure 10.4 does not emphasise evacuation routes or the characteristics of those living in particular storm zones. Rather, its purpose is to encourage preparation and awareness in the general population.

Similar to Sandy, Hurricane Katrina, which struck the city of New Orleans, Louisiana, in late August 2005, was a historic example of a confluence of hurricane-related weather events, including wind and especially flooding, affecting a poorly prepared, densely populated area. The historic nature of the storm and its impacts has resulted in an unprecedented amount of research in the social sciences on inequality and how, at virtually every stage – hazard mitigation, preparation, response and recovery – impact and handling of the emergency varied by socio-economic group. Vulnerability within the context of New Orleans and Hurricane Katrina is discussed in more detail below.

Table 10.3 Useful GIS techniques and methods

As made (hopefully) amply clear above, GIS is used at every stage of emergency planning, from evaluating potential hazard impacts and vulnerability to evacuation plans to response management. Most applications will combine data sources discussed above and will apply a wide range of GIS techniques. A few relevant methods are highlighted here, alongside details of where to look in this book for more information.

Cartography and visualisation (Chapter 3) is simple but powerful. Seeing the location of hazards, the areal extent of vulnerability and even evacuation routes is among the most powerful modes of communication to the general public but also to policy makers and emergency responders. Visualisation of emergencies at every stage, from preparation to recovery, can help provide common ground for multiple stakeholders with a range of knowledge about a subject and a range of perspectives. It can also help focus attention on priority areas.

Network analysis (Chapter 4), which can be used to find the best route out of a place or the closest emergency shelter, is used not only during the emergency preparedness and response phases of planning, but also in estimating vulnerability. Just as individuals might use a network to leave an area, some locations are more vulnerable to impacts of hazards, given their connectivity to a network (think, for example, of international transmission of disease, via aeroplane routes or waterborne illness via water mains or fresh water systems).

Spatial analysis (Chapter 2) covers an enormous number of methods, but can be summarised as tools that consider the overlay of layers as well as those that are concerned with proximity to a feature or set of features. Together it is these types of tools that enable researchers to estimate the spatial impact of hazards, exposure to them and underlying vulnerability.

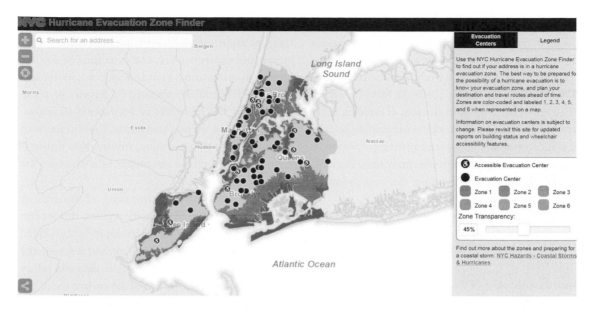

Figure 10.4 Map showing storm vulnerability and location of evacuation centres in New York City

Source: http://maps.nyc.gov/hurricane/

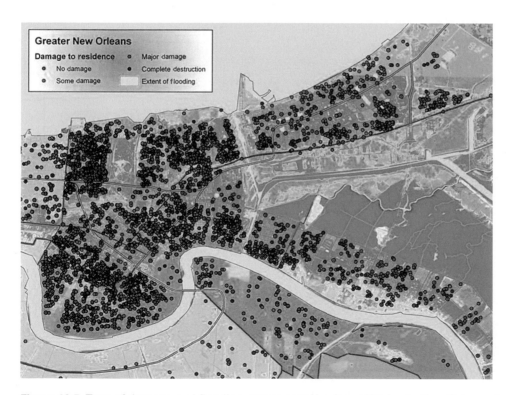

Figure 10.5 Type of damage and flooding extent post-Hurricane Katrina in New Orleans, Louisiana, US

Source: Frederick Weil's research on recovery from Hurricane Katrina. Available from: http://lsu.edu/fweil/KatrinaMaps/index.htm

This section highlights a different aspect of Katrina: the use of maps and cartographic visualisation to assess the extent and type of damage, as well as the unequal impacts. Figure 10.5 shows one element of research done by Frederick Weil at Louisiana State University. This particular map shows the area of flooding in the city and compares it to the extent of resulting housing damage. Other ways of presenting the information, as tables or figures, might also effectively show the magnitude of damage and flooding; the map shows the reader how location mattered.

GIS for evacuation

A core component of disaster or emergency response is evacuation – the process of removing individuals from affected areas. Evacuation planning requires a GIS, not only for planning of efficient routes, but also for locating evacuation centres or temporary hous-ing and assessing how the emergency might impact transportation routes. Underlying population distribution and characteristics, and even time of day, are important components to include in evacuation planning. Efficient routes, measured by distance or travel time, might not be nearly as desirable if all households in an area employ that route. Thus, in terms of effective evacuation planning, information dissemination and beforehand assessment of alternatives are important.

Rather than provide a summary of research or real-world examples of evacuation plans, this section provides an example of a tool developed by the United States Geological Survey (USGS). A federal agency, the USGS is responsible for information, data and policy relating to natural resources and the environment, but also natural hazards. Within this latter area, the agency has developed a tool – essentially an add-on or extension for ESRI's ArcGIS software – that assists with the development of pedestrian evacuation plans for natural

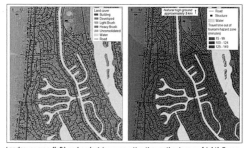

Landcover map (left) and pedestrian evacuation time estimate map (right) Ocean Shores, WA.

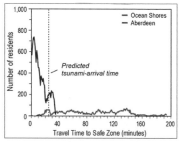

A graph comparing pedestrian evacuation time estimate for Ocean Shores and Aberdeen, WA

How do I use it?

Download the user's guide for details on the steps necessary to run the modeling. The workflow chart shown identifies the basic steps in the modeling process. The tool was used to generate the maps on the Tsunami and In the News tabs of this website.

How was the tool developed?

The tool is an ArcGIS extension that combines a Python geoprocessing toolbox with a C#/.NET 4.0 user interface and is designed for ArcGIS 10.1 SP1, Advanced version. The tool also runs on versions 9.3 and 10.2.

Figure 10.6

Snapshot of USGS's pedestrian evacuation tool capability

Source: Jones *et al.* (2014) and Wood and Schmidtlein (2013)

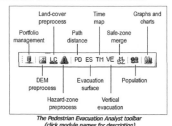

The Pedestrian Evacuation Analyst toolbar
(click module names for description)

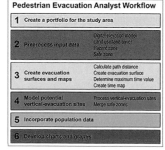

Pedestrian Evacuation Analyst workflow

hazards such as volcanoes or floods. Figure 10.6 provides an overview of the tool's capabilities. From a pedagogical standpoint, the USGS's tool is useful; it provides a sense of how a variety of spatial data is integrated, including hazard areas, elevation, land use and networks, to find paths (and distances) to areas of sufficiently high elevation. Tool inputs make clear that evacuation plans are not solely least-cost routes, but must also acknowledge hazard-specific constraints. In the case of flash flooding, for example, evacuation must be speedy but also from lower to higher ground. The USGS extension can also incorporate population data, which can be used to estimate how many people are in various time bands for evacuation.

GIS for vulnerability assessment

Vulnerability measures the 'combination of physical, social, economic, and political components that influence the degree to which an individual, community, or system is threatened by a particular event' (Wood and Schmidtlein, 2013: 1604). The measurement of vulnerability is important: not only is it used to estimate impacts of a hazard, but also to focus scarce resources for education or awareness, as well as potential mitigation efforts. Vulnerability can also affect a group's ability to evacuate an area (Wood and Schmidtlein, 2013) as well as recover from a disaster (Flanagan *et al.*, 2011). Vulnerability assessments, like the two other examples of emergency planning applications above, form their own, large literature. This section focuses on two aspects of vulnerability assessment that are important for any researcher to consider. The first is the measurement of the socio-economic aspects of vulnerability. The second is the spatial – how will individuals with a given set of characteristics respond to an emergency? Specifically, how will their characteristics affect the ability to evacuate?

Vulnerability is typically measured along socio-economic and demographic lines by developing composite indices that weight a variety of factors, including age, race/ethnicity and income. A recent paper by Flanagan *et al.* (2011) uses US census data at the tract level (small administrative units that typically contain around 4,000 people) to develop an index that

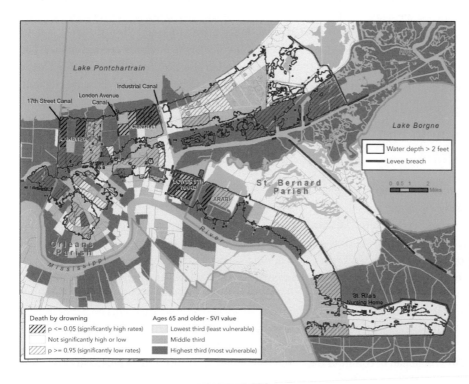

Figure 10.7
Vulnerability and drowning deaths from Hurricane Katrina

Source: Flanagan *et al.* (2011)

considers, simultaneously, socio-economic status, household characteristics, minority status and language spoken, housing and transportation. The development of the index is straightforward and *aspatial*. That is, there is no need for a GIS to calculate the measure, although the units themselves are spatial. The results, though, can be mapped to show variation in vulnerability and how one location compares to others. Moreover, locations of other variables can be compared to vulnerability scores. Figure 10.7, taken from Flanagan *et al.* (2011), shows the elderly vulnerability at the tract level in New Orleans and compares it to drowning deaths that occurred. This information would not be helpful as an emergency was under way, but provides baseline information for future emergencies and an assessment of why outcomes (e.g. drownings) were shown to vary over space.

A recent paper by Wood and Schmidtlein (2013) shows how the spatial variation in vulnerability, however measured, also plays into 'evacuation potential'. Their paper, which uses two counties in Washington – areas prone to tsunamis – uses evacuation times to help estimate exposure to potential tsunamis. The argument here is that characteristics might matter, but ability to leave an area, which is mediated by characteristics but also distance to safety, is an important component of exposure. A nice aspect of this study is that it emphasises the importance of both spatial and aspatial data in evaluating exposure and vulnerability.

GIS and social media and crowd-sourced data

A small but growing source of information in emergency planning, especially response and recovery, comes from social media and crowd-sourced data. Social media data typically refer to posts and tweets coming from Facebook or Twitter, for example. Crowd-sourced data can come from social media sources but are also the result of volunteers creating useful data products from raw information. Recent large-scale natural disasters – the Haiti earthquake (2010), the Japan earthquake and subsequent tsunami (2011) and Hurricane Sandy (2011) – all had in common the

increased use of these non-traditional data types. The Haiti earthquake is often cited as the first major disaster for which these sorts of data took on prominence and this is partly due to timing (Katrina was pre-Facebook, pre-Twitter) and also due to characteristics specific to Haiti, specifically the lack of existing spatial data to aid the response and recovery efforts (Zook *et al.*, 2010). This section is brief and its goal is simply to draw attention to a new source of information that is likely to take on a more central role in the future. To date, applied examples of Twitter's effectiveness in disaster response is unproven. Likewise, although technological capability and knowledge are increasing, crowd-sourced data are becoming increasingly important but are still secondary to traditional data sources.

Concluding comments

Concepts such as emergency, vulnerability, resilience and risk can all be viewed and studied through a spatial lens. Indeed, some aspects of each of these are inescapably spatial. Whether assessing or responding to disasters that have already occurred, or planning for response to future emergencies, GIS has much to contribute. A key aspect of research in this area is data input requirements; the methods and techniques that are used are often also employed in other subject areas and you are likely to be exposed to those throughout this textbook. Emergency planning itself is a broad topic. Those working in this area are advised to seek out further resources in their particular subfield (see below for some suggestions and a practical exercise).

Accompanying practical

This chapter is accompanied by Practical 7: Emergency planning – fire station location, which provides an opportunity to work with a road network data set in the context of fire station coverage in the Australian state of Victoria. The practical builds on previous network analysis practicals and walks you through the process of identifying areas of 'risk' of domestic fires. You will import and geocode data related to fire station locations and work with the road network to assess the provision of fire services and their accessibility to the population 'at risk'. Practical D (Part I) is also related to the content of this chapter and uses social media data from Hurricane Sandy to pre-process, import and visualise social media data related to a natural disaster.

References

All website URLs accessed 30 May 2017.

Emrich, C. T., Cutter, S. L., & Weschler, P. J. (2011) GIS and emergency management, in T. Nyerges, R. Macmaster, & H. Couclelis (eds) *The SAGE Handbook of GIS and Society*, SAGE Publications Ltd, London, 321–343.

Flanagan, B. E., Gregory, E. W., Hallisey, E. J., Heitgerd, J. L., & Lewis, B. (2011) A social vulnerability index for dis-aster management. *Journal of Homeland Security and Emergency Management.* DOI: https://doi.org/10.2202/1547-7355.1792.

Jones, J. M., Ng, P., & Wood, N. J. (2014) The pedestrian evacuation analyst: geographic information systems software for modeling hazard evacuation potential: U.S. Geological Survey Techniques and Methods, book 11, chap. C9, 25pp.

van Westen, C. J. (2013) Remote sensing and GIS for natural hazards assessment and disaster risk management, in J. F. Schroder & M. P. Bishop (eds) *Treatise on Geomorphology* (Remote Sensing and GIScience in Geomorphology Vol. 3), Academic Press, Elsevier, San Diego, CA, 259–298.

Wood, N. J., & Schmidtlein, M. C. (2013) Community variations in population exposure to near-field tsunami hazards as a function of pedestrian travel time to safety. *Natural Hazards*, 65(3), 1603–1628.

Zook, M., Graham, M., Shelton, T., & Gorman, S. (2010) Volunteered geographic information and crowdsourcing disaster relief: a case study of the Haitian earthquake. *World Medical & Health Policy*, 2(2), 7–33.

GIS and education planning

Introduction

In this chapter we look specifically at GIS for education planning, in particular at the following topics:

■ demography and pupil forecasting;
■ school catchment area analysis and the dynamics of provision;
■ education planning in an era of market systems;
■ GIS and its potential use in the analysis of geographical issues in higher education.

The chapter will primarily address GIS applications in the context of school education in urban and regional environments. A main issue to note is that around the world more and more education systems are becoming market oriented, in which parental choice and school performance become the most important drivers of change, not the decisions of the local planning agencies. As Stillwell and Langley (1999) comment, planning in the UK education sector, for example, developed far greater complexity following the introduction of the 1988 Education Reform Act and the creation of a state education system in which parents were given free choice in deciding the schools to which they would send their children. The requirement for secondary schools to publish achievement information in the form of examination results, coupled with a geography of non-fixed catchment boundaries, has resulted in a market system in which schools must effectively compete for pupils. That said, in some countries the state or local planning agencies do maintain a higher level of jurisdiction. This chapter will demonstrate the methods through which GIS and spatial modelling can be used effectively by planners to monitor the education environment and to advise on decision-making processes within whichever of the education systems are in operation.

Demography and pupil forecasting

Pupil forecasting is undertaken around the world to ensure that sufficient accommodation exists for the students within a particular area. If a new housing development is built, or current pupil levels increase/decrease due to demographics, the local education planning agency needs to be aware of the likely impacts for education provision as a whole. These factors have an obvious impact on school location and catchment area policy, hence the long tradition and importance of using spatial analytical techniques to predict changes in pupil numbers through population data, geodemographics and population forecasting techniques.

At the most basic level, schools are able to roll their attendance figures forward year on year in a straightforward manner. In a secondary school (in the UK this is typically children aged 11–16/18), this operation would enable a rough estimate of year group size to be followed through for up to seven years. Primary intake (in the UK, children aged 5–11 typically) can be simulated using birth rate data (if available) providing a guide for pupil numbers within a four-year prediction period. A slightly more committed approach to the analysis of current roll data might involve a study of student population trends and a comparison of patterns in neighbouring authorities and on a national scale. For example, while nationally pupil numbers might be increasing, regionally there is likely to be a very marked difference between the number of pupils of primary and secondary school age and the magnitude of change over time. In the UK, for example, by using sub-national population projections produced by the ONS, the Department for Education reported that between 2012 and 2017 the primary aged population was forecast to increase by between 9% and 15% depending on region, while the population of secondary school aged pupils was set to decline.

Cohort-survival techniques are widely used in population forecasting generally. Cohorts in this context refer to age groups: 0–4, 5–11, 12–18, etc. In relation to school planning, Simpson (1987) argues survival rates work in the following way:

Compared to the births in area X, how many children will appear five years later in school Y; compared to the five-year-olds now in school Y, how many six-year-olds will be there next year; and so on. The survival rate would not usually be exactly 100 per cent because some children move home and school, some children attend private schools and after 16 children leave school. The birth-to-school survival rate is often called an 'arrival rate', while the post-15+ survival rates are usually called 'staying-on rates' . . . It represents the net effect of migration, dropping-out, mortality and transfers between the state system and other schools: it estimates the balance of children who are lost and gained to the system at each age.
(Simpson, 1987: 67–68; also see Simpson, 1988)

A greater level of understanding can be achieved by incorporating other influences on student numbers, in particular migration (where the primary factor is the development of new housing). The process uses the same current school attendee information rolled forward by year as the basic approach for calculating the number of pupils by postcode-based catchment area. This number is then supplemented by forecasting for new housing developments from the first 'future allocation' of land to housing – this gives an approximate number of houses likely to come on stream around five years later. Exact numbers and house types to be built are only known when final planning permission is granted, and are used to provide a last revision to earlier indicators. It is this number, and the nature of the new housing, which are the greatest influences on likely pupil numbers. For example, a new development of three-bed semi-detached housing is likely to generate far greater pupil increases than a block of sheltered accommodation for the elderly. Table 11.1 shows the general rules for predicting student numbers adopted by one UK local authority: Norfolk County Council (in East England). The overall pupil number forecasting approach of Norfolk County Council is illustrated in Figure 11.1.

It is interesting to note that new housing developments will not necessarily result in increased pupil numbers in local schools. This is especially the case in

Table 11.1 Norfolk County Council, UK, predicted pupil number multipliers for catchment areas with new housing developments

School type	First	Infant	Junior	Primary	Middle	11–16+ High
Development <20 houses	5.8	4.3	5.3	9.6	4.8	10.9
Development >20 houses	15.1	11.3	14.1	25.4	13.7	17.4

Source: Norfolk County Council

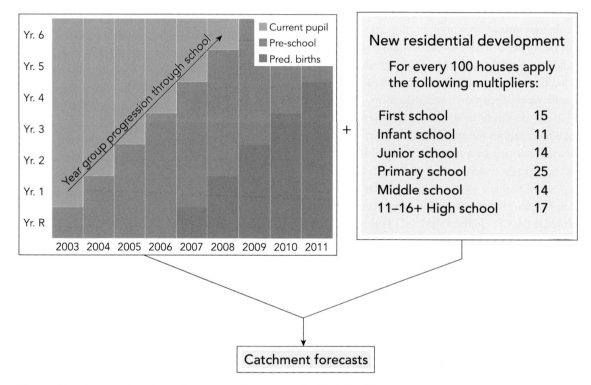

Figure 11.1 Generation of catchment pupil forecasts in Norfolk, UK

Source: Norfolk County Council

countries that have a market system in education (see below for further discussion). Gulson (2007) gives a very good example of this in relation to urban renewal in inner-city Sydney in the 2000s. Such renewal brought new houses and a greater population to many inner-city Sydney communities (a process of regeneration and gentrification), but as these new residents were more middle class there was a tendency not to use the local schools with poorer exam scores and reputations, but to send children across the city to the better performing schools.

GIS can be used to aid the process of pupil number forecasting at all levels. The most basic application might involve a map of current population figures overlaid by school and catchment locations. This simple presentation would afford planners a greater understanding of pupil distribution in relation to current catchments. GIS could also be used to locate a new housing development (position and size) and allocate developments and their pupils to current catchments and schools. In a council or authority situation, a cross platform GIS might aid the better integration between departments to give a more unified approach to planning issues through the encouragement of data compatibility and sharing. A bespoke system could also offer better facility for data entry and display with standard automated features to deal with common inputs such as term or annual pupil counts, class sizes, current roll and leavers' roll.

A broader approach to geographic information in a computing environment might use further data input to aid understanding of flows and patterns and cater more efficiently and effectively for pupils' needs. GIS might be used to illustrate geodemographic information as a basis for a modelling environment. This approach might incorporate birth rate patterns and data on parents or those of parental age to monitor and predict what demographic are likely to purchase which property type and have children of a certain age. Such a system could also analyse and forecast the likelihood of pupils being drafted into private schooling or those continuing to higher education.

School catchment areas and the dynamics of provision

As detailed above, school networks and service provision are inextricably linked to the location and dimension of the student population. Due to the investment in schools often over tens or hundreds of years, and the long-established relationships between populations and primary and secondary schools, it is not feasible to completely redesign a school network on the basis of an optimisation approach (unless we are considering an unlikely new town scenario). However, such an exercise might be appropriate in developing countries, especially as plans to expand access to education come to fruition. The Kenyan School Mapping Project has used GIS to create a database of school locations, resources, teachers and pupil numbers across Kenya (Mulaku and Nyadimo, 2011). This baseline information captured in a GIS provides decision makers with the ability to answer questions such as: where should new schools be placed? Where are teachers required? Are the school resources adequate? Is current provision adequate? Hite (2008) looks at the

potential for using a GIS system for educational micro planning in Nepal and Uganda, arguing that, given the increasing use of tools to visually represent the world, there will be

> a move in educational micro-planning efforts toward more visually-conceptualised and oriented planning at all levels of decentralisation. In this regard, it is reasonable to expect that GIS and other user-controlled visualisation techniques and solutions should become more prominent in education micro-planning efforts in the near future.
>
> (Hite, 2008: 16)

As noted above, any changes to future provision must be based on the current school network and are driven by population changes and school performance indicators. Potential interactions in this respect include expansion, downsizing, openings, closures, mergers and divisions, all of which will have implications for demand and supply functions in other education facilities or catchments. GIS could be applied in the approach to education planning by offering a better understanding of relationships between pupil populations and school locations through visualisation and analytical techniques. An example of this application might be best realised in the planning of a new school location or the selection of an existing location most suitable for extension or investment, a situation considered by Cropper (2003). This commentary reviews potential uses of GIS facilities, with suggestions including location of student populations, and analysis of available and affordable land through aerial photos and data on acreage and current appraised values. As the author rightly declares, all of these operations minimise searching and planning time while ensuring an extremely high rate of accuracy. In addition, Cropper (2003) proposes GIS for the investigation of:

- potential causes of special education population;
- integration plans for de-segregation orders;
- plans for future school sites;
- transportation routes;
- student/school interactions.

The fundamental concept when assessing the need for new schools is that of capacity. This refers to the number of pupils that can be accommodated at individual schools and within catchments and the region as a whole. This information is vital as decisions regarding locations, networks and resource allocation must be based on a sound understanding of the relationship between demand and supply (i.e. pupil population and available accommodation). An interesting investigation has been undertaken across the city of Chicago in response to concerns over disparities between education supply and demand. Capacity is calculated by the Chicago school board as a formula accounting for the number of classrooms, courses scheduled and the maximum class size recommended by teachers' union contracts. Comparing information on capacity vs. enrolment provides an index of resource utilisation, revealing schools that enrol greater than 100% of design capacity as overcrowded. Further investigation has shown that several of these schools are over attended by choice and tend to attract students from outside of their attendance boundaries. Results of this trend have become apparent in other sectors whose schools are now suffering under-utilisation and forced closure, which is a pressing concern for the city's education management authorities who must ensure that all areas and pupil populations are adequately served by educational facilities. Fortunately the phenomenon has now been identified and recognised as a geographical issue, offering great potential for solutions by GIS application. Information on student population, capacity and location, as well as data on school closures, renovations and upgrades can be used to better manage attendance catchments and future planning of facilities and resource management. The first stage might be to map existing data and recent shifts in attendance patterns to improve understanding of the changes in pupil/parental preference or forced movements due to closures/openings. Scenario applications could also be used to interactively manipulate school networks, considering situations for more favourable provision through adjustment of school location and facility or catchment boundaries.

Of course population change can involve massive population decline, putting great pressure on the future of schools in certain locations. Lubienski and Dougherty (2009) use GIS to plot population loss in New Orleans after the devastating Hurricane Katrina in 2005. Figure 11.2 shows how many schools (of different types) are located in the areas of high population loss, putting great strain on the future viability of each one.

When considering school performance and catchment areas GIS can offer useful additional insights (although see the following section for a more detailed discussion of GIS and school performance in market systems). Peters and Hall (2004), for example, demonstrate the usefulness of GIS for representing and improving the quality of education provision in Peru. They provide an analysis of education quality and what they refer to as neighbourhood well-being. They first use a series of different indicators such as number of computers, student-to-teacher ratio, presence of a library, etc. to produce an education quality indicator for each school. They produced a similar indicator for neighbourhood quality based on 13 socio-economic inputs. Plotted together these indicators show clear spatial inequalities. When correlated, they demonstrate that a statistically significant relationship exists between the two. The analysis also allows for the identification of individual schools that urgently require more resources.

An interesting aspect of school catchment area analysis is planning transportation. In England, local education authorities are obliged to provide public transport to pupils from outside a two- to three-mile radius of the school (depending on pupil age). All pupils within this radius are expected to find their own way to school. GIS could be used here to:

- create spatial buffers at various distances from the school and to generate a list of pupil names and addresses for students entitled to transport services;
- work out the shortest walking route from one point to another (using network analysis described in Chapter 4);
- analyse the most efficient routes for buses to take in order to collect pupils who qualify for free transport.

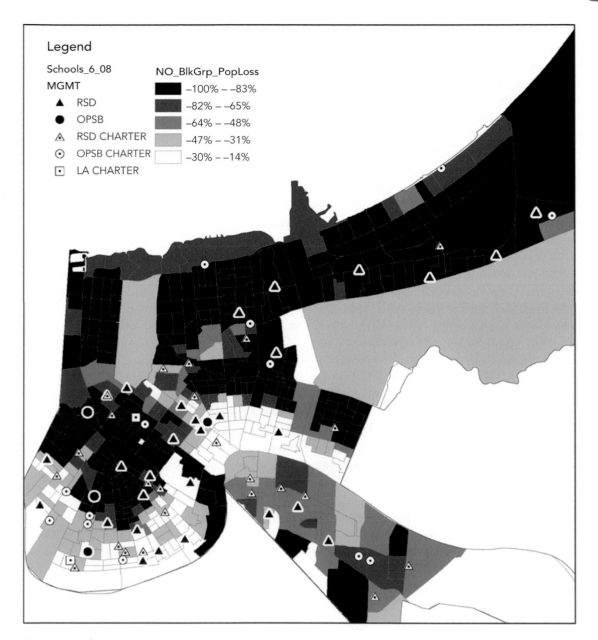

Figure 11.2 School locations plotted against population decline in New Orleans, Louisiana, US

Source: Lubienski and Dougherty (2009)

In the US, many schools run their own student transport (where suitable public transport does not exist), or fund provision of pupil transport services directly from other providers. The ArcNews website features a 2001 article on GIS adoption by school districts. The article introduces the application of GIS products by the Central Valley School District (Spokane, Washington) in student routing and management of transportation information. School facility information, route maps and geocoded student data are the main inputs to the system, enabling a variety of analyses to be reviewed. One of the claims of this site suggests that adoption of GIS reduced the time taken for report production by a week and revealed that the elimination of five or six bus routes, through more efficient planning of routes, could realise savings of up to $125,000.

Individual postcode-level data are not always available for each pupil's home address, although if it is then plotting the precise geographical location of pupil addresses will permit a disaggregate analysis of patterns and provide some insight into the true nature of school catchment boundaries and transport mechanisms. Such a system has been implemented in Forsyth County, Atlanta. When a new student enrols with a school, or a student moves address, the details are geocoded and the system assigns the student to a specific bus route and stop. Details of the system are available at Directions Magazine (2015) (see also Nayati (2008) for an interesting application to school bus routing in Hyderabad in India, and a neat overview of studies in this area by Park and Kim (2010)).

A further advanced approach to GIS and spatial analysis application in the education planning field involves the use of gravity or spatial interaction models to analyse the impact of school openings and closures, and also in modelling optimum sizes, numbers and location of schools (see Chapter 8 for more details on these models in a retail context). York Local Education Authority in the UK has used such spatial models and GIS in consultation exercises with the public when revising an old system of catchments. In this project, four proposals were analysed for their impacts using GIS. The first stage in this process was to derive a set of mathematical functions, coefficients and constraints to effectively model the current situation of flows or trips from pupils' homes to school facilities. This was best achieved by sub-dividing a study region into suitable geographical boundaries and considering flows to and from zone centroids. Following calibration and testing with observed data, such a model can be used to run scenarios or impact assessments to study how the network would respond to change, most likely in the form of adjustment to demand (e.g. new housing development) or supply (e.g. opening/closure of a school facility). Such a model could also react to changes in destination site attractiveness, for example, if a school

were to improve performance and hence increase the attendance demand. The simplest form of attractiveness indicator might be school size or roll, which does act as a proxy for other school features such as facilities and funding. Another useful variable in this context would be a performance figure such as proportion of students achieving more than five A*–C grades at GCSE level. Stillwell and Langley (1999) also examined a spatial interaction approach to modelling the education system in Leeds. Having calibrated a model with existing flow data, they studied the effects of several scenarios:

- Removing 200 pupils from the population of one central city postal district revealed a significant impact on the city as a whole with declining student numbers in most facilities.
- A single large school closure in a suburban ward resulted in attendance increases across all city zones.

The authors recognise the unlikely situation of such large-scale events, but claim that the model evidence is useful to planners in providing confirmation that a number of small local shifts could lead to significant alterations to the system as a whole.

The remaining examples of geographical analysis in education planning refer to more advanced approaches of spatial modelling, scenarios and optimisation, and aim to broaden the understanding of the concerns and complexity of school planning issues. Pacione (1989) demonstrated the relative accessibility of secondary schools to the residents of 104 postcode sectors in Glasgow using a variation of the SIM (again see Chapter 8 for more details on these models). An index of accessibility extended the model by accounting for the relative access to cars and public transport, and the population of the neighbourhoods. Impacts of projected school closures on the resulting *accessibility surface* were evaluated by comparing provision prior to and after school closure. This research highlights the need to recognise the social as well as the spatial dimensions of accessibility, through the fact that two neighbours within the same locality can experience different levels of access depending on social grouping.

Add to this the need for consideration of changing pupil populations and school capacities and the traditional element of school relationships that were mentioned in an earlier section, and the true complexity of the school planning strategy can be realised. Only with the aid of computational functions, data manipulation and mapping facilities can we realistically hope to account for all these potentially important factors in the management of educational planning.

Finally, work by Harland and Stillwell (2007) was concerned with analysis of the PLASC (Pupil Level Annual School Census) database, in particular to look at the movements of pupils between their homes and schools by ethnic group and over time. They note a number of challenges in adapting SIMs to account for the specific characteristics of the education sector. These include school capacity constraints, school admissions criteria and complex drivers of school choice decision making by pupils and parents. Nevertheless, they build a 'Spatial Education Model' which they demonstrate can be used to predict or forecast future school roll and capacity constraints (Harland and Stillwell, 2008).

Educational performance in market systems

As noted in the introduction, many countries now have a more market-oriented education system where schools compete for pupils based on parental choice. At the heart of such systems are school performance indicators. These can include discipline record, extra-curricular activities, truancy rates, ambience, etc., but most important are league tables of school success (in many countries the outcomes of exams at various key stages in academic development). League tables are intended to consolidate information on school performance and education quality, and a web-based GIS system is used by the UK Department for Education, for example, to display this information geographically (Department for Education, undated).

In many countries the introduction of market systems seems to have led to increasing social segregation as the higher income groups choose the better schools (or are chosen by the schools themselves if they have too many applications for limited places). Waslander and Thrupp (1995) were among the first to demonstrate this following the introduction of markets in New Zealand education in 1988 (also see an update in Thrupp, 2007). Since then a wealth of UK studies have shown the same (Ball *et al.*, 1995; Ball, 2003, 2013; Taylor, 2009; Butler et al., 2007; for US studies see Bell, 2009). In spatial terms, it is not surprising perhaps that the schools that do the best in educational attainment terms are those in middle and higher income areas where students are more motivated and have had a more advantageous upbringing, including better parental support, better knowledge of the education system by those parents in general, access to more financial resources to attend better schools and perhaps more chance to attend pre-school educational establishments. In contrast, such market systems tend to produce a set of schools that struggle to attract high performing pupils, and are normally in working-class areas, which lack many of the attributes described above for the wealthier areas. These schools have less impressive exam success and often face a spiral of decline as resources are shifted to the better schools, as resources normally follow pupils. Clarke and Langley (1995) show an example of the catchment areas for two schools in Leeds, UK, following the full market system coming into operation in the UK in 1988.

Figure 11.3 shows that the school in North Leeds (top diagram), which has high attainment scores, is attractive to the higher income residents right across the north of the city. The bottom diagram shows the opposite is the case for a poor performing school in working-class inner-city Leeds – this has pupils drawn from only the immediate catchment, pupils perhaps trapped into having very few effective alternative choices. Parsons *et al.* (2000) also use GIS to plot inflows and outflows in an anonymised UK local authority area and show the vast number of non-local flows to school each day, especially to better performing schools.

A number of other studies have linked GIS to regression models (also see Chapters 5 and 8) in order to try to show a more robust statistical relationship between school attainment (success) levels and various

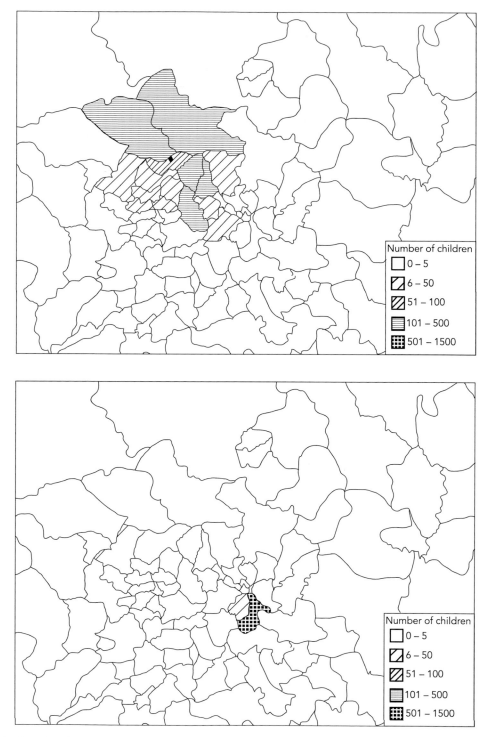

Figure 11.3 Mapping flows of pupils to two schools in Leeds, UK, following the introduction of the full market system

Source: Clarke and Langley (1995)

socio-economic small-area indicators (i.e. Gordon and Monastiriotis, 2007). A difficulty with such analysis is that it is hard to estimate the nature of catchment area boundaries without actual pupil address data. Yet, it is important to try this:

> The performance indicators cannot be considered in isolation from the social characteristics of its local area because the background of the pupils has a clear influence upon how well a school performs. It is, therefore, important to develop methods that link together the different geographies of the school performance tables and the census data.
>
> (Pearce, 2000: 302)

Pearce (2000) showed one way GIS could help with this issue – he drew weighted Thiessen polygons (see Chapter 8 for more details on this technique) around schools in his study area of Lancashire, UK. Figure 11.4 shows the results.

Using these polygons he was able to link census data to school catchment areas and undertake regression analysis (see also Flowerdew and Pearce, 2001). While the use of global regression analysis is common, Fotheringham *et al.* (2001) applied their GWR technique to school performance data in north-east England. They showed that the power of individual key socio-economic variables used to 'explain' poorer exam success (low income, unemployment, no cars, etc.) could vary spatially, thus producing a more powerful and interesting local set of regression models.

A number of other studies have examined school performance by geodemographics. One of the more powerful is Butler *et al.* (2007), in East London, UK (though see Hamnett *et al.* (2007) and Webber and Butler (2007) for similar discussion). They used individual pupil addresses from the six top- and bottom-performing schools and profiled those students by the Mosaic geodemographic classification (which contains categorisations such as 'symbols of success' for higher income households and 'blue-collar enterprise' to represent working-class households). The results showed that generally school results were better for those schools closer to the wealthier suburban edge, as we might expect, but that a number of schools seemed to be perhaps punching above their weight in less well-off areas of East London (showing of course that school teachers, school policy, etc. can make a difference and improve performance in working-class environments).

In such a market environment, and given the arguments of how the market is playing out in many countries outlined above, Mayston (2003) argued that the assessment of performance needs to adequately account for the complexities of the real world and that the simple performance league tables published by countries such as the UK apply pressure to individual

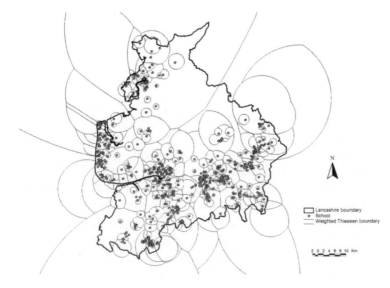

Figure 11.4 Weighted Thiessen polygon boundaries around schools in Lancashire, UK

Source: Pearce (2000)

schools, as slipping within the national tables can have adverse effects on their ability to generate future funding (again, note that normally resources follow pupils in such quasi-market systems). GIS can provide a more complete review by adjusting data for these catchment area factors. The aim with such work is to try to show these poorer performing schools in a better light which may help their future attractiveness – so the exam scores could be repackaged for example as 'actual performance against expected performance'. The higher the number of low-income residents in a catchment, then the lower expectations of exam success probably are – so if a school can show a better actual performance against expected outcome then this will offer a more positive light for marketing purposes. This exercise can be facilitated by incorporating more data into a model environment and might use, for example, deprivation indicators to give greater insight to real conditions. Eligibility for free school meals is another useful variable for assessing socio-economic structure (in the UK) in this context.

A good illustration of this kind of work is provided by Higgs *et al.* (1997) in Wales. They used a multi-regression approach to estimate what they felt a benchmark exam score was for each school in Glamorgan, South Wales. They could then compare with actual results to see if the school was in effect over- or under-performing. For example, a school in the most deprived community might only be predicted to have 20% of its pupils attain five or more GSCEs grade A*–C (the standard UK performance indicator). If, in reality the school achieves a 30% pass rate then it is doing much better than predicted – this type of analysis could then produce an alternative performance list based on 'distance from expected' (much like the example with crime lists presented in Chapter 7). Brimicombe (2000) also produces a new index of contextualisation for school exam scores in north-east England. Instead of using Thiessen or Voroni polygons for defining school catchment areas he uses a process called kriging. The net result is similar to Higgs *et al.* (1997) – a set of schools doing better than 'expected' and a set performing worse than 'expected'.

For other interesting attempts to model school catchment areas in market systems of education see Harris and Johnston (2008) and Singleton *et al.* (2011).

GIS in higher education

As mentioned at the beginning of the chapter, the issue of planning for higher education is associated more with regional and national student flows than with the local residential catchment area. This seems especially so in the UK where students are more unlikely to stay at home and attend higher education establishments in comparison to other European countries and the US and Australia. In terms of GIS analysis, many issues for higher education establishments are the same as for primary and secondary schools. For example, higher education establishments follow a similar approach for student forecasting to those adopted by schools and local planners. The University of Leeds, for example, recognises the importance of understanding the student population to support planning across departments and campuses. Estimates are based on current student counts and historical projection rates as well as forecast intakes for each individual course. These forecasts reflect market intelligence, growth targets, specific initiatives and promotional activities, and are used to predict fee incomes and financial margins as well as academic teaching space and accommodation needs. The university is determined to maintain its commitment to offering accommodation for all first-year students and projections are essential to meet these requirements.

In addition to roll forecasting, as with primary and secondary schools, many higher education establishments might be interested to use GIS to plot their catchment areas. In the UK, as noted above, these are likely to extend across the country. Figure 11.5 maps the home locations of students at the University of Manchester in 2005–2006.

Once the data are mapped as in Figure 11.5, the marketing team at the university can perhaps try to figure out why flows are high from certain feeder locations compared to others. This might relate to the spatial distribution of students with the highest educational attainment rates at secondary schools. It is not

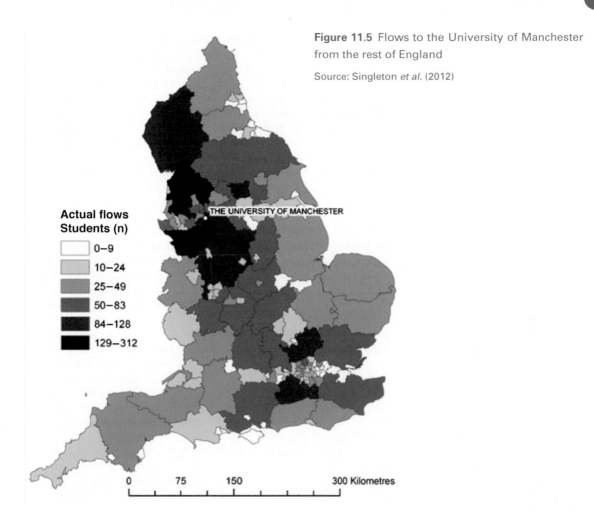

Figure 11.5 Flows to the University of Manchester from the rest of England

Source: Singleton *et al.* (2012)

Actual flows Students (n)

	0–9
	10–24
	25–49
	50–83
	84–128
	129–312

THE UNIVERSITY OF MANCHESTER

0   75   150   300 Kilometres

surprising therefore that there are a number of studies linking GIS with geodemographic profiling (Singleton, 2010; Brunsden *et al.*, 2011). They could also calculate indicators such as market shares (as the original totals or outflows will be known) and thus build decision support systems similar to those found in retailing (see Chapter 8).

increase understanding of past trends and current patterns, as well as the potential for modelling of future scenarios. To this end, we have considered how spatial analysis, modelling and GIS can be applied to analyse pupil populations and forecasts, school networks and facilities to replicate interactions of demand and supply in primary, secondary and higher education.

Concluding comments

Despite the challenges faced by education and planning departments, this chapter has illustrated a strong case for the implementation of GIS and spatial analysis in the education sector. Such technologies, when combined with the appropriate data, offer advantages to

Accompanying practical

This chapter is accompanied by Practical 7: Education provision and planning, giving you the opportunity to apply your spatial analysis skills to a series of data sets related to education provision. We guide you through the process of handling spatial data related to school

catchment areas, school performance and access to educational opportunities, using data related to a specific administrative area within the UK. The data sets used contain more complex variables than some of our previous practicals, giving you the opportunity to develop your spatial and attribute data handling skills.

References

All website URLs accessed 30 May 2017.

Ball, S. J. (2003) *Class Strategies and the Education Market: The Middle Classes and Social Advantage*, Routledge, London.

Ball, S. J. (2013) *The Education Debate*, Policy Press, London.

Ball, S. J., Bowe, R., & Gewirtz, S. (1995) Circuits of schooling: a sociological exploration of parental choice of school in social class contexts. *The Sociological Review*, 43(1), 52–78.

Bell, C. (2009) Geography in parental choice. *American Journal of Education*, 115(4), 493–521.

Brimicombe, A. (2000) Constructing and evaluating contextual indices using GIS: a case of primary school performance tables. *Environment and Planning A*, 32(11), 1909–1934.

Brunsdon, C., Longley, P., Singleton, A., & Ashby, D. (2011) Predicting participation in higher education: a comparative evaluation of the performance of geodemographic classifications. *Journal of the Royal Statistical Society: Series A (Statistics in Society)*, 174(1), 17–30.

Butler, T., Hamnett, C., Ramsden, M., & Webber, R. (2007) The best, the worst and the average: secondary school choice and education performance in East London. *Journal of Education Policy*, 22(1), 7–29.

Clarke, G. P., & Langley, R. (1995) A review of the potential of GIS and spatial modelling for planning in the new education market. *Environment and Planning C*, 14, 301–323.

Cropper, M. (2003) GIS in School Facility Planning, Peter Li Education Group. Available from: https://webspm.com/articles/2003/02/01/gis-in-school-facility-planning.aspx.

Department for Education (undated) www.education.gov.uk/schools/performance.

Directions Magazine (2015) www.directionsmag.com/pressreleases/onegis-implements-school-bus-route-website-with-forsyth-county-georgia-boar/109482.

Flowerdew, R., & Pearce, J. (2001) Linking point and area data to model primary school performance indicators. *Geographical and Environmental Modelling*, 5(1), 23–41.

Fotheringham, A. S., Charlton, M. E., & Brunsdon, C. (2001) Spatial variations in school performance: a local analysis using geographically weighted regression. *Geographical and Environmental Modelling*, 5(1), 43–66.

Gordon, I., & Monastiriotis, V. (2007) Education, location, education: a spatial analysis of English secondary school public examination results. *Urban Studies*, 44(7), 1203–1228.

Gulson, K. N. (2007) Repositioning schooling in inner Sydney: urban renewal, an education market and the 'absent presence' of the 'middle classes'. *Urban Studies*, 44(7), 1377–1391.

Hamnett, C., Ramsden, M., & Butler, T. (2007) Social background, ethnicity, school composition and educational attainment in East London. *Urban Studies*, 44(7), 1255–1280.

Harland, K., & Stillwell, J. (2007) Using PLASC data to identify patterns of commuting to school, residential migration and movement between schools in Leeds. *Working Paper 07/03*, School of Geography, University of Leeds. Available from: www.geog.leeds.ac.uk/fileadmin/documents/research/csap/working_papers_new/2007-03.pdf.

Harland, K., & Stillwell, J. (2008) *Commuting to School: A New Spatial Interaction Modelling Framework for the Education Sector*. NCRM Research Methods Festival 2008, 30 June–3 July, St Catherine's College, Oxford. (Unpublished.)

Harris, R., & Johnston, R. J. (2008) Primary schools, markets and choice: studying polarization and the core catchment areas of schools. *Applied Spatial Analysis and Policy*, 1(1), 59–84.

Higgs, G., Bellin, W., Farrell, S., & White, S. (1997) Educational attainment and social disadvantage: contextualising school league tables. *Regional Studies*, 31, 775–789.

Hite, S. (2008) School mapping and geographical information systems in education micro-planning. Paper presented at the Symposium on Directions in Educational Planning, UNESCO International Institute of Educational Planning (IIEP), Paris, 3–4 July.

Lubienski, C., & Dougherty, J. (2009) Mapping educational opportunity: spatial analysis and school choices. *American Journal of Education*, 115(4), 485–491.

Mayston, D. (2003) Measuring and managing educational performance. *Journal of the Operational Research Society*, 54(7), 679–691.

Mulaku, G., & Nyadimo, E. (2011) GIS in education planning: The Kenyan School Mapping Project. *Survey Review*, 43, 567–578.

Nayati, M. A. K. (2008) School bus routing and scheduling using GIS. Available from: diva-portal.org.

Pacione, M. (1989) Access to urban services: the case of secondary schools in Glasgow. *Scottish Geographical Magazine*, 105, 12–18.

Park, J., & Kim, B. I. (2010) The school bus routing problem: a review. *European Journal of Operational Research*, 202(2), 311–319.

Parsons, E., Chalkley, B., & Jones, A. (2000) School catchments and pupil movements: a case study in parental choice. *Educational Studies*, 26(1), 33–48.

Pearce, J. (2000) Techniques for defining school catchment areas for comparison with census data. *Computers, Environment and Urban Systems*, 24(4), 283–303.

Peters, P. A., & Hall, G. B. (2004) Evaluation of education quality and neighbourhood well-being: a case study of Independencia, Peru. *International Journal of Educational Development*, 24, 85–102.

Simpson, S. (1987) School roll forecasting methods: a review. *Research Papers in Education*, 2, 63–77.

Simpson, S. (1988) The use of school roll forecasts in LEA administration: the allocation of resources to schools. *Cambridge Journal of Education*, 18(1), 89–98.

Singleton, A. D. (2010) The geodemographics of educational progression and their implications for widening participation in higher education. *Environment and Planning A*, 42(11), 2560–2580.

Singleton, A. D., Longley, P. A., Allen, R., & O'Brien, O. (2011) Estimating secondary school catchment areas and the spatial equity of access. *Computers, Environment and Urban Systems*, 35(3), 241–249.

Singleton, A. D., Wilson, A. G., & O'Brien, O. (2012) Geodemographics and spatial interaction: an integrated model for higher education. *Journal of Geographical Systems*, 14(2), 223–241.

Stillwell, J. C. H., & Langley, R. (1999) Information and planning in the education sector, in J. C. H. Stillwell, S. Geertman, & S. Openshaw (eds) *Geographical Information and Planning*, Springer, Heidelberg, 316–333.

Taylor, C. (2009) Choice, competition, and segregation in a United Kingdom urban education market. *American Journal of Education*, 115(4), 549–568.

Thrupp, M. (2007) School admissions and the segregation of school intakes in New Zealand cities. *Urban Studies*, 44(7), 1393–1404.

Waslander, S., & Thrupp, M. (1995) Choice, competition and segregation: an empirical analysis of a New Zealand secondary school market, 1990–93. *Journal of Education Policy*, 10(1), 1–26.

Webber, R., & Butler, T. (2007) Classifying pupils by where they live: how well does this predict variations in their GCSE results? *Urban Studies*, 44(7), 1229–1253.

12 GIS and transport analysis and planning

Introduction

Transport plays an undeniably vital role in contemporary society as among some of the most commonly used services provided by both national and local governments. Most modern cities are a result of some historic transport feature, for example river crossings and road or rail intersections, and must consider transport provision for development and planning at every level including the provision and maintenance of pavements (sidewalks) and local roads, traffic signals and signs, pedestrian crossings, car parks, cycle lanes, transport interchanges, local air quality, road safety and bus services. The provision of, and rights to, private and public transport, and easy access to goods and services, are often taken for granted, but have realised greater concern in recent times with increasing issues of congestion and pollution.

Transport systems are inherently spatial and are particularly well suited to planning, analysis and management through developing GIS and spatial analysis, a relationship that forms the content for this chapter. We should first, though, acknowledge that transport has been a core component of many previous chapters, especially when we have discussed accessibility to goods and services which often consider some kind of transport network. We first look at the fundamental elements of movement, and more specifically the reasons that trips are made and the networks that facilitate these movements. The chapter will then draw this information together in the investigation of a range of GIS applications to control and manage the many facets of transport planning in the urban or regional environment.

As GIS has become more widespread in transport planning so we have witnessed a number of useful reviews of the history and development of applications in this field which make useful complementary reading to this chapter (i.e. Goodchild, 2000; Thill, 2000; Nyerges, 2004).

Flows on networks

The understanding of movements or flows is a concept fundamental to the field of transport planning, but is a subject that will never be static or completely controlled due to the random nature of human factors. The best representation that we can hope to achieve is by making generalisations about the volume, frequency, timing and purpose of trips. Fortunately for planners, this method is fairly consistent and reliable. For example, at the simplest level, we know that road and rail links into and out of city centres are likely to host major flow events at 'rush hours' on weekdays due to worker commuting patterns. Table 12.1 illustrates an attempt to classify personal trips by destination and mode. Although somewhat dated, it is still very relevant and suggests that a large proportion of personal journeys consist of local and suburban movements, usually for the purpose of everyday activities such as work, school, shopping and visiting friends and relatives. With increased access to cars and improved provision of public transport, populations have become more mobile, facilitating greater ease and frequency of trips.

The controlling factor to movement will continue to be cost and, as we will see later in this chapter, there is a function of cost termed a 'distance decay' factor which determines how far people are prepared to pay to travel for a particular good, service or function. As an extreme example, employees could not justify living over a certain distance to work if their travel costs exceeded income or if time taken for the trip was unfeasible. This parameter provides a guide for flow modelling on a regional or national basis but might be of less importance in an urban environment where goods, services and consumers are fairly well concentrated and connected. Despite this assertion, the distance decay factor could be argued to work at all levels depending on trip purpose; for example, a consumer is unlikely to drive or take a bus across a metropolitan landscape for a pint of milk, but would probably make this journey for work or leisure purposes.

Unfortunately for the planner, information on personal movements can be difficult to come across or costly to acquire. This could be a blessing in disguise when considering the complexity and volume of information that such a data set would generate,

Table 12.1 Characteristic residentially based intra urban personal movements

Spatial zone	Daily journeys	Weekly journeys	Infrequent journeys	Modes
Local	Convenience shop	Food shops	Clinics and doctors	Walk
	Primary school	Clubs	Parks	Car
	Friends and relatives	Pubs		Bike
	Pubs	Places of worship		
	Work			
Suburban	Secondary school	Food shops	Durable goods shops	Bus
	Work	Personal business	Personal business	Car
		Friends and relatives	Friends and relatives	Bike
		Sports	Entertainment	Walk
Urban	Work	Friends and relatives	Specialist shops	Car
	Higher education	Sports	Entertainment	Bus
				Rail
Extra-urban	Work		Specialist shops	Rail
			Entertainment	Car

Source: Adapted from Daniels and Warnes (1980)

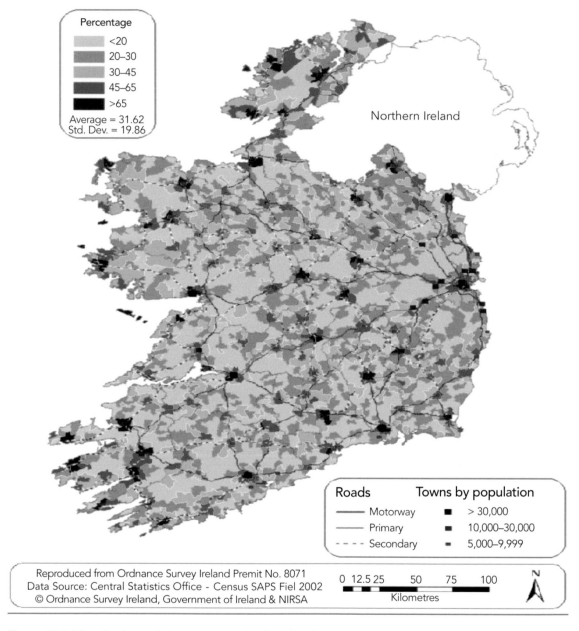

Persons travelling 0–4 miles to work

as a percentage of total at work, 2002

Figure 12.1 Mapping long-distance commuting in Ireland

Source: Walsh *et al.* (2006)

and the realisation that such data would require spatial aggregation before any meaningful analysis could be undertaken (although the increasing amount of data becoming available from media sources such as mobile phone or 'twitter' data should provide a new era of 'big data' analysis – see Birkin *et al.* (2017)). The standard format for flow data operates under geographic subdivisions such as census wards or tracts or some zoning system based on postal geography as origins and destinations.

A primary source of this type of flow data in many countries is the journey to work data set collected by the various census agencies. In the UK, for example, it is possible to get workplace and travel statistics for commuting flows from census captures in 1981, 1991, 2001 and 2011. These data have the potential to address many issues regarding commuter flows and hence the situations that normally generate maximum demand on transport networks. One such example is described by Horner (2007), who used data from the 1990 and 2000 US Census Transportation Planning Package (CTPP) to examine the flows of commuters in the Tallahassee region of the US. He looked at the change in minimum, average and maximum commute distances over the ten years, all of which increased. Horner (2007) interpreted this change as an increase in the level of dispersal of jobs and housing over the decade. He also carried out analyses at the local level and found that areas within a city that have a higher jobs–housing balance generally have lower average commute times (where this is defined as zones having a smaller out commute). In terms of policy, it is therefore worthwhile trying to improve the jobs–housing balance of areas, possibly through a better mix of housing and employment land use (see also Wang (2000) for his analysis of commuting and the housing market in Chicago, and Horner's (2004) review of this research area in general).

There have been many studies of commuting patterns in cities and regions. Li *et al.* (2012) provide a good case study of the analysis of changing commuting patterns over time and space in Brisbane using data from various census periods. Walsh *et al.* (2006) provide a comprehensive analysis of commuting patterns in 2002 for small areas throughout Ireland. Figure 12.1

shows the GIS map of those travelling for longer than 90 minutes each day for their daily commute – the dominance of the Dublin labour market in Ireland is obvious from this map (see also the use of GIS to examine long-distance commuting from sparsely populated regions in Sweden provided by Sandow (2008)).

Other flow data can be collected by various interested parties in fields such as health and retail. Health care planners are particularly concerned with the pattern of service uptake in the form of GP and hospital catchment area analysis. For the effective provision of services it is paramount that trip distances to facilities are controlled and minimised, and data to illustrate trends can be taken from patient registration information (see Chapter 9 for more details).

Retailers often use their loyalty cards to gather information on customer shopping trends. From basic address and purchase information, a retailer can study patterns of who is travelling from where (and to buy what) in order to aid their understanding of local retail environments, understand market shares and facilitate customer targeting and branch/outlet location reviews. In the UK, the 'Nectar' loyalty card scheme contains details on the shopping behaviour of the holders of over 18 million cards, around 12 million of which are actively used, enabling participating retailers to understand more about the trip-making behaviour of their customers (for example, understanding flows of consumers from home postcodes to specific stores). Chapter 8 explores the use of GIS in retail analysis in more detail.

Around the world many transport companies are themselves increasingly using technology in addition to paper ticketing. These smart card systems are providing new sources of data to understand how people travel around on urban transit systems. Tao *et al.* (2014) show how these data can supplement census data and offer new insights into commuting – by linking space and time to show movements on a 24/7 basis. Figure 12.2 shows the data for the Brisbane Rapid Transit (BRT) system using this smart card data. Tao *et al.* summarise the main patterns:

> For BRT trips, a number of large volume pathways
> from the northern, southern and western outer

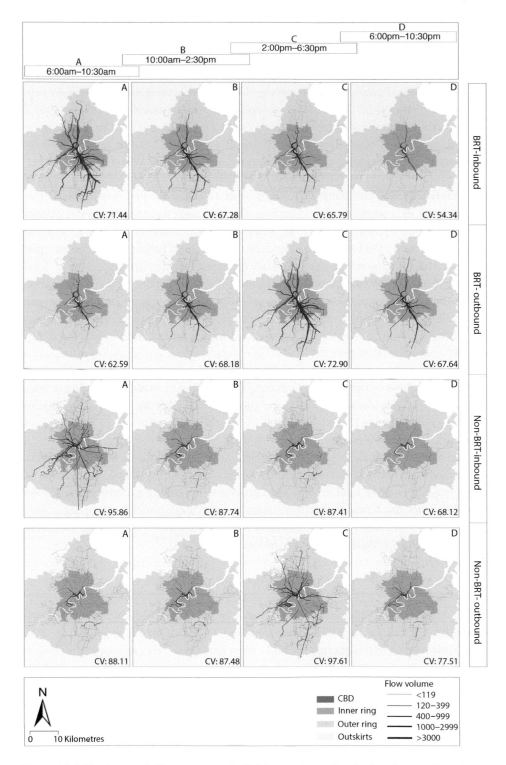

Figure 12.2 Rapid transit flow patterns in Brisbane, Australia, during the working day

Source: Tao *et al.* (2014)

rings formed and persisted throughout the two days, suggesting that the BRT busway serves as a backbone corridor in channelling trips from outer suburbs (more notably in the north–south direction) to the CBD and surrounding areas within the Brisbane bus network.

(Tao *et al.*, 2014: 96)

Monitoring flows and providing suitable capacity for these movements is a simplified summary of the role undertaken by a transport planner. In UK cities we have a well-developed network of access routes and flows, but these are under constant pressure from increasing volumes and can be subject to rapid changes in demand (where the concern is most likely to involve issues of congestion, rather than under use). This scenario might arise, for example, with the introduction of a new business park or shopping centre as a flow destination, or a new housing development as an origin.

SIMs (introduced in Chapter 8) are worthy of a mention here due to their use of flow information to understand and simulate population movements across a spatially subdivided surface. Indeed, some of the earliest uses of these models were for transport analysis and planning (see the examples in Foot (1981)). A matrix of volumetric measurements is derived for each possible eventuality of origin – destination flow (usually between centroids of the aforementioned spatial units), and these interactions are modelled by characteristics of source zone (possibly population affluence or socioeconomic condition) and destination zone (usually an attractiveness feature such as perceived popularity of a shop or service), constrained by a function of the distance between these two points (a distance decay factor). Depending on the size of spatial unit chosen for such a model, it might be useful to a transport planner in terms of understanding trends of population movement.

SIMs have been covered in detail, in a retail context, in Chapter 8 and are therefore not explored further in this chapter. Instead, we focus on GIS-based network analysis for modelling and evaluating routing through a transport system.

Networks and examples of GIS in use

Transport networks are the channels that facilitate the flow of people and goods and can be represented within a GIS by points (nodes of intersection or access) and vector lines. Despite the earlier discussion of flow patterns generalised to area-based interaction, these networks form the real connections between populations, origins and destinations, and can take a variety of forms such as roads, rails, footpaths, canals, etc.

Network links and nodes can be associated with a range of attributes that determine their function. The fundamental properties when considering routes through a transport network must be length and capacity. The length of a road or rail link could be allocated in terms of distance or time taken to negotiate at a set speed. The capacity feature might be extremely difficult to ascribe figures to, but could be represented in the example of roads by some function of road width, road lane count and speed limit (factors that could, in turn, be represented by a proxy such as road classification, e.g. motorway, A-road, B-road, etc.). For the example of public services such as rail, an analyst might consider the capacity of carriages and frequency of service as indicators of capability.

There is a vast range of applications in transport networking to which GIS and spatial analysis can be adapted. These operations lend themselves most readily to functions of network management and adjustments to current layouts, offering the additional advantage of analysing implications for network changes. Applications relating specifically to networks will be discussed in the following sections, which include a range of illustrative examples relating to the main transport networks of roads, rail systems and air traffic spaces.

Road networks

The planning of roads in highly populated developed countries presents a difficult balance of conflicting interests. From the perspective of most end users the objective is a straightforward package of high capacity and minimised travel time. The planner, however, must account for factors such as sustainability, economics and environmental and social interests. A study by Rapaport and Snickars (1999) integrated all of these concerns in the application of GIS to road location for minimised environmental damage, building costs and travel time in Sweden. The environmental variables considered included salt damage, interruptions to surface water flow, greenbelt incursions, visual impacts and noise disturbance (some of which are subjectively determined or have complex relationships with traffic volume). Road building costs were calculated as a function of terrain (geological conditions and slope), current land-use type and level of disturbance to existing traffic during construction. Finally, travel time indicators used a simple application of distance and time which could be derived quickly and accurately in GIS software. These factors were analysed independently to identify possible routes for a new road system, revealing that each constraint produced an alternative optimum pathway. The constraints were then combined in a weighted linear function to derive the most suitable route under consideration of all factors.

At a more localised level, a GIS system might also prove useful for the management of network repairs which could be recorded in a database as attributes to a mapped system. Information along road networks such as power lines, water mains and other utilities information could all be recorded as separate layers with attributes to advise when maintenance is due, and what effects the changes might have on other network properties. This subject was researched by Khan (2000) in the design of a GIS for street lighting and traffic signal maintenance in Leeds, UK. In an ideal management system, such an advisory scheme might enable planners to manage repairs more efficiently in a spatial sense, whereby a maintenance team could be directed to fix other problems nearby to an ongoing work site (also see Meehan, 2007; Liu and Issa, 2012).

Websites such as the UK's 'Fix my street' (www.fixmystreet.com) allow local residents to use a web-based GIS interface to report transport network repairs needed within their neighbourhood, such as potholes, broken traffic lights and illegal parking. Users can view the status of other problems reported in their area or submit their own report, which is passed instantly to the relevant authority. This is an excellent example of the use of GIS interfaces to promote public participation and community engagement within local issues influencing the transport network.

We will now turn to the issue of pollution and a potential use of GIS and spatial analysis in its management. In a simple application of GIS, Mavroulidou *et al.* (2004) used mapping software to visualise simulated patterns of pollution vulnerability and contributory factors in an area surrounding Guildford, UK. Data including traffic volume, wind, topography and building density were combined through the overlay procedure to examine how air quality was susceptible to change. These maps identify the areas adjacent to the main 'A3' road and in the centre of town as most vulnerable to poor air quality, plus a hot spot to the east of town where high traffic volume and complex terrain combine to increase pollution. The paper highlights the potential for planners to use this tool in highlighting areas with adverse conditions for pollution dispersal, and to identify routes where factors favouring pollutant dilution are greater.

In another application to the same field, Guldmann and Kim (2001) used statistical regression models to explain observed variations across urban areas in concentrations of ozone and carbon monoxide. Using a cell-based raster approach, their model considered pollutant concentration as a balance between inputs and outputs, using GIS and databases to spatially relate pollution measurements, meteorological factors, land-use characteristics, socio-economic data and major highway networks. The GIS hosted model allowed the researchers to assess impacts of land-use categories or major highways on particular emission concentrations by studying samples at buffered distances from these suspected sources, and comparing figures with background pollutant levels. All of the explanatory variables relating to transport were highly significant and suggested improvements of network capacity and flow at intersections as a vital means to reduce emissions.

GIS can also be used as an important tool for network planning in the context of transporting hazardous material. In one such example, Monprapussorn *et al.* (2009) use a GIS to assess the risks associated with transporting hazardous materials and hazardous waste by road in Thailand. Specifically, they considered the potential impact of a truck being involved in an accident while carrying these materials, identifying a number of potential social, economic and environmental impacts, in order to determine the optimum routing for these vehicles to minimise potential risk and impact (also see Lovett *et al.*, 1997; Verter and Kara, 2001).

Route planning is an issue that is implicitly linked to networks. Devlin *et al.* (2008), for example, compared the actual routes taken to transport timber across Ireland with the 'optimal' routes suggested by GIS. The range of data that is required to consider options and plan events is well suited to a GIS, which can facilitate a variety of spatial queries. An excellent example of this application is in GPS & GIS routing of emergency service vehicles such as ambulances. In many countries, ambulances are now fitted with GPS navigation as standard and increasingly detailed mapping products can be used to route ambulances to specific addresses or other locations not covered by a geographic address. In the UK, Ordnance Survey's Master Map Integrated Transport Network (ITN) data set can be used in conjunction with OS Address Point to route vehicles effectively to required locations. A study by Ota *et al.* (2001) confirmed the benefits of this approach through an analysis of emergency service response times in the urban environment. The first test in this research was to compare the progress of two ambulance teams, one with standard street maps and another with GPS navigator, in locating and routing to randomly selected residential addresses. With results taken from a total of 29 situations, the mean distances covered were 8.7 kilometres (GPS) and 9.0 kilometres (standard maps) with a mean travel time of 12.5 (GPS) versus 14.6 minutes (standard maps). The GPS & GIS equipped teams were faster in 72% of responses. The second test involved a qualitative assessment by emergency teams who were allowed to combine GPS & GIS resources with standard methods. Most providers noted that such technology enhanced their ability to navigate to destinations in areas that were unfamiliar. Such systems can be operated from a central control, where GIS systems can be used to relay maps and information on emergency conditions to response vehicles. These systems have the potential to receive and process more detailed information on traffic conditions to ensure responses are optimised.

GIS & GPS combinations can also be used by logistics companies to manage fleet operations, ensuring that delivery routes through networks and conditions of cargos are optimised. Toll Tranzlink, a New Zealand distributor, use an application developed by Vodafone to track and control their trucks in relation to 39 bases and 400 delivery locations. From a central control unit, the management team are able to monitor live information on location and time to destination, as well as on-board conditions such as capacity and temperature which are relayed by on-screen maps and attribute information. The claimed benefits of this system include a 55% improvement in on-time undamaged deliveries, more effective use of carrying capacity due to live monitoring, and greater efficiency due to real-time information provision and instant decision making. In 2011 it was widely reported in the media that US parcel delivery company UPS had implemented a new GIS-based algorithm to determine optimum routes for their fleet of delivery vehicles in the US. Delivery routes were optimised in order to minimise the number of left-turns made by drivers. Unlike right-turns, left-turns involve waiting to cross oncoming traffic – wasting time and fuel. This approach thus saves money and reduces emissions.

Many web-based services provide real-time traffic information alongside route-planning functions available free-of-charge to users. These services make use of an extensive road network database, coupled with real-time information on delays, traffic flows, roadworks, weather conditions and current traffic speeds in order to identify the most appropriate route to a destination. For the primary route network across much of the EU, the US and parts of China, Japan, Russia and Australia, Google Maps displays real-time flow speeds, along with the opportunity to identify average flow speeds at any given date or time, making use of a large spatially referenced database of historic traffic conditions.

Satellite navigation devices (e.g. TomTom) have become an important in-car accessory and almost all smartphones also feature some form of mapping and navigation device. These applications use route-planning software to help you navigate a particular route, taking account of real-time traffic flows. Traffic flow information is delivered to these devices in the UK using the 'Radio Data System-Traffic Message Channel' – this is an FM channel through which traffic information is broadcast to an agreed format which can be decoded by a range of devices. In the case of a satellite navigation system, the receiving unit will decode key information such as the nature and location of the incident, provide a visual/audible alert to the user and, most importantly, attempt to recalculate the route to find an alternative quicker route to the destination (see also the discussion on evacuation planning in Chapter 10).

In relation to the road network, GIS can be used to find the optimal location of support services. We have already seen this in relation to siting fire stations

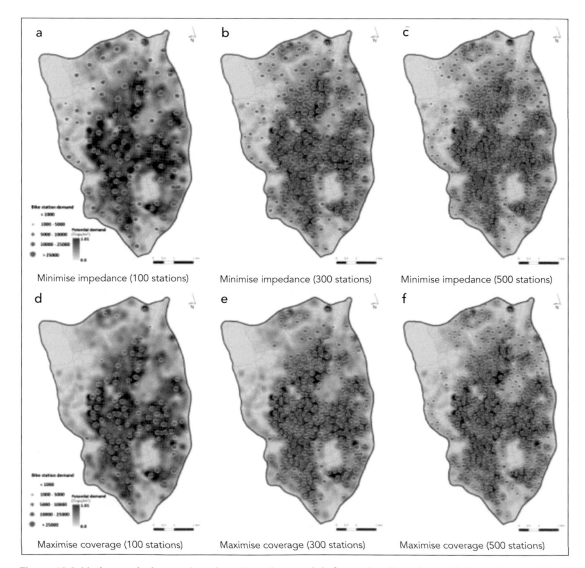

Figure 12.3 Various solutions to location-allocation models for optimal locations of bike stations in Madrid, Spain

Source: García-Palomares *et al.* (2012)

(Chapter 7) and hospitals (Chapter 9). In relation to road networks this could entail finding the best location for car parks based on current network usage and trip destination modelling or perhaps park and ride schemes. A nice example of the use of location-allocation models (discussed widely in Chapter 9) in transport GIS is provided by García-Palomares *et al.* (2012). They built a location-allocation model to find the best location for providing 'bike stations' (key locations where people can hire a bike or leave it once having finished their journey), based on the movement of bike trips through the city of Madrid. Figure 12.3 shows the results of the model for various alternative numbers of bike stations.

Accident research is a particularly interesting field to which GIS have been applied in relation to road networks. Accident data are routinely recorded in many countries for legal requirements. Such data enable the identification of 'blackspot' locations where remedial action might be possible. In addition to event location and time, the information might include additional attribute data, such as a classification of seriousness (slight, serious or fatal), circumstances of the accident (weather and traffic conditions), vehicles involved, casualties and any special circumstances. Unfortunately, records will often be incomplete as some non-fatal accidents do not make it into the accident database.

Nevertheless, these statistics play a major role in the definition of priorities for road improvements and financial justification for road construction. The rapid data set query and visualisation capabilities of GIS allow accident data to be studied and better understood in terms of particular spatial and attribute trends. The *Guardian* newspaper in the UK has compiled UK road accident and injury data and produced an excellent visualisation tool, available at (www.theguardian.com/news/datablog/interactive/2011/nov/18/road-casualty-uk-map). Identification of specific risks or potential risks resulting from network or flow changes is the area of greatest interest here. However, these patterns are extremely complex due to countless variations in accident cause and the need to consider the rate of accidents relative to the numbers of road users. The best approach must be to review geospatial and temporal incidence of events through what data are available, before identifying specific locations or occurrences worthy of further investigation.

Lovelace *et al.* (2016) have been able to reveal the spatial patterns in cycling accident data in West Yorkshire in the UK, which are often masked by limitations of raster format. Of particular interest is the study of changes in accident occurrence and location over time, which has enabled the identification of possible causal factors; for example, a change in road or junction layout which might have positive or negative implications for accident probability. Figure 12.4 shows hot spots of accident data in West Yorkshire

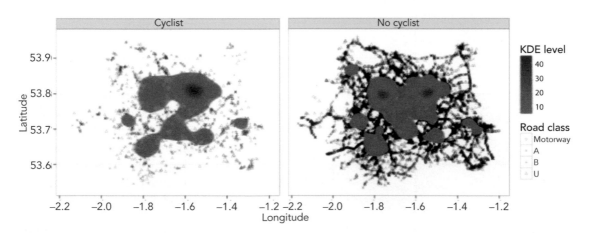

Figure 12.4 Accident hot spot analysis for cyclists (left) and other road users (right), West Yorkshire, UK

Source: Lovelace *et al.* (2016)

which helps to plan for change (note that this study has actually been undertaken using the computing package R and its analysis and GIS-style mapping capabilities).

A second example is taken from the work of Young and Park (2014). Again they use kernel density hot spot mapping, in this case for cycling injuries in Regina, Canada (Figure 12.5; also see Li *et al.*, 2007; Yiannakoulias *et al.*, 2012; Dai, 2012).

Rodrigues *et al.* (2015) used a GIS-based framework in a case study in the municipality of Barcelos, Portugal. A set of road-related variables was defined to obtain a road safety index. The indicators chosen were: severity, property damage and accident costs. In addition to the road network classification, the application of the model allowed them to analyse the spatial coverage of accidents in order to determine the centrality

and dispersion of the locations with the highest incidence of road accidents.

The broad range of examples presented in this sub-section has highlighted the range of uses of GIS-based network analysis to optimise routing and to minimise accidents. In the following sub-section we present examples to illustrate that a similar range of network-based tools can support the evaluation of public transport networks, with a particular focus on accessibility.

Public transport networks and accessibility

Many of the applications of GIS in relation to public transport networks come under two main topics – accessibility analysis and the planning (and impact) of designing new routes. We have seen accessibility as a crucial element of GIS throughout this book, and indeed many of the examples used elsewhere in the book use network analysis as the basis for calculating accessibility indicators of various types, especially Chapter 9, in relation to health services. The Tyne and Wear (UK) Accessibility Planning Project forms part of the requirements of the local transport plan in monitoring the performance of the conurbation's public transport network. Focusing on issues of social exclusion, which arise primarily through lack of vehicular access, the project looks to advance the UK Government's standard indicator of rural accessibility (the number of houses within a 13-minute walk of a bus service that runs at least hourly), by studying the real patterns of access between origins and destinations and incorporating real timetable information to calibrate accessibility functions. The software that facilitates this analysis is called AutoAccess, a collection of software engines, databases and GIS that enable comprehensive analysis of networks in the production of two main forms of statistic: the time taken to travel between any two pairs of public transport stops in the area (of which there are approximately 7,000) and a full breakdown of how this journey would be made for a given time and date (and including walking, waiting, interchanges and bus/rail travel time).

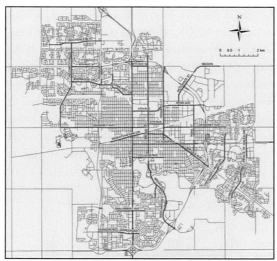

Figure 12.5 Top four hot spots for bike accidents in Regina, Canada

Source: Young and Park (2014)

The key inputs for this product are the Post Office Address File (PAF), the Ordnance Survey's definitive road survey product OSCAR and a detailed timetable of services that operate on these routes. Proprietary GIS systems such as ArcMap can be used to host the models' output information, illustrating how the public transport network operates spatially to serve access to work, education, health and other facilities. Applications have been identified as follows: regular long-term monitoring of accessibility, guidance for land-use planning and development control, definition of the sustainable public transportation network and short-term accessibility audits. The data set and GIS can be queried for specific timetables and origin–destination flows. This enables users to identify optimum routes and a duration guide for journeys to work, trips to the nearest pharmacy, etc. The system can also be updated for changes to the network and could run scenarios to examine effects of these adjustments on travel time and accessibility. Table 12.2 identifies a series of accessibility targets set by Tyne and Wear Local Transport Plan (LTP) for the period 2006–2011 identifying, for example, that in 2004, 87% of households were within a 30-minute journey by public transport to their nearest hospital – with a target of raising this to 91% by 2010.

Around the world, there are a number of accessibility packages that can be used in conjunction with GIS to produce standard access scores (e.g. see Liu and Zhu, 2004; Miller and Wu, 2000). Liu and Zhu show their outputs for Singapore: powerful 3-D maps representing accessibility surfaces in this instance (Figure 12.6).

Finally, in relation to accessibility and transport GIS, we should note an interesting set of papers that estimate accessibility to jobs in various locations for various modes of travel: car, bicycle, walking, etc. (Mavoa *et al.*, 2012; Wang and Chen, 2015).

As noted above, the second major use of GIS in public sector transport management is to help in the design of new systems: whether they be new bus routes, bus lanes or light transit systems. Often these studies are part of much wider urban and regional planning projects. The building of a new light rail system, for example, is not simply about route planning. Planners and engineers would normally consider the demand and route planning first, but only as a component of a much larger study around economic costs and impacts. This might involve a systematic appraisal of current road and public transport use, current journey to work patterns and land-use costs and availability. There is certainly a wealth of studies in economics and regional science which have explored the impacts of new transit routes not only on traffic movements, accessibility patterns and congestion, etc., but also on job creation, house/office prices and revised commuting patterns.

Table 12.2 Tyne and Wear LTP: public transport access thresholds from the LTP

Definition	Threshold	Base year		Base	2006	2007	2008	2009	2010
Secondary education	20 mins	2004	Actual	98%	99%	99%	99%		
			Target		98%	98%	99%	99%	99%
Hospitals	30 mins	2004	Actual	87%	86%	88%	87%		
			Target		88%	89%	90%	90%	91%
Employment (Newcastle city centre)	40 mins	2004	Actual	72%	74%	74%	72%		
			Target		73%	73%	74%	74%	75%
Employment (Great Park)	40 mins	2004	Actual	25%	25%	28%	28%		
			Target		26%	27%	27%	28%	29%
Employment (Newburn)	40 mins	2004	Actual	29%	32%	25%	30%		
			Target		29%	30%	30%	31%	31%
Employment (Team Valley)	40 mins	2004	Actual	24%	37%	37%	36%		
			Target		24%	25%	25%	26%	26%

Source: www.tyneandwearltp.gov.uk/wp-content/uploads/2010/09/DRPt2.pdf

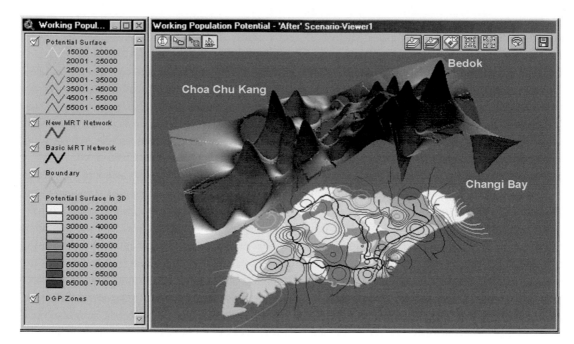

Figure 12.6 Accessibility surface for public transport in Singapore

Source: Liu and Zhu (2004)

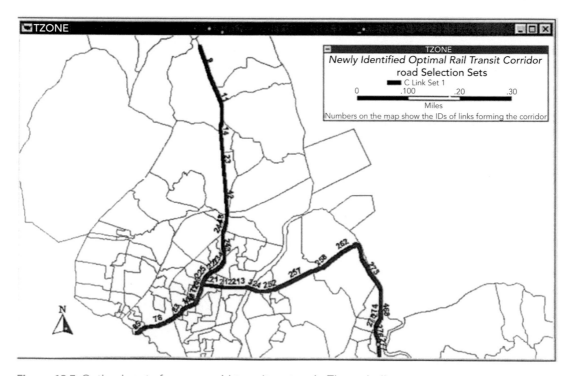

Figure 12.7 Optimal route for new rapid transit system in Thane, India

Source: Verma and Dhingra (2005)

Some, or all, could be done within a GIS environment but the detailed land-use/transportation modelling takes us beyond the scope of this book (although the interested reader could take a look at Wilson (1998) for a general review and Vuchic (2005) in relation particularly to urban transit developments).

That said, there are a number of interesting studies that have focused principally on the transport aspects of new transit system planning: namely the planning of new routes and stations/stops. Most applications use a two-stage process (Samanta and Jha, 2008). First, is the estimation of demand. This might involve GIS in its own right – to identify areas where many persons are using the road network (especially for commuting) who might be persuaded to switch to public transport if a better system is in operation. Second, is then the identification of suitable corridors or networks to locate the route itself. Station planning along such a corridor then needs to be undertaken so that there are not too many stations (too costly and slow in operation) and not too few (thus missing opportunities to increase access and remove cars from the road). GIS could again be used here – looking, for example, at the amount of demand within, say, two to five kilometres of potential stations. Those locations that maximise accessibility could then be chosen. Verma and Dhingra (2005) show how such a combination of demand estimation techniques overlaid onto a network of possible routeways for Thane in India can lead to the development of an optimal route (see Figure 12.7).

This desire to produce an optimal mix of lower costs and maximum accessibility has also led to the design of many sophisticated optimisation techniques, especially in the field of operational research. Such optimisation procedures might use genetic algorithms for example to find optimal station locations (for a good introduction to these approaches see Laporte *et al.* (2011) and García-Palomares *et al.* (2012)). Having helped to design new routes, GIS can also help add new infrastructure around these routes. Horner and Grubesic (2001), for example, used GIS to calculate an index of demand within the catchment area of the rail stations in Columbus, Ohio, US. Each station was a potential candidate for a park and ride scheme. They produced a type of suitability index to rank and compare potential park and ride locations. Of course, the evaluation of transformations in public transport networks does not always shine a positive light on such changes. Blair *et al.* (2013) used a variety of GIS access type indicators to evaluate the major revision to bus routes and times in Belfast in Northern Ireland during the mid-2000s. They showed the negative impacts on a variety of user groups, especially those in low-income areas.

As with the road network there are many studies of GIS in relation to freight movements, pollution and carbon footprint analysis. Zuo *et al.* (2013), for example, built a spatial flow model of the transportation of aggregates across the UK in a GIS, for both road and rail. They were then able to estimate the carbon footprint associated with travel by both modes and test a series of what-if? scenarios for trying to reduce those carbon footprints (i.e. impacts of building new high-technology quarries or increasing the amount of aggregates transported by rail).

Also, as with the examples of roads above, other examples of real-time management for flows and networks can be seen in the case of bus and train/tram services in many cities, where computer aided dispatch and passenger information systems have been introduced. Mobile technology and GPS tracking enable synchronised tracking of vehicle location, electronic displays at bus stops which indicate when the next bus will arrive and the relaying of service information to customers by mobile telephone text messaging. In many cases, intelligent bus priority traffic information systems are also used, whereby bus location and distance to the next signalised junction will be used to control traffic-light sequences for quicker public transport flow. A range of examples of these projects from around the world can be found on the 'Innovation in Traffic Systems' website (www.initusa.com/en/index.php).

Finally in this section we look at how GIS has been used in a more limited number of studies relating to air travel. First there are a number of studies that provide a similar decision support mechanism to the facilities management systems described for roads. Using GIS and GPS a number of so-called airport pavement systems have been created. McNerney (2000) reported

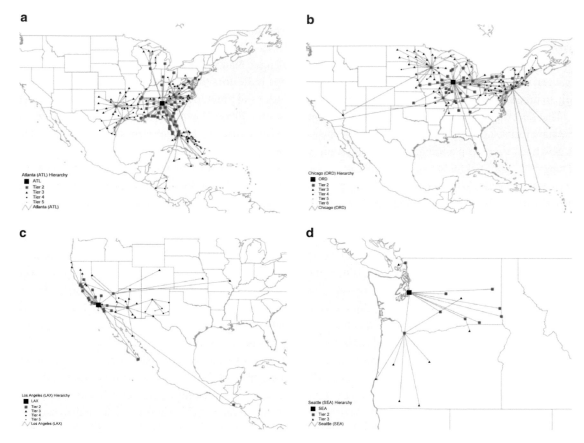

Figure 12.8 Airport connectivity in the US: (a) Atlanta, (b) Chicago, (c) Los Angeles, (d) Seattle

Source: Grubesic *et al.* (2008)

that 60% of US airports use these systems to keep track of the condition of runways and taxi areas, often replacing the old vehicle-based inspection services (also see Shwartz *et al.*, 1991; Chen *et al.*, 2012).

On a broader geographical scale, GIS has been used to plot connectivity at different airports in terms of the spatial variations in locations of potential destinations available. Grubesic *et al.* (2008) examined the emerging global hierarchy of airline network connectivity using data from 900 airline carrier schedules between 4,600 worldwide destinations. Figure 12.8 shows the connectivity from four major US hubs – showing Atlanta as the most connected with 163 destinations (2006 data).

GIS and models for transport-based location analysis

In this section we explore the use of GIS and location models for the optimal locations for transport-based services such as ambulance or fire stations. These services are called 'outreach' services, in the sense that they come to their clients, rather than the other way round. The models are location-allocation models, introduced in Chapters 4 and 9. Figure 12.9 plots the location of ambulance and fire stations in London in 2012. It is immediately apparent that there are 'open points': areas that are not as well served by these services. Thus it is not difficult to estimate where the worst place to live in London is, should there be a house fire (i.e. the location which is furthest from the nearest fire station).

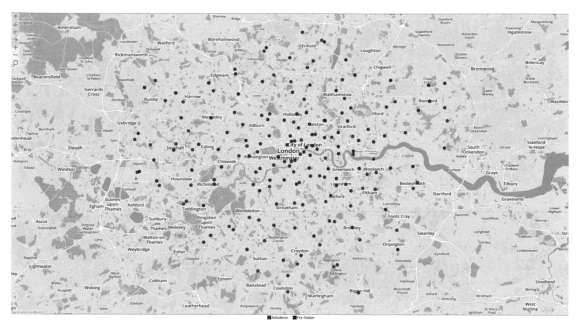

Figure 12.9 Spatial variations in the location of ambulance stations (blue dots) and fire stations (red dots) in London, 2012

Source: Spatiametrics (2013). Background mapping source: https://www.mapbox.com/

Concluding comments

As emphasised throughout this chapter, flows as demand structures and networks as movement facilitators are the underlying components of transportation. In the urban environment these factors are subject to continuous change due to variations in population pressure and the erratic nature of personal journeys. The examples of GIS in this chapter have demonstrated, however, that transport systems can be understood and effectively managed by thoughtful consideration of relevant data sources and the understanding of interactions that occur in our cities and regions. Moreover, the application of suitable software and modelling environments can be used to run controlled simulations of scenario changes such as new origins, destinations and network links or changes to access policy. These opportunities should serve the transport planner well in the ongoing pursuit of more efficient, reliable and serviceable urban transport networks.

We should finish with recognition that many applications of transport planning involve large-scale research projects which might actually link many of the components we have discussed above. The final example in this section is a research project that attempts to bring together many of the elements discussed in this section through the development of a GIS-based decision support system for planning urban transport policy. Arampatzis *et al.* (2004) developed a tool to assist transport administrators in attempts to improve transport efficiency while monitoring environmental and energy indicators. Transport network analysis is facilitated by a road traffic simulation using traffic flow patterns (based on known network characteristics and traffic demand) for each link in the system. This information is supplemented by energy consumption and pollutant emission calculations when combined for data hosting and visualisation in a GIS environment. The main analytical use of GIS comes in the form of case studies for policy change, regarding vehicle access and parking restrictions for the centre of Athens. More specifically, the scenarios study example implementations of an area traffic restriction on the use of private cars in the Municipality of Athens, and a simulated 50% reduction of parking places in the central town. The model outputs for these scenarios include implications for other transport modes (taxis and other public transport), estimated average vehicle speeds, estimated fuel consumption and air pollutant emission levels. The tool is heralded as 'a GIS-based decision support system that involves realistic representation of the multi-modal transportation network and efficient implementation of network equilibrium solutions for problems related to the application of urban transportation policies' (Arampatzis *et al.*, 2004: 475).

We looked at some of these bigger decision support systems for transport planning in relation to GIS for emergency planning in Chapter 10. We hope that this chapter has served as an introduction and overview of the literature and range of GIS-based applications in the modelling, planning and evaluation of transport systems.

Accompanying practical

This chapter is accompanied by Practical 8: Transport planning. The practical gives you the opportunity to assess the provision of public transport in relation to commuter flows (journey to work) in the city of Chicago. Specifically you consider the opportunities to introduce new cycle hire stations to promote commuting by bike. Practicals D: Network analysis and 6: Emergency planning are also related to the network approach outlined in this chapter.

References

All website URLs accessed 30 May 2017.

Arampatzis, G., Kiranoudis, C. T., Scaloubacas, P., & Assimacopoulos, D. (2004) A GIS-based decision support system for planning urban transportation policies. *European Journal of Operational Research, 152*(2), 465–475.

Birkin, M., Clarke, G. P., & Clarke, M. (2017) *Retail Location Planning in an Era of Multi-Channel Growth*, Routledge, London.

Blair, N., Hine, J., & Bukhari, S. M. A. (2013) Analysing the impact of network change on transport disadvantage: a GIS-based case study of Belfast. *Journal of Transport Geography, 31*, 192–200.

Chen, W., Yuan, J., & Li, M. (2012) Application of GIS/GPS in Shanghai Airport pavement management system. *Procedia Engineering, 29*, 2322–2326.

Dai, D. (2012) Identifying clusters and risk factors of injuries in pedestrian–vehicle crashes in a GIS environment. *Journal of Transport Geography, 24*, 206–214.

Daniels, P. W., & Warnes, A. M. (1980) *Movement in Cities: Spatial Perspectives on Urban Transport and Travel*, Methuen, London.

Devlin, G. J., McDonnell, K., & Ward, S. (2008) Timber haulage routing in Ireland: an analysis using GIS and GPS. *Journal of Transport Geography, 16*(1), 63–72.

Foot, D. (1981) *Operational Urban Models: An Introduction*, Taylor & Francis, London.

García-Palomares, J. C., Gutiérrez, J., & Latorre, M. (2012) Optimizing the location of stations in bike-sharing programs: a GIS approach. *Applied Geography, 35*(1), 235–246.

Goodchild, M. F. (2000) GIS and transportation: status and challenges. *GeoInformatica, 4*(2), 127–139.

Grubesic, T. H., Matisziw, T. C., & Zook, M. A. (2008) Global airline networks and nodal regions, *GeoJournal, 71*(1), 53–66.

Guldmann, J. M., & Kim, H. Y. (2001) Modelling air quality in urban areas: a cell based statistical approach. *Geographical Analysis, 33*(2), 156–180.

Horner, M. W. (2004) Spatial dimensions of urban commuting: a review of major issues and their implications for future geographic research. *The Professional Geographer, 56*(2), 160–173.

Horner, M. W. (2007) A multi-scale analysis of urban form and commuting change in a small metropolitan area (1990–2000). *Annals of Regional Science, 41*, 315–332.

Horner, M. W., & Grubesic, T. H. (2001) A GIS-based planning approach to locating urban rail terminals. *Transportation, 28*(1), 55–77.

Khan, M. Z. (2000) The design and application of a GIS for street lighting and traffic signal maintenance. Unpublished PhD Thesis, University of Leeds.

Laporte, G., Mesa, J. A., Ortega, F. A., & Perea, F. (2011) Planning rapid transit networks. *Socio-Economic Planning Sciences, 45*(3), 95–104.

Li, L., Zhu, L., & Sui, D. Z. (2007) A GIS-based Bayesian approach for analyzing spatial–temporal patterns of intra-city motor vehicle crashes. *Journal of Transport Geography, 15*(4), 274–285.

Li, T., Corcoran, J., & Burke, M. (2012) Disaggregate GIS modelling to track spatial change: exploring a decade of commuting in South East Queensland, Australia. *Journal of Transport Geography, 24*, 306–314.

Liu, R., & Issa, R. (2012) 3D visualization of sub-surface pipelines in connection with the building utilities: integrating GIS and BIM for facility management. *Computing in Civil Engineering, 2012,* 341–348.

Liu, S., & Zhu, X. (2004) Accessibility analyst: an integrated GIS tool for accessibility analysis in urban transportation planning. *Environment and Planning B, 31*(1), 105–124.

Lovelace, R., Roberts, H., & Kellar, I. (2016) Who, where, when: the demographic and geographic distribution of bicycle crashes in West Yorkshire. *Transportation Research Part F: Traffic Psychology and Behaviour, 41*(B), 277–293.

Lovett, A. A., Parfitt, J. P., & Brainard, J. S. (1997) Using GIS in risk analysis: a case study of hazardous waste transport. *Risk Analysis, 17*(5), 625–633.

Mavoa, S., Witten, K., McCreanor, T., & O'Sullivan, D. (2012) GIS based destination accessibility via public transit and walking in Auckland, New Zealand. *Journal of Transport Geography, 20*(1), 15–22.

Mavroulidou, M., Hughes, S. J., & Hellawell, E. E. (2004) A qualitative tool combining an interaction matrix and a GIS to map vulnerability to traffic induced air pollution. *Journal of Environmental Management, 70*(4), 283–289.

McNerney, M. (2000) Airport infrastructure management with geographic information systems: state of the art. *Transportation Research Record: Journal of the Transportation Research Board, 1703,* 58–64.

Meehan, B. (2007) *Empowering Electric and Gas Utilities with GIS,* ESRI Press, Redlands, CA.

Miller, H. J., & Wu, Y. H. (2000) GIS software for measuring space-time accessibility in transportation planning and analysis. *GeoInformatica, 4*(2), 141–159.

Monprapussorn, S., Thaitakoo, D., Watts, D. J., & Banomyong, R. (2009) Multi criteria decision analysis and geographic information system framework for hazardous waste transport sustainability. *Journal of Applied Sciences, 9*(2), 268–277.

Nyerges, T. L. (2004) GIS in urban-regional transportation planning, in S. Hanson & G. Giuliano (eds) *The Geography of Urban Transportation,* Guilford Press, London, 163–198.

Ota, F. S., Muramatsu, R. S., Yoshida, B. H., & Yamamoto, L. G. (2001) GPS computer navigators to shorten EMS response and transport times. *American Journal of Emergency Medicine, 19*(3), 204–205.

Rapaport, E., & Snickars, F. (1999) GIS-based road location in Sweden, in J. Stillwell, S. Geertman, & S. Openshaw (eds) *Geographical Information and Planning,* Springer, Berlin, 135–153.

Rodrigues, D. S., Ribeiro, P. J. G., & da Silva Nogueira, I. C. (2015) Safety classification using GIS in decision-making process to define priority road interventions. *Journal of Transport Geography, 43,* 101–110.

Samanta, S., & Jha, M. (2008) Identifying feasible locations for rail transit stations: two-stage analytical model.

Transportation Research Record: Journal of the Transportation Research Board, 2063, 81–88.

Sandow, E. (2008) Commuting behaviour in sparsely populated areas: evidence from northern Sweden. *Journal of Transport Geography, 16*(1), 14–27.

Schwartz, C. W., Rada, G. R., Witczak, M. W., & Rabinow, S. D. (1991) GIS applications in airfield pavement management. *Transportation Research Record: Journal of the Transportation Research Board, 1311,* 267–276.

Spatiametrics (2013) Simple Spatial Analysis of Fire & Ambulance Services in London. Available from: https://spatiametrics.wordpress.com/2013/01/16/simple-spatial-analysis-of-fire-ambulance-services-in-london/.

Tao, S., Corcoran, J., Mateo-Babiano, I., & Rohde, D. (2014) Exploring bus rapid transit passenger travel behaviour using big data. *Applied Geography, 53,* 90–104.

Thill, J. C. (2000) Geographic information systems for transportation in perspective. *Transportation Research Part C: Emerging Technologies, 8*(1), 3–12.

Verma, A., & Dhingra, S. L. (2005) Optimal urban rail transit corridor identification within integrated framework using geographical information system. *Journal of Urban Planning and Development, 131*(2), 98–111.

Verter, V., & Kara, B. Y. (2001) A GIS-based framework for hazardous materials transport risk assessment. *Risk Analysis, 21*(6), 1109–1120.

Vuchic, V. R. (2005) *Urban Transit: Operations, Planning, and Economics,* Wiley, Hoboken, NJ.

Walsh, J., Foley, R., Kavanagh, A., & McElwain, A. (2006) Origins, destinations and catchments: mapping travel to work in Ireland in 2002. *Journal of the Statistical and Social Inquiry Society of Ireland, XXXV,* 1–55.

Wang, C. H., & Chen, N. (2015) A GIS-based spatial statistical approach to modeling job accessibility by transportation mode: case study of Columbus, Ohio. *Journal of Transport Geography, 45,* 1–11.

Wang, F. (2000) Modeling commuting patterns in Chicago in a GIS environment: a job accessibility perspective. *The Professional Geographer, 52*(1), 120–133.

Wilson, A. G. (1998) Land-use/transport interaction models: past and future. *Journal of Transport Economics and Policy, 31*(1), 3–26.

Yiannakoulias, N., Bennet, S. A., & Scott, D. M. (2012) Mapping commuter cycling risk in urban areas. *Accident Analysis & Prevention, 45,* 164–172.

Young, J., & Park, P. Y. (2014) Hotzone identification with GIS-based post-network screening analysis. *Journal of Transport Geography, 34,* 106–120.

Zuo, C., Birkin, M., Clarke, G., McEvoy, F., & Bloodworth, A. (2013) Modelling the transportation of primary aggregates in England and Wales: exploring initiatives to reduce CO_2 emissions. *Land Use Policy, 34,* 112–124.

13 GIS for environmental justice and policy evaluation

LEARNING OBJECTIVES

- Define environmental justice and explain how it is spatial
- Aspects of the environment often studied using GIS
- Common GIS concepts and tools used to study environmental justice
- How tools can be combined to answer spatial research questions

Introduction

The environment can be thought of in terms of the natural world: water, green space and forest, or dumps and hazardous waste sites. It can also be conceptualised in more personal terms; our environment is the day-to-day spatial context of our lived experience. In both senses of the word, geography, and by extension GIS methods, have much to contribute. Both definitions of the 'environment' assume spatial variations, patterning and processes to underlie the outcomes we perceive. Environmental justice – as well as environmental law and policy appraisal – are areas of research and social action that aim to measure and demonstrate how one aspect of spatial inequality operates. Typical questions that are addressed in the environmental justice literature have to do with basic fairness: does one group have better access to environmental amenities or goods – parks or open space, for example – than others? Do some groups tend to bear the burden of

'disamenities', such as hazardous waste dumps or poor air quality, disproportionately? Often research focuses on differential outcomes by race or ethnicity, but income or even age might also be factors considered. These questions are complex and are conceptualised and analysed in a range of different fashions. In common to virtually all is a dependence on the tools of GIS, both visualisation and methods. We cannot state with confidence that burdens are unequally distributed if we cannot measure *what* (amenity or disamenity) is close to *whom*.

Much of the attention in environmental justice and policy evaluation (henceforth referred to simply as environmental justice) is devoted to distributional questions, such as those listed above. The field is larger than the measurement of outcomes, however. What we observe on the ground, fair or unfair, is the result of processes that might privilege one group over another. As Maantay (2007) observes, environmental justice also focuses on unequal protection from environmental ills or 'disamenities', as well as unequal representation when policies and laws are being developed. Often, observed outcomes are a combination of unequal representation and unintended consequences. New York City's recent growth, for example, necessitated conversion of land from industrial to residential purposes. This, inadvertently, led to a concentration of manufacturing activity in the Bronx (a lower income, high minority borough of the city where land was cheaper and residential demand lower), subsequently leading to even poorer air quality and increased adverse health effects for many in the area (Maantay, 2007).

Research and applications in environmental justice range in their use of GIS methods, but virtually all employ at least some. The use of GIS is arguably stronger in those studies that emphasise outcomes. In such cases, GIS is indispensable, as distance, proximity and containment are the fundamental spatial concepts used to judge environmental burden and access. In the case of air quality, especially, GIS is occasionally used to model dispersion of pollutants from a point source and thus estimate burden. Such applications require collaboration between GIS and climate experts. And, of course, GIS lends itself well to visualising the locations of population groups and environmental elements. GIS can also be used to develop solutions to unjust situations. If some groups suffer lack of access to parks, for example, where should new parks be located (Sister *et al.*, 2010)? Or if truck routes through urban neighbourhoods are linked to poor air quality, how might routes be redesigned to minimise impact on neighbourhoods but still meet truck drivers' requirements (Fisher *et al.*, 2006)? Increasingly, research aims not only to show correlation between population characteristics and locations of environmental hazards and goods but also to show causation. In short, in this field, GIS is used to calculate spatial measures of burden or distribution, but also for visualisation and modelling.

Because a strength of GIS's contribution to the field of environmental justice lies in burden estimation, and because a range of estimation strategies has been developed, this chapter focuses on how GIS can best be used for outcome studies. The goal of the chapter is to illustrate how conceptualisation of burden can be matched to GIS methods and to provide examples of applications. Along the way, the chapter covers some specific challenges that have been identified with regard to spatial analysis and environmental justice research. It provides examples of common GIS methods introduced in Part I. The chapter also complements material covered in the other chapters in Part II of the book, such as health, crime and emergency planning, and also provides concrete examples of how methods can be combined to address methodological and topical issues related to environmental justice and policy appraisal.

Data and conceptual challenges

The environmental justice and spatial analysis/GIS literature is large. It is common in these published studies, which often highlight a theoretical or conceptual point using a case study, to note challenges associated with measuring environmental injustice (see e.g. Cutter, 1995; Maantay, 2007; Maroko *et al.*, 2009). Some challenges are aspatial. For instance, when measuring burden, how do we identify different groups – by income, by race or by age? Others are clearly spatial and it is these that this section focuses on.

Two main challenges arise where environmental data are concerned. First, data on environmental hazards, such as air or water pollutants, might not be complete or accurate. Knowledge about dangerous substances and their behaviour in nature is also changing. Because the research needs to be done, the analyst likely needs to make assumptions and might need to narrow the scope of the project to match the type of data available. For example, data on measured air quality are spotty in many locations, making it difficult to use actual measures of air pollutants to assess whether, for example, those living close to highways or factories are worse off than those living further away (the first step in studying whether different groups have different levels of exposure being to measure the exposure).

This leads to the second challenge. Our understanding of how people are exposed to substances likely requires assumptions. How far does an individual need to be from an incinerator before its impacts are null? In the air quality example above, what is often done is to use known facts about levels of air pollutants that are dangerous and levels of pollution emitted by various sources to develop estimates of pollution at various distances from the source. This could be done formally, using models of air dispersion. It can also be done using buffers (Chapter 2). Buffers of different radii can be employed to check the sensitivity of results to the size of buffer. Distances might also be used – how far away do different groups live on average from a noxious site? In this case, the challenge is deciding whether straight-line (or Euclidian) or network distance is more appropriate (Maroko *et al.*, 2009). The decision is not a matter of convenience! Ideally, it should be justified by the spatial process at work. Air pollution, for example, does not disperse on a network, so a buffer around a manufacturing plant might be appropriate (it is certainly more straightforward than modelling the actual dispersion in continuous space). Pollutants emitted from trucks, though, are likely emitted along a network, so that it might be best to buffer the route being used. Another example is from the perspective of access to amenities. Being closer to a park is likely better in terms of access, but what is the appropriate distance, whether straight-line or network?

A final challenge is a pan-social science issue – establishing causality. Using GIS, a researcher might establish that one group has more polluting facilities in their neighbourhoods than others or that some groups have more trees or parks, but this does not establish that the location of any of these was because of group characteristics. This leads to explanatory models, using neighbourhoods or similar as units of observation, that attempt to control for other factors that might lead to differential locations. More complicated, though, is the fact that neighbourhoods evolve. Today's demographic characteristics might not reflect those of the time period when the rubbish dump or incinerator was initially sited. This leads to the difficult but important question whether disamenities are located in poorer/minority neighbourhoods or whether the poor/minorities are constrained to locate in areas containing disamenities (see Cutter (1995) for good background). Neither is good but the appropriate policy response might depend on the answer.

What do we study when we study GIS and environmental justice?

So far this chapter has made ample reference to the idea of environmental 'amenities' and 'disamenities'. An amenity is a good, some aspect of the landscape that is preferred or desired. A disamenity is the opposite: a feature or characteristic in the landscape that is harmful, unpleasant or dangerous. The origins of the environmental justice movement and field of study lie in studying locations of disamenities in relation to population characteristics, but increasingly researchers have expanded their focus to include amenity distribution (there are links here to the geography of happiness and well-being discussed in Chapter 6). In both cases, it is also common to study how some outcome results from the unequal distribution – health and air pollution or health and park access, for example. This section briefly discusses some of the more typical choices for investigation. This is by no means intended to be an exhaustive list, but rather to give a sense of how various elements of the environment are treated.

Air pollution

Poor air quality is an important public health concern that has been associated with a range of illnesses, respiratory and otherwise. Pollutants could be heavy metals, sulfur or nitrogen oxides, or chemical pollutants. They can also be particulate matter, which refers to particles – smoke, debris or chemical compounds – that are very small (10 micrometres or less) and easily inhaled. The United States Environmental Protection Agency distinguishes between 'criteria pollutants' and 'hazardous air pollutants' or HAPs. The latter, which include benzene and mercury among others, are chemicals known to cause cancer, birth defects or other serious illnesses (United States EPA, 2017). Criteria pollutants are particulate matter, ground-level ozone, nitrogen dioxide, carbon monoxide, sulfur dioxide and lead. Air quality standards in the US are based on measurements of these compounds. Sources of pollution can be point-based and stationary, like factories or incinerators, or mobile as from automobile traffic.

Regardless of pollutant type, air quality stations generally provide insufficient information about air quality across a study area. Instead, many point sources must provide information about pollutants released from the facility and this information can be used in analysis. These data might be more or less complete, requiring interpolation or exclusion of observations. In other cases, information about facilities and mobile sources of air pollution are combined with estimates of pollution that have been calculated elsewhere. These studies tell us, for instance, that highways with a given amount of traffic produce a certain amount of pollution that disperses over some area. This estimate can then be applied to all highways.

Published research on air pollution and environmental justice has investigated neighbourhood characteristics of areas containing both point and mobile pollution sources (e.g. Fisher *et al.*, 2006; Maantay, 2007). Applications focusing on health outcomes might also consider school or park locations, as well as other outdoor spaces where air quality might be important.

Hazardous waste

Land can also be contaminated or polluted. Unlike air, sources of land pollution can be contemporary or historical in nature. Governments track locations of ongoing pollution, such as factories, landfills or toxic waste dumps, but also areas where contamination might have occurred in the past, or was stored in a fashion that the hazard today presents a threat. Research in this area often considers some type of land pollution, such as brownfield or landfill sites and compares these locations to demographic characteristics of the surrounding area. The resulting statistics can then, in turn, be placed aside average characteristics of those areas without the hazardous land use. In areas with a long history of industrial activity, long-term impacts on health (from activities and decisions from decades ago) could be substantial. In the US, for example, underground storage tanks containing hazardous waste might leak over time, such that the burden faces today's population, rather than that of the past (e.g.

Wilson *et al.*, 2012). As in the case of air pollution, measuring the population's exposure to any type of land contamination often requires extrapolating from other research findings, as the areal extent of pollution is unlikely to have been thoroughly documented for every site.

Green space

Access to parks, forest or other public green space is considered an amenity. Green space allows for recreation and exercise but is also associated with cleaner air and a higher quality of life. Unlike land or air pollution, measuring green space is relatively straightforward. Many urban areas maintain data about parks – their size, as well as availability of recreational facilities. Even in the event of no data, however, satellite imagery can often be used to develop estimates of green space or tree canopy across different areas.

As with 'disamenity' studies, the challenge is to estimate exposure (or accessibility) to green space. Some studies approximate access via distance; for instance, if a park is within half a mile, children or adults might be expected to walk to it. Other studies use containment as a metric: which neighbourhoods have parks and how many? At other times, benchmark metrics have been established, against which distribution can be measured. Comber *et al.* (2008) cite UK guidelines that recommend at least two hectares of green space per 1,000 inhabitants. The guidelines also suggest that each person should have small green space (of at least two hectares/five acres) within 300 metres of home, and larger space (at least 20 hectares/49 acres) within two kilometres. In a similar vein for the US, Sister *et al.* (2010) cite a guideline of 6–10 acres (2.4–4 hectares) per 1,000 people.

Water

Clean water is an amenity, but also a fundamental requirement to support life. Environmental justice could therefore treat water as a public good that all should have equal access to, for drinking, washing, swimming, fishing or simply enjoying. It could also assess water quality in ways similar to those employed when studying air or land pollution. In the latter case, surface and groundwater are both important and quality will be affected by point source polluters such as factories, power plants or mines, but also by storage sites, which might date from previous eras. In areas of drought and water shortage, access to water itself could become an environmental justice issue.

Common GIS methods employed in environmental justice and policy research

Researchers benefit from a host of GIS and spatial analysis tools that can be used in environmental justice

research. This section re-introduces many tools covered in Part I of the book, providing short explanations and examples of research that have employed the method. Remember that methods are often used in conjunction with each other.

Containment or 'point-in-polygon'

One easy way to estimate community exposure or access is to rule that all areas – often neighbourhoods or their statistical proxies – that possess a feature are exposed to it (or have access). So areas with landfills are exposed, while those without are not. Neighbourhoods with parks have access; those without do not. This is an admittedly simple way of gauging exposure but it is also frequently employed, given its straightforwardness.

Figure 13.1 Number of points lying within each polygon, which can be used to determine community accessibility to an amenity (or exposure to a hazard)

Figure 13.1 shows areas and parks, with the number of parks counted for each area. The containment method uses a point-in-polygon algorithm (see Chapter 2 for more detail) to count the number of points (parks) in a polygon and generate a variable therefrom. This is what Maroko *et al.* (2009) refer to as the 'container method'. In GIS software it is implemented as a spatial join (see Maroko *et al.*, 2009; Davidson and Anderton, 2000). You will have the opportunity to make use of these techniques to explore access to green space in the associated practical. Although quite simple to operationalise, the container method is quite vulnerable to the MAUP, discussed in Chapter 2. In its most basic interpretation, MAUP suggests that results – here, exposure – might be dependent on the size and boundary locations of units. This is almost certainly the case. Moreover, it is quite likely that the units employed, whether census tracts, wards or enumeration districts, bear little resemblance to real neighbourhoods or activity spaces – they are simply administrative or statistical units. Thus the container method, while straightforward and easy to implement, risks generating spurious, useless or uninterpretable results and should be used with caution.

Buffers

Another way to estimate exposure is to use a buffer, covered in Chapter 2 in more detail, which creates polygons of certain radii around points (e.g. stationary sources of pollution or parks) or lines (e.g. highways). Buffers allow for pollutant impact that extends beyond predefined statistical area boundaries. They can also be used to estimate areas affected by different pollution sources, such as highways or incinerators. Maantay (2007) uses buffers of 140 metres from highways and 800 metres from stationary polluters to evaluate whether proximity to these air pollution sources is related to asthma hospitalisations. They can also be used in conjunction with a spatial join to capture how many polluting features lie within a given distance of a neighbourhood. Buffers are an exceedingly versatile and useful tool.

Distance

Another option is to calculate the distance between pollution source and neighbourhood (or park and neighbourhood, etc.). Average distances can then be calculated for each type of neighbourhood to see if differences are detectable. Or, if impact from a pollutant is thought to be limited to some area around it, neighbourhoods can be classified as 'affected' or 'unaffected' using a distance measure. Distances can be calculated as straight-line or network (see examples in Davidson and Anderton (2000) or Talen and Anselin (1998)).

Cartography

The power of visualising differences in access or exposure should not be underestimated. Maps that show distribution of population by some characteristic (e.g. race/ethnicity or income) and distribution of some environmental amenity or disamenity can help motivate a study or help select locations for more in-depth research (Maroko *et al.*, 2009).

Geometric centroids

Calculating distance or containment with polygons can be tricky, so researchers sometimes collapse polygons down to their geometric centroid, the 'middle' of the polygon (Figure 13.2). Point to point distances are then more straightforward to calculate and spatial joins (the 'point-in-polygon' measure discussed above) are simpler (Wilson *et al.*, 2012).

Areal interpolation

In order to calculate the number and types of people exposed to some hazard, buffered features are often compared to population in census areas such as tracts or output areas. Because boundaries of these two sets of areas are unlikely to be coincident, it is necessary to estimate the share of population lying only in the portion of the census area exposed to the pollutant.

Figure 13.2 Collapsing polygon features (and their attendant attributes) to representative points, or centroids

Figure 13.3 Example of partial coverage of administrative areas by buffers, where areal interpolation can be helpful

Because we do not generally know how population is distributed within the polygon, we do what is termed an areal interpolation (Flowerdew and Green, 1994; Goodchild and Lam, 1980; also see the discussion in Chapter 8) and assume that the share of area affected is equal to the share of population. So, if 30% of a tract falls within a pollutant buffer, we will assume that 30% of the population does, as well. Figure 13.3 shows, visually, how buffers might only partially cover administrative areas for which we have data. Where buffers cover areas completely, all population numbers and characteristics can be preserved. In those cases where buffers cover only a portion of the area, we can recalculate counts and characteristics using areal interpolation (Maantay, 2007). We provide a practical example of areal interpolation in the practical activities linked to this chapter.

Thiessen polygons

If accessibility is the topic of interest, Thiessen polygons offer a different evaluation perspective. Starting from a set of points, the GIS can generate a set of polygons such that borders demarcate areas closest to each point. If we assume that individuals will use the park closest to them, as opposed to employing some threshold distance, we can use Thiessen polygons to determine 'service areas' for parks. This is the approach used by Sister *et al.* (2010) in their analysis of park access in Los Angeles (again see Chapter 9). They then relate numbers of people (and their characteristics) to these service areas to determine where there is more competition for park space.

Kernel density

An alternative to distance or containment is to use feature locations to generate a density surface that reflects access or exposure to the feature from every location in the area. This kernel density estimation technique produces a raster that sums point densities for predetermined cell sizes. The density surface can then be summarised to neighbourhood levels, giving a measure of overall exposure/access for each neighbourhood. Although more complicated than other methods discussed here, kernel density avoids having communities assigned one measure of exposure based on containment or distance from a centroid (for more details see Maroko *et al.*, 2009; Moore *et al.*, 2008).

(Geographically Weighted) Regression

To isolate the link between pollutant burden (or access to environmental amenities), researchers might estimate explanatory models that seek to control for other factors that could explain exposure/access. An issue that might arise is spatial non-stationarity in the relationship between the dependent variable and the explanatory variables. For example, in their study of park access in New York City, Maroko *et al.* (2009) suggest that minority group share might have different predictive power for park size and characteristics, depending on location in the city. They could have used a global regression model but instead use a technique called Geographically Weighted Regression (GWR), which estimates a series of local regression models that allow relationships between the Xs and Ys to vary (Fotheringham *et al.*, 2002). Results can be difficult to interpret and are often best assessed visually on a map. They can help point researchers to unusual locations, however, where predicted relationships are less strong or different.

Applications of GIS in environmental justice

Research that evaluates outcomes – or the differential burden or access of groups to environmental goods and 'bads' (and uses these results to make statements about potential environmental injustice) often proceeds in a similar fashion. The distribution of the population is compared to locations of amenities or 'disamenities'. Where research varies is in terms of the population groups studied, the type of environmental justice issue, and especially how nearness or access is conceptualised and measured. In addition, some studies go beyond exposure or accessibility to compare some health or education outcome to disproportionate exposure to some type of hazard.

Much of the richness in GIS applications comes from the variety of ways in which exposure/accessibility is measured. Once this has been estimated, it can be compared to the spatial distribution of race/ethnic or income groups (or some composite index that captures vulnerability) in an area. Then maps, descriptive statistics or explanatory models can be created that evaluate the hypothesis that some groups are advantaged/disadvantaged over others. Two examples are presented below and each utilises a range of GIS techniques. The case study city is Washington, DC, chosen for a variety of reasons. The capital city of the United States, the District of Columbia is comparable to a city-state or county. It is fairly uniformly urban, with no rural land. It is also known as a racially diverse but very residentially segregated city, which would lead us *a priori* to expect differential outcomes for a range of variables, including education, access to transportation and employment, and exposure to environmental amenities and disamenities. Last, but most certainly not least, the city government has a strong data provision and transparency policy, such that GIS (and tabular) data for a wide variety of topics are freely available to the public. Table 13.1 provides some basic demographic information for the city as a whole.

The first example relates to accessibility of parks. As discussed above, access to green space is important for a variety of reasons. Parks, generally the result of public provision of green space and recreational facilities, fulfil not only health and quality of life goals for the population they serve, but their location can also be taken as an indication of the extent of public investment in the communities in which they are located. Decision about park location could also reflect neighbourhood representation at the municipal level. Figure 13.4 shows the distribution of the white, non-Hispanic (WNH) population across census block groups in Washington (a statistical area composed of blocks of groups containing approximately 600–3,000 inhabitants) and the location of parks, classified by size. From the map, it is difficult (if not impossible) to draw conclusions about any relationship between park and population distribution. While perhaps more parks appear in the darker red areas (those that are more occupied by WNH), this could also be a trick of the eye and certainly does not provide any indication of park accessibility for different areas.

Measuring accessibility, however, requires decisions about what constitutes 'access'. One option is to consider which areas have parks and which do not – the so-called container methods. An alternative is to measure the number of parks within some distance of the neighbourhood. This example uses a one-mile threshold for this distance, using the neighbourhood centroid. Yet another choice, not covered here, would be to use the Washington, DC, road network to see how many parks are within a mile of neighbourhood centroids (using actual streets).

Many block groups in Washington, DC, do not have a park (Figure 13.5, left-hand map). Visual inspection suggests that these areas are located throughout the city and there is no obvious relationship to the distribution of the WNH population. Many of these areas are small, however, and appear to be in proximity to areas containing parks. If we assume that parks within a mile of block group centroids are accessible to those living there, a different picture emerges (Figure 13.5, right-hand map). Now, most block groups have access to at least one park within a mile but two distinct clusters of areas have a very great number of parks to choose from. Checking the correlation between the proportion of the block group population that is WNH and the number of parks within a mile, we find a slight positive correlation (0.19), but Table 13.2 confirms the

Table 13.1 Population characteristics, Washington, DC, 2010

Variable	Statistics
Total population	601,723
Total block groups (n)	450
Percentage white, non-Hispanic	34.8
Percentage African-American, non-Hispanic	50.0
Percentage Hispanic	9.1

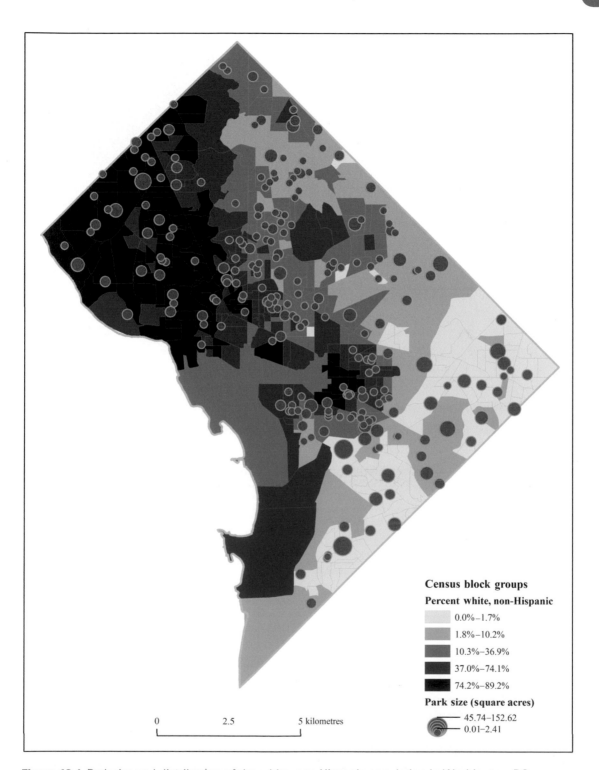

Figure 13.4 Park size and distribution of the white, non-Hispanic population in Washington, DC

Source: Authors' calculations from data retrieved from the DC Data Catalog (http://data.dc.gov/)

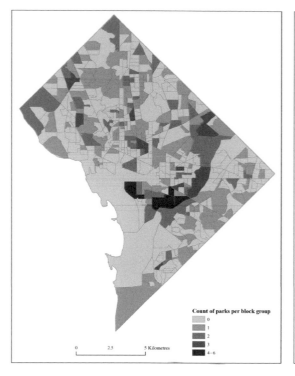
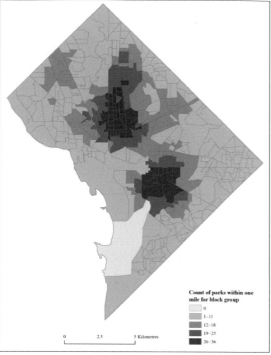

Figure 13.5 Assessing park accessibility via containment (left) and distance (right)

Source: Authors' calculations from data retrieved from the DC Data Catalog (http://data.dc.gov/)

Table 13.2 Demographic characteristics of population with access to parks

		Total	Hispanic	White, NH	Black, NH
Percentage of population in block groups	Containing at least one park	37.3	37.3	37.2	37.5
	Containing no park	62.7	62.7	62.8	62.5
	At least one park within a mile	99.5	99.3	99.2	99.8
	No park within a mile	0.5	0.7	0.8	0.2

Source: Authors' calculations from data retrieved from the DC Data Catalog (http://data.dc.gov/)

visual impression that although most inhabitants of the city do not have a park within their block group, almost everyone has at least one park within a mile – and this statistic holds for each demographic group in the table.

A difficulty with the container and distance methods is that access is treated as binary: one either has parks nearby or not. The kernel density method used in Maroko *et al.* (2009) gets around this challenge by generating a density surface from the distribution of parks. To understand how this method works, imagine moving point by point across a study area and, at each location, counting how many parks, weighted by size, can be reached within a given distance (in this case, one mile). This number is then summed for each cell of a grid, in this case cell size is 50 metres, and the resulting surface gives an idea of the density, or access, to parks from each cell of the raster. Summing cell values returns the total park acreage in the city; that is, the kernel density method has distributed park access across the city, depending on distance and size of park.

Park density for Washington, DC, is shown in Figure 13.6. The left-hand map shows raster values, density with 50-metre grid cells. The density surface suggests that much of the city has relatively low access to parks, with several clusters of high park density in the south-east and north-west. By using zonal statistics, the average park density can be calculated for each block group. This is shown on the right-hand map of Figure 13.6. Summing or averaging values by statistics area is helpful as it brings all measures back into a common unit of analysis. For example, the correlation between the new mean park density variable and the proportion of the block group population that is WNH is negative (−0.40), suggesting that the more WNH the area, the lower the average access to parks.

A final method to assess equitable distribution of parks in the city is to look at access in terms of competition for individual parks, as well as park 'service areas'. The Thiessen polygon technique explained above is applied here. The assumption is that individuals will go to the park closest to them. If this is the case, polygons can be drawn around each park centroid that delineates territory closest to that park. All things being equal, smaller polygons indicate parks that do not have to serve as much territory. Conversely, larger polygons mean more territory served, but also longer distances to reach the park for those on the fringes. Measured this way, park accessibility might be within a mile for some, but further for others.

The visual results indicate a range of size of service area for each park (Figure 13.7). The next questions are: how many people are 'served' by each park and what are their characteristics? A preliminary determination of unjust distribution of parks would be indicated

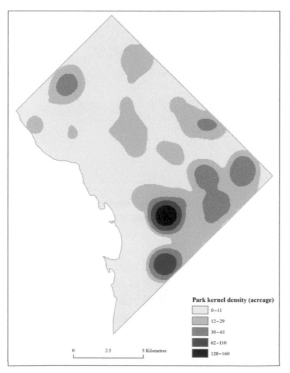
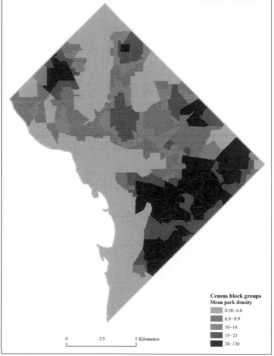

Figure 13.6 A park density surface (left) using kernel density estimation techniques and average density by block group (right)

Source: Authors' calculations from data retrieved from the DC Data Catalog (http://data.dc.gov/)

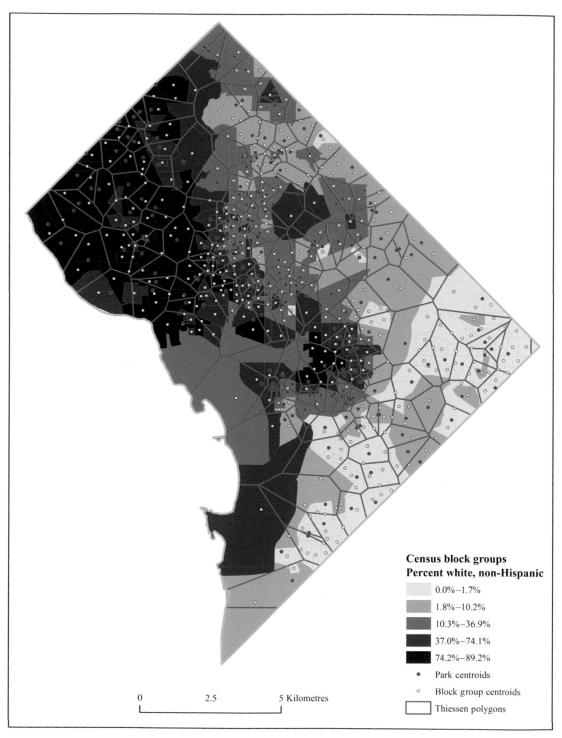

Figure 13.7 Thiessen polygons for parks, paired with distribution of the white, non-Hispanic population and block group centroids

Source: Authors' calculations from data retrieved from the DC Data Catalog (http://data.dc.gov/)

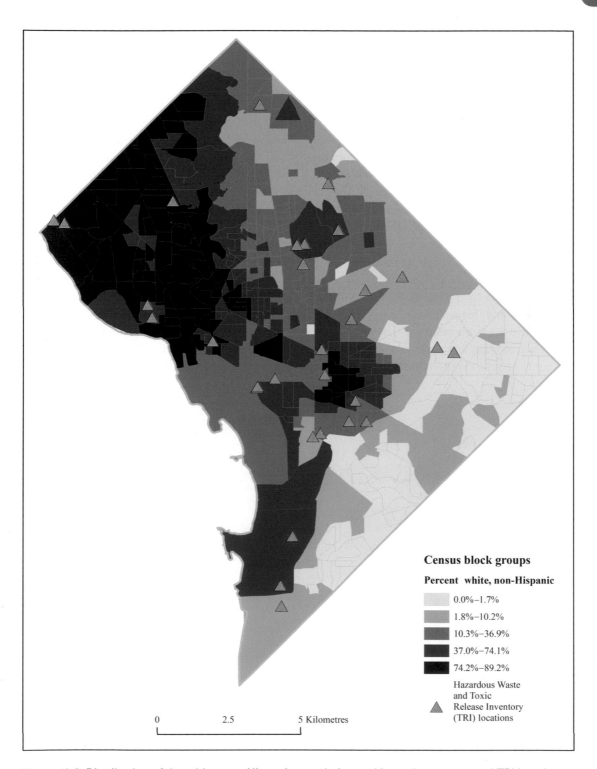

Figure 13.8 Distribution of the white, non-Hispanic population and hazardous waste and TRI locations

Source: Authors' calculations from data retrieved from the DC Data Catalog (http://data.dc.gov/)

by larger areas for minority areas, as well as potential park crowding resulting from parks in minority areas serving larger numbers of people than parks in other locations in the city. To be able to draw these conclusions, the block group population data should be combined (or intersected) with the new Thiessen polygons. As Figure 13.7 shows, intersection of polygons will result in a number of new, smaller polygons for which updated population information will need to be calculated (see areal interpolation, above). Another alternative is to use block group centroids rather than polygons and to allocate population to parks when the centroid falls within the park's service area.

We now consider a second application of GIS for environmental justice in Washington, DC. This example considers the spatial distribution of hazardous waste and Toxic Release Inventory (TRI) locations. TRI locations are those that must report their release of toxic substances to the federal government. Here, power stations, military bases, and a number of manufacturing plants are included. The question is, who lives near these potential hazards? Are all groups equally likely to live nearby? Visual inspection of the distribution of the WNH population and hazardous waste producers seems to suggest that, if anything, many producers are located in or near predominantly WNH areas (Figure 13.8).

As with the park example above, there are two basic approaches to further spatial analysis. The first calculates the number of facilities located in each block group. The second uses a one-mile buffer around block group centroids to capture the number of facilities within that distance. In the case of block group-level exposure to polluting facilities, it appears that very few areas are directly affected (Figure 13.9, left). Of course, the environmental justice question is whether inhabitants of these areas are different from those elsewhere, poorer or belonging to minority groups. If the definition of exposure is expanded, however, to include all areas within a mile of a polluter, the impact appears

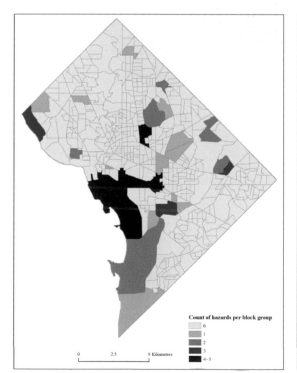

Figure 13.9 Hazard exposure via containment (left) and distance (right)

Source: Authors' calculations from data retrieved from the DC Data Catalog (http://data.dc.gov/)

Table 13.3 Demographic characteristics of population exposed to hazards

		Total	Hispanic	White, NH	Black, NH
Percentage of population in block groups	Containing at least one hazard	6.6	4.4	7.7	6.1
	Containing no hazard	93.4	95.6	92.3	93.9
	At least one hazard within a mile	60.2	54.2	64.8	57.4
	No hazard within a mile	39.8	45.8	35.2	42.6

Source: Authors' calculations from data retrieved from the DC Data Catalog (http://data.dc.gov/)

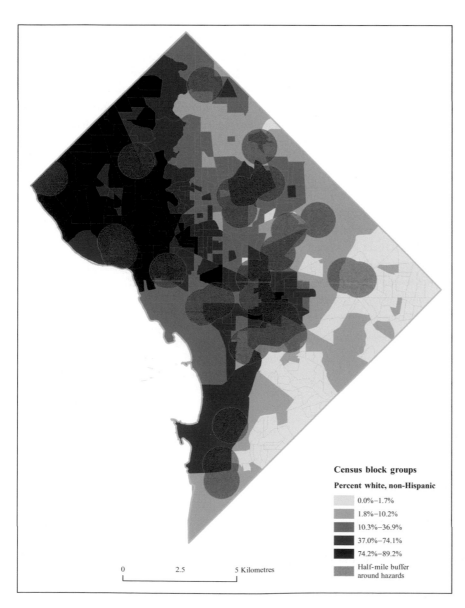

Census block groups

Percent white, non-Hispanic

- 0.0%–1.7%
- 1.8%–10.2%
- 10.3%–36.9%
- 37.0%–74.1%
- 74.2%–89.2%
- Half-mile buffer around hazards

0   2.5   5 Kilometres

Figure 13.10
A half-mile buffer around hazard locations permits the research to compare characteristics of those living close by and further away

Source: Authors' calculations from data retrieved from the DC Data Catalog (http://data.dc.gov/)

267

much greater. A large portion of the city is exposed to at least five polluters. Conversely, clusters of zero impact (from these facilities using the one-mile buffer) areas are also apparent: one in a predominantly WNH section of the city and one with large shares of minority populations. Table 13.3 presents the information in tabular form. Here, we see that while the bulk of the population lives in a block group containing no polluter, over 60% live within a mile of one. For every racial/ethnic category, the majority of the group lives within a mile of a hazardous waste/TRI source. WNHs are more likely than the overall population to have at least one polluter within a mile of their home. While these results give no preliminary evidence of environmental injustice where these pollution sources are considered, the two block groups with eight pollution sources within a mile are both majority minority (only 13% and 43% WNH). Also, this summary approach does not permit the researcher to assess the impact of any one pollution source individually.

Another approach, also using buffers, would be to buffer the facilities themselves and then analyse the population characteristics close by and compare them to characteristics of those living further away. This strategy is shown visually in Figure 13.10. By dissolving overlapping buffers, the researcher loses the ability to compare impacts of one facility to another. However, comparing exposed to non-exposed is more straightforward. As with Figure 13.9, buffering facilities seems to indicate impacts in largely WNH areas.

Concluding comments

In both examples discussed above – parks and pollution – the emphasis has been on providing visual examples of the techniques introduced earlier in the chapter. While a picture can be worth a thousand words, a strength of the GIS is the generation of numbers. Virtually all of the methods shown above produce data, numbers that can be used to calculate descriptive statistics or estimate models to find the relationship between some aspect of the environment and some set of population characteristics. Articles cited in the chapter are an excellent way to learn more about how

GIS output can be used to generate actual findings on environmental justice issues.

Accompanying practical

This chapter is accompanied by a practical activity (Practical 10: Environmental justice – access to green space in Washington, DC) which makes use of the Washington, DC examples presented throughout this chapter. We give you the opportunity to evaluate access to green spaces (parks), using skills previously introduced, including point-in-polygon and buffer and overlay. You will gain further hands-on experience using areal interpolation, an important tool when working with population counts within artificial spatial units. We also introduce a new technique, Thiessen polygons, in order to delineate park 'catchment areas'.

References

All website URLs accessed 30 May 2017.

Comber, A., Brunsdon, C., & Green, E. (2008) Using a GIS-based network analysis to determine urban greenspace accessibility for different ethnic and religious groups. *Landscape and Urban Planning*, *86*(1), 103–113.

Cutter, S. L. (1995) Race, class and environmental justice. *Progress in Human Geography*, *19*(1), 111–122.

Davidson, P., & Anderton, D. L. (2000) Demographics of dumping II: a national environmental equity survey and the distribution of hazardous materials handlers. *Demography*, *37*(4), 461–466.

Fisher, J. B., Kelly, M., & Romm, J. (2006) Scales of environmental justice: combining GIS and spatial analysis for air toxics in West Oakland, California. *Health & Place*, *12*(4), 701–713.

Flowerdew, R., & Green, M. (1994) Areal interpolation and types of data, in A. Fotheringham & P. Rogerson (eds) *Spatial Analysis and GIS*, Taylor & Francis, Bristol, 121–145.

Fotheringham, A. S., Brunsdon, C., & Charlton, M. (2002) *Geographically Weighted Regression: The Analysis of Spatially Varying Relationships*, Wiley, London.

Goodchild, M., & Lam, N. (1980) Areal interpolation: a variant of the traditional spatial problem. *Geo-Processing*, *1*, 297–312.

Maantay, J. (2007) Asthma and air pollution in the Bronx: methodological and data considerations in using GIS for environmental justice and health research. *Health & Place*, *13*(1), 32–56.

Maroko, A. R., Maantay, J. A., Sohler, N. L., Grady, K. L., & Arno, P. S. (2009) The complexities of measuring access to parks and physical activity sites in New York City: a quantitative and qualitative approach. *International Journal of Health Geographics*, *8*(34). DOI: https://doi.org/10.1186/1476-072X-8-34.

Moore, L. V., Diez Roux, A. V., Evenson, K. R., McGinn, A. P., & Brines, S. J. (2008) Availability of recreational resources in minority and low socioeconomic status areas. *American Journal of Preventive Medicine*, *34*(1), 16–22.

Sister, C., Wolch, J., & Wilson, J. (2010) Got green? Addressing environmental justice in park provision. *GeoJournal*, *75*(3), 229–248.

Talen, E., & Anselin, L. (1998) Assessing spatial equity: an evaluation of measures of accessibility to public play-grounds. *Environment and Planning A*, *30*(4), 595–613.

United States EPA (2017) Pollutant Types. Available from: www.epa.gov/air-quality-management-process/pollutant-types.

Wilson, S. M., Fraser-Rahim, H., Zhang, H., Williams, E. M., Samantapudi, A. V., Ortiz, K., *et al.* (2012) The spatial distribution of leaking underground storage tanks in Charleston, South Carolina: an environmental justice analysis. *Environmental Justice*, *5*(4), 198–205.

14 Conclusions

GIS, social media and the future of GIS applications

The chapters in this book have presented a wide range of application areas in GIS for human geography and related social sciences. We hope we have illustrated how GIS can help shed light on important spatial variations in the incidence of activities and how additional spatial analysis can help understand those patterns further. The fact that we have included many examples of work being undertaken for, or on behalf of, key public and private sector organisations, we believe helps to drive home the fact that GIS represents a very public facing, applied perspective for the discipline as a whole. In an era of increasing academic accountability, this part of geography can at least point to many successes in forging partnerships with both business and planning organisations to help solve so-called real-world problems (see Birkin *et al.* (2014a) for more discussion on their perceived benefits of applied GIS and spatial analysis in geography and academia).

However, we recognise that GIS technology moves on and new data sets are appearing with the potential to transform some of the application areas we have reviewed, especially round the area of social media. Indeed, some might argue that this should now be an important application area in its own right. In making some observations about the future of GIS we briefly explore some key issues around the area of GIS and social media in this final chapter.

The first interpretation of GIS and social media is the increasing use of GIS by organisations involved in journalism. The news reports seen on television, plus many newspaper articles, are increasingly using GIS to help present those news stories to the general public. Sometimes these maps are provided in an interactive framework, with accompanying data sets (on related websites) for viewers to plot variables relating, perhaps, to their home locations or own personal interests (see Herzog (2003) for a good review and Murray and Tong (2009) for a broader examination of GIS applications seen in the media). ESRI (2007) give the example of the partnership between CBS in the US and themselves, with the former increasingly using a version of ArcGIS on their regular news and current affairs bulletins. For example, they quote Al Oritz of CBS in relation to their coverage of the US elections in 2004: 'The GIS mapping and data system allowed us, for the first time, to show county-level election results integrated with demographic data using 3D digital maps . . . The system enriched the presentation of results for our viewers.'

Second, is the increasing use of online mapping tools to provide the general public with more interactive GIS-style mapping. Hudson-Smith *et al.* (2009) review a range of online mapping products (Figure 14.1).

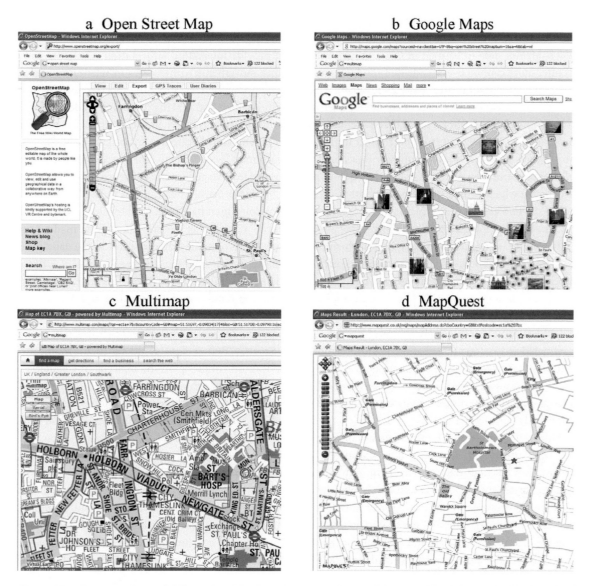

Figure 14.1 A comparison of different public domain map systems with respect to detail and added content

Source: Hudson-Smith *et al.* (2009)

As Hudson-Smith *et al.* (2009) and Batty *et al.* (2010) explain, many of these mapping systems allow users to tag new information to their maps. They note, for example, that Google have made available an application program interface (API) that lets users embed their maps into third-party applications. Through such interactive mapping interested parties can communicate with each other, often helping to add to maps using so-called 'volunteered information' (Goodchild, 2007). Sui and Goodchild (2011), for example, give the examples of people adding their maps of the 'location of Bin Laden's death, Google Earth mashups of critical sites using data posted on WikiLeaks and tracking the diffusion of BP's oil spill in the Gulf of Mexico' (see also the excellent text of Sui *et al.* (2012) which explores the notion of volunteered information and online mapping in far more detail). Such volunteered information is seen by many as an important new source of data to be used in GIS although others offer more caution over its reliability and therefore use (Goodchild and Li, 2012; Batty *et al.*, 2010). This type of information has also been referred to as 'crowdsourcing'. This phrase has different definitions in different disciplines but in a GIS context it means information provided by groups of individuals (initial information and then response data to key questions or debates) which is then shared with all other users within the GIS portal. Collecting data through maps has also been supplemented by the increasing availability of GIS on tablets and smartphones which can make them ideal also for data collection. In the past, for increased accuracy, smartphone users could connect their phone to an accurate GPS device using applications such as Bluetooth. Nowadays, most smartphones and tablets contain increasingly accurate GPS recorders which help to collect information on location more routinely (and can be used for analysis in the field also – for example, retail location planners exploring demographic or drive times via GIS on a tablet while in the field).

Third, in addition to the increasing use of online mapping, the public is increasingly volunteering information through data provided through social media websites such as Twitter and Facebook. These social media sites are now generating enormous levels of traffic, and hence potential new data. Twitter, for example, claimed to have around 500 million tweets per day globally in 2014. Again these media sites allow interested parties to communicate with others, and show the interactions between key events of the day and the reactions of the public. These networks can also be used by media or government agencies to relay vital information back to the public, especially in the event of emergencies (see discussion below). Sadly, not all of these data are georeferenced. However, Stefanidis *et al.* (2013) note that much of Twitter data, for example, can provide 'geospatial footprints' in the form of where the tweet originates or references in the context to geographic entities.

Some brief illustrations of the use of Twitter data are useful here and can be related back to our earlier chapters. First let us examine their potential use in retail analysis, discussed in Chapter 8. Twitter data can provide interesting information relating to consumer movements within cities in relation to choice of shopping centres and offer greater insight into multi-purpose trip making. Location models in retail planning often rely on limited data on consumer movements to calibrate trip-making models. GIS can incorporate Twitter data to plot where consumers have travelled from to get to a shopping centre and help that calibration process. They can also be used to examine where consumers have been prior to visiting a shopping centre – i.e. whether the trips have been made from home, work or some other location in the daily cycle of movement around the city (see Birkin *et al.* (2014b, 2017) and Lovelace *et al.* (2016) for more details in relation to retail use of Twitter data and Arribas-Bel *et al.* (2015) for the implications of Twitter data for understanding general movements across cities).

Second, Twitter data could be useful for health applications (building on the material of Chapter 9). Many public health organisations are already monitoring tweets for any mention of illness or disease (and the spread of such disease in a geographical context). Paul and Drezde (2011), for example, consider a broad range of public health applications seen on Twitter, examining over one and a half million health-related tweets, and discover mentions of over a dozen ailments, including allergies, obesity and insomnia. An interesting application has recently begun at the University of Leeds to look at the geography of food poisoning through the discussion on Twitter/Facebook of ill-health induced by bad experiences of meals in cafés, restaurants or bars (Oldroyd, forthcoming).

Third, many applications of GIS and social media have used Twitter data to explore reactions to natural disasters (relevant especially to emergency planning, discussed in Chapter 10). The common methodology here is to use Twitter data to identify and localise the impact areas of major events (event monitoring). This information could also have implications for emergency planning and real-time evacuation procedures (see interesting applications in Crooks *et al.*, 2013; Crooks and Wise, 2013; Stefanidis *et al.*, 2013; Goodchild and Glennon, 2010; Middleton *et al.*, 2014).

Not surprisingly, technical advances have allowed the development of new GIS and related software packages to handle the increase in social media data. Tsou *et al.* (2013), for example, have designed new mapping software to analyse Twitter data by searching on key words and then producing 'geospatial fingerprints' of matters related to the variables associated with keywords (see also Batty *et al.*, 2010). Their application area is based around political geography. They showed that it was possible to map the popularity of different electoral candidates across the US by an exploration of complimentary tweets towards those candidates. These geospatial fingerprints showed a strong correlation to the geography of the final election results. Thus, this type of analysis could be useful for forecasting the results of future elections, especially at the small-area level.

The increased availability of social media has also been associated with the development of a new information era in GIS and spatial analysis – that of 'big data'. Big data are not only social media data: they can also come from public and private organisations – perhaps not data typically in the public domain in the past. So retailers, for example, might make data on sales available for academics to explore spatial variations in diet and nutrition, as seen through the bundles of foods bought by customers in different parts of cities or regions. We have already seen some examples of this type of data in the earlier chapters. For example, in Chapter 12, we referred to the data being provided by the transit companies in Brisbane, Australia, allowing researchers to map and analyse vast quantities of (hourly) public transport data across the city.

Big data (and especially social media data) have inevitably raised new and additional concerns over accuracy and ethics. These concerns are not new in the history of GIS. Pickles (1995, 1999) was one of the first to look at the ethical use of data in GIS and geodemographics. For big data there are now a plethora of studies re-exploring these issues, along with important concerns over data accuracy (e.g. Kitchin, 2014; Goodchild and Li, 2012; Graham and Shelton, 2013; Xu *et al.*, 2013; Dalton and Thatcher, 2015; Thatcher, 2014). All users of big data should at least be aware of these wider debates.

All of these developments are likely to increase the ease of access to GIS technology and perhaps increase the number of ways in which data can be captured and shared. However, wherever the data come from in the future we hope that the presentation of what GIS can do in different areas of the social sciences that we have offered in this book (with the related practical examples) will still be a valuable learning tool for students and practitioners alike, especially those new to the discipline.

References

All website URLs accessed 30 May 2017.

Arribas-Bel, D., Kourtit, K., Nijkamp, P., & Steenbruggen, J. (2015) Cyber cities: social media as a tool for understanding cities. *Applied Spatial Analysis and Policy*, 8(3), 231–247.

Batty, M., Hudson-Smith, A., Milton, R., & Crooks, A. (2010) Map mashups, Web 2.0 and the GIS revolution. *Annals of GIS*, 16(1), 1–13.

Birkin, M., Clarke, G. P., Clarke, M., & Wilson, A. (2014a) The achievements and future potential of applied quantitative geography: a case study. *Geographia Polonica*, 87(2), 179–202.

Birkin, M., Harland, K., Malleson, N., Cross, P., & Clarke, M. (2014b) An examination of personal mobility patterns in space and time using Twitter. *International Journal of Agricultural and Environmental Information Systems* (*IJAEIS*), 5(3), 55–72.

Birkin, M., Clarke, G. P., & Clarke, M. (2017) *Retail Location Planning in an Era of Multi-Channel Growth*, Routledge, London.

Crooks, A. T., & Wise, S. (2013) GIS and agent-based models for humanitarian assistance. *Computers, Environment and Urban Systems*, 41, 100–111.

Crooks, A. T., Croitoru, A., Stefanidis, A., & Radzikowski, J. (2013) Earthquake: Twitter as a distributed sensor system. *Transactions in GIS*, 17(1), 124–147.

Dalton, C. M., & Thatcher, J. (2015) Inflated granularity: spatial 'Big Data' and geodemographics. *Big Data & Society*, 2(2), 2053951715601144.

ESRI (2007) *GIS for Media*, ESRI, Oaklands, CA. Available from: www.esri.com/library/bestpractices/media.pdf.

Goodchild, M. F. (2007) Citizens as sensors: the world of volunteered geography. *GeoJournal*, 69(4), 211–221.

Goodchild, M. F., & Glennon, J. A. (2010) Crowdsourcing geographic information for disaster response: a research frontier. *International Journal of Digital Earth*, 3(3), 231–241.

Goodchild, M. F., & Li, L. (2012) Assuring the quality of volunteered geographic information. *Spatial Statistics*, *1*, 110–120.

Graham, M., & Shelton, T. (2013) Geography and the future of big data, big data and the future of geography. *Dialogues in Human Geography*, *3*(3), 255–261.

Herzog, D. (2003) *Mapping the News: Case Studies in GIS and Journalism*, ESRI Press, Redlands, CA.

Hudson-Smith, A., Batty, M., Crooks, A., & Milton, R. (2009) Mapping for the masses: accessing Web 2.0 through crowdsourcing. *Social Science Computer Review*, *27*(4), 524–538.

Kitchin, R. (2014) Big Data, new epistemologies and paradigm shifts. *Big Data & Society*, *1*(1), 2053951714528481.

Lovelace, R., Birkin, M., Cross, P., & Clarke, M. (2016) From big noise to big data: toward the verification of large data sets for understanding regional retail flows. *Geographical Analysis*, *48*(1), 59–81.

Middleton, S. E., Middleton, L., & Modafferi, S. (2014) Real-time crisis mapping of natural disasters using social media. *IEEE Intelligent Systems*, *29*(2), 9–17.

Murray, A. T., & Tong, D. (2009) GIS and spatial analysis in the media. *Applied Geography*, *29*(2), 250–259.

Oldroyd, R. (forthcoming) Outbreaks of food-borne illness through syndromic surveillance using online consumer data. Unpublished PhD thesis, School of Geography, University of Leeds.

Paul, M. J., & Dredze, M. (2011) You are what you tweet: analyzing Twitter for public health. *ICWSM*, *20*, 265–272.

Pickles, J. (ed.) (1995) *Ground Truth: The Social Implications of Geographic Information Systems*, Guilford Press, New York.

Pickles, J. (1999) Arguments, debates, and dialogues: the GIS-social theory debate and the concern for alternatives. *Geographical Information Systems*, *1*, 49–60.

Stefanidis, A., Crooks, A. T., & Radzikowski, J. (2013) Harvesting ambient geospatial information from social media feeds. *GeoJournal*, *78*(2), 319–338.

Sui, D., & Goodchild, M. (2011) The convergence of GIS and social media: challenges for GIScience. *International Journal of Geographical Information Science*, *25*(11), 1737–1748.

Sui, D., Elwood, S., & Goodchild, M. (eds) (2012) *Crowdsourcing Geographic Knowledge: Volunteered Geographic Information (VGI) in Theory and Practice.* Springer Science & Business Media, Heidelberg.

Thatcher, J. (2014) Big data, big questions. Living on fumes: digital footprints, data fumes, and the limitations of spatial big data. *International Journal of Communication*, *8*, 1765–1783.

Tsou, M. H., Yang, J. A., Lusher, D., Han, S., Spitzberg, B., Gawron, J. M., & An, L. (2013) Mapping social activities and concepts with social media (Twitter) and web search engines (Yahoo and Bing): a case study in 2012 US Presidential Election. *Cartography and Geographic Information Science*, *40*(4), 337–348.

Xu, C., Wong, D. W., & Yang, C. (2013) Evaluating the 'geographical awareness' of individuals: an exploratory analysis of Twitter data. *Cartography and Geographic Information Science*, *40*(2), 103–115.

Index